## About the Author

Martin Heidegger (1889–1976) was born in Messkirch, in Baden-Württemberg, Germany. He studied theology and philosophy at the University of Freiburg and lectured there as an assistant to Edmund Husserl in the early 1920s. He then taught at the University of Marburg and wrote his magnum opus, *Being and Time* (1927), before returning to Freiburg to assume the chair of philosophy upon Husserl's retirement. He was rector of the University of Freiburg from 1933 to 1934. An official member of the Nazi Party, he was banned from teaching from 1945 to 1951, when he officially retired, after which he continued to lecture. He died in 1976 and is buried in Messkirch.

# MARTIN HEIDEGGER

# BASIC WRITINGS

# MARTIN HEIDEGGER

# BASIC WRITINGS

## from *Being and Time* (1927) to *The Task of Thinking* (1964)

REVISED AND EXPANDED EDITION

EDITED, WITH GENERAL INTRODUCTION AND INTRODUCTIONS TO EACH SELECTION, BY

DAVID FARRELL KRELL

**Foreword by Taylor Carman**

HARPERPERENNIAL MODERNTHOUGHT

LONDON • TORONTO • SYDNEY • NEW DELHI • AUCKLAND

HARPER**PERENNIAL** MODERN**THOUGHT**

The selections in this volume are translated from the following German books: *Sein und Zeit*, © Max Niemeyer Verlag, 1972; *Wegmarken*, © Vittorio Klostermann, 1976; *Holzwege*, © Vittorio Klostermann, 1972; *Die Frage nach dem Ding*, © Max Niemeyer Verlag, 1962; *Vorträge und Aufsätze*, © Verlag Günther Neske, 1954; *Was heisst Denken?* © Max Niemeyer Verlag, 1971; *Unterwegs zur Sprache*, © Verlag Günther Neske, 1959; *Zur Sache des Denkens*, © Max Niemeyer Verlag, 1969. Acknowledgment is made to Henry Regnery Company, Publishers, for permission to reprint "Modern Science, Metaphysics, and Mathematics" from *What Is a Thing?* © 1967 by Henry Regnery Company. Portions of this work originally appeared in somewhat different form in *What Is Called Thinking?* © 1968 in the English translation by Harper & Row, Publishers, Inc.; *Poetry, Language, Thought*, © 1971 by Martin Heidegger; *On Time and Being*, © 1972 by Harper & Row, Inc.



FIRST HARPER PERENNIAL MODERN THOUGHT EDITION PUBLISHED 2008.

Library of Congress Cataloging-in-Publication Data is available upon request.

ISBN 978-0-06-162701-9

23 24 25 26 27 LBC 26 25 24 23 22

TO Hannah Arendt

J. Glenn Gray

Joan Stambaugh

*It is proper to every gathering that the gatherers assemble to coordinate their efforts to the sheltering; only when they have gathered together with that end in view do they begin to gather.*

—Martin Heidegger, *Logos*

# CONTENTS

# FOREWORD

## *to the Harper Perennial Modern Thought Edition*

In his lecture course on Nietzsche in 1936, Heidegger said, "All great thinkers think the same. Yet this 'same' is so essential and so rich that no single thinker exhausts it."[1] The same (so to speak) could be said of Heidegger himself, though he claimed to have made a decisive break from the metaphysical thinking that had dominated Western philosophy for some 2,300 years. Metaphysical thinkers from Plato to Nietzsche had thought widely and deeply about *entities*, indeed the totality of entities, entities as a whole, the entirety of *what there is*. That sounds utterly comprehensive, yet Heidegger noticed that it leaves out something crucial, namely that *in virtue of which* entities *are* entities, what "makes" them entities, as it were. What makes entities entities cannot itself be another entity—Platonic forms, say, or substance, or God, or human experience—since that would beg the question, What makes *that* entity an entity? Heidegger's new question, the question the metaphysical tradition since Plato had neglected and forgotten, is what he calls the question of *being*. The question of being is not, *What is there*? but, *What does it mean to be*? This was the question that gripped Heidegger throughout his life, his one great thought, the "same" that he thought from the beginning to the end of his intellectual career.

And yet, just as Plato, Descartes, and Nietzsche said profoundly different things about the most general structure and character of entities, so too, at different stages along what he called his "path" of thinking, did Heidegger say very different things about being. Consequently, just as we now distinguish the early Wittgenstein of the *Tractatus* from the later Wittgenstein of *Philosophical Investigations*, it has long been customary to distin-

1. *Nietzsche*, vol. 1, *The Will to Power as Art*, D. F. Krell, trans. (San Francisco: Harper & Row, 1979), 36.

guish the early Heidegger of the 1920s from the later Heidegger of the 1930s, '40s, and '50s. Heidegger referred to the radical change in the style and orientation in his thinking—from the quasi-systematic scholarly treatise *Being and Time* (1927) to the emphatically unsystematic essays on art, science, poetry, language, technology, and the history of metaphysics (written from about 1930 onward)—as "the turn" (*die Kehre*). The writings collected in the present volume trace Heidegger's path of thinking and the turn it took from its earlier to its later incarnations, and in this way they offer a glimpse of a good part, though by no means all or even most, of his philosophical work.

As he makes clear in the introduction to *Being and Time*, Heidegger conceived of his early project of fundamental ontology as a kind of "transcendental" inquiry. Since Kant's *Critique of Pure Reason*, transcendental philosophy has been concerned with the *a priori* conditions for the possibility of various manifest phenomena, for example our knowledge of objects and our awareness of ourselves as subjects. But whereas Kant either simply states or tries to provide formal arguments for what he considers the necessary conditions of objective experience—namely, space, time, and the categories of the understanding (for example, *cause, substance*, and *number*)—Heidegger undertakes a concrete phenomenological description of being as such, which he calls "*the transcendens pure and simple*," so that "Every disclosure of Being as the *transcendens* is *transcendental* knowledge. *Phenomenological truth* (*disclosedness of Being*) *is veritas transcendentalis*."[2]

But how is it possible for *being* as such—as opposed to *entities*—to be concretely manifest to us, hence susceptible to phenomenological description? Surely experience puts us in touch with entities only, not abstract metaphysical categories like being. Isn't being itself in effect nothing at all, a kind of mirage, as Nietzsche put it, one of "the

2. *Being and Time*, J. Macquarrie and E. Robinson, trans. (New York: Harper & Row, 1962), 62; page 85 (cf. 240) of the present volume, to which all further references will be made in parentheses in the body of the text.

most general, emptiest concepts, the last wisp of evaporating reality"?[3] Heidegger thinks not. Or rather, as he says, being is "nothing" only in the literal sense that it is *no thing,* i.e. not something, not an entity. This is why he concedes, '"Pure Being and pure Nothing are . . . the same.' This proposition of Hegel's (*Science of Logic,* vol. I, *Werke* III, 74) is correct" (108).

Correct but misleading, for far from rendering being and nothing equally empty and devoid of interest, the equation points us beyond our ordinary and scientific experience of entities to our underlying affective apprehension that they *are,* that might *not* have been or might not continue to be. Hence "the fundamental question of metaphysics": *Why are there entities at all, and not nothing?*[4] Again, this is not Heidegger's question, for it is still a question about entities, not being as such. What it asks is why there are entities, not what it *means* to be. Still, the question of metaphysics is heuristically crucial, for it points us *toward* the question of being. When we ask—or better yet, *encounter*—the metaphysical question in the right mood, it stirs up in us a sense of astonishment, awe, perhaps a vague sense of dread.

This experience of wonder just *is* the question of being. The question of being, as Heidegger comes to understand it, is not strictly speaking an interrogative, a "question" in a grammatical or illocutionary sense of the word, a question with an answer. It is an experience, both a mood and an understanding, an *apprehension* (in both senses of that word) of the sheer *being* of what is. Moods thus have a significance for Heidegger that they have had for very few other philosophers. In boredom we feel the dull weight of things as a whole, as it were, quite apart from whatever else we happen to think or feel about them. In joy we sense their presence as a kind of abundance or blessing. In anxiety, by contrast, we feel things—and ourselves—"slipping away" into nothing

3. Nietschze, *Twilight of the Idols,* D. Large, trans. (Oxford: Oxford University Press, 1998), III, §4. Heidegger quotes Nietzsche's remark in *Introduction to Metaphysics*, G. Fried and R. Polt, trans. (New Haven: Yale University Press, 2000), 38.

4. *Introduction to Metaphysics*, chapter 1.

(101). Indeed, in anxiety, Heidegger famously says, we encounter "the nothing" itself (95ff.).

Beginning with Rudolf Carnap, critics have often objected to Heidegger's rhetorical use of the nominalizing definite article—"*the* nothing"—as if nothing were a substantive something, after all.[5] Some have been driven especially crazy by Heidegger's (not quite translatable) reflexive formula, *das Nichts selbst nichtet*—"the nothing itself nihilates" (or, it is tempting to say, "noths"). But Heidegger's rhetoric is not just a product of confusion or cynicism. His point is that our understanding of entities rests on a subrational apprehension of things as a whole existing and hanging together, more or less coherently, in a definite way. Anxious apprehension of "nothing" is a perfectly familiar experience, after all, albeit no doubt more intensely felt by some than by others, and what it reveals is our primitive sense of the being of things, a sense we simply find ourselves with, and feel that we have had all along, though we normally don't give it a thought.

That primitive sense of being is also what underlies our "usual concept of truth" (116), a concept that comprises both the *correctness* of factual and evaluative propositions and the *realness* or genuineness of things ("true" gold as opposed to fool's gold). Since Aristotle, philosophers have conceived of truth as a kind of correspondence or "accordance" (120) between the mind and the world, or between concrete particulars and ideas in Platonic heaven or the mind of God. Heidegger points out, though, that that image of fit or agreement presupposes the prior "unconcealment" of a *world* (124–6), a phenomenon the philosophical tradition has consistently overlooked in its "oblivion" or "forgetfulness" of being (232).

The worldly unconcealment in which we live—and which constitutes the transcendental condition of the possibility of true and false, right and wrong, real and sham—can itself be brought out into the open in a number of ways, most importantly for Heidegger in the creation and institution of great works of art, such as the

5. Carnap, "The Elimination of Metaphysics Through Logical Analysis of Language," in *Logical Positivism*, A. J. Ayer, ed. (New York: Free Press, 1959).

Greek temples of Athena and Hera in Paestum or the Romanesque cathedral in Bamberg (166). In these and other paradigmatic works, an entire world emerges—as if out of the darkness, into the light—for the human beings inhabiting it. A great work of art is itself an occasion for wonder, for we pause in the face of "the simple *factum est* . . . that such a work *is* at all rather than is not . . . this '*that* it is' of createdness" (190).

What is genuinely important about a work of art, however, according to Heidegger, is not that it supplies us with pleasurable subjective experiences, as the subdiscipline of philosophical aesthetics has assumed for more than two centuries, but that it discloses the world itself: "The work holds open the open region of the world" (170). More specifically, a great work of art opens the world in such a way as to reveal, however obliquely, something that *resists* unconcealment. That opaque, resistant, recalcitrant, anomalous accompaniment to the transparent worldliness of the world is what Heidegger calls *earth* (168). A great work of art shows us our world and ourselves, it "puts up for decision what is holy and what unholy, what great and what small, what brave and what cowardly, what lofty and what flighty, what master and what slave" (169). But it does this by as it were *unearthing* something that stubbornly hides itself, something that refuses to be drawn out fully into the light of day, something that matters to us profoundly, but is so close to us that we can barely recognize or comprehend it: "*The work lets the earth be an earth*" (172).

It is tempting to fall back on the comfortable idea that what Heidegger is talking about, both in *Being and Time* and in the later essays, is something merely "subjective" in contrast to the "objective" world described by the sciences. This is a mistake, but not because Heidegger thinks "the open region of the world" is something objective or independent of human being (*Dasein*). Far from it. As Heidegger says in *Being and Time*, and then reiterates by quoting himself twenty years later, "Only so long as Dasein is, is there [*gibt es*] Being" (240). Being is not independent of man, but neither can human understanding be said to "create Being" (241). To say that the peculiar locality and orientation and significance of *our* world, the human world—society, history, civilization, culture—is merely "subjective" would be to say that it is *grounded* in the human subject, the mind as the true underlying *subiectum* or foundation of the world (cf. 296–304).

That radical idea of subjectivity, Heidegger suggests, is the metaphysical meaning of modern humanism, and it is perhaps the single idea he was most deeply resolved to question, to resist, and to think around and against. The "Letter on Humanism" is a response to Jean-Paul Sartre's famous definition of existentialism as the doctrine that, for human beings, "existence precedes essence," that "man is nothing other than what he makes of himself."[6] Sartre's inspiration had been Heidegger's own reference in *Being and Time* to "the priority of '*existentia*' over *essentia*" as a basic ontological characteristic of Dasein.

Crucial to that formulation, however, are the scare quotes around the term *existentia*, and so too those around the word "essence" on the previous page: "The 'essence' ["Wesen"] of this entity lies in its "to be" [Zu-sein]."[7] The scare quotes are crucial in both instances because the traditional metaphysical concepts named by the Latin terms *essentia* and *existentia* do *not* coincide with *what* (or rather *who*) Dasein is and the peculiar fact *that* it is (its *Existenz* or being-in-the-world). By failing to question the meanings of those terms, Heidegger argues, Sartre merely trades one dubious metaphysical proposition for another. The tradition since Plato "has said that *essentia* precedes *existentia*. Sartre reverses this statement. But the reversal of a metaphysical statement remains a metaphysical statement" (232).

Heidegger concedes, however, that even in *Being and Time*, whatever steps that book might have taken beyond tradition ontology, "Being is thought on the basis of beings, a consequence of the approach—at first unavoidable—within a metaphysics that is still dominant" (240). In the third, unpublished division of *Being and Time*, he later confesses—in a strangely impersonal tone—that "thinking failed in

6. Sartre, *Existentialism Is a Humanism*, C. Macomber, trans. (New Haven: Yale University Press, 2007), 22.
7. *Being and Time*, 68, 67.

the adequate saying of this turning [*Kehre*]" away from metaphysics, toward "this other thinking that abandons subjectivity" (231).

Some time after the Second World War Heidegger came to believe that Western metaphysics had moved beyond the Cartesian-Kantian subjectivism that projects the world as an objective "picture" (*Bild*) over against itself. Modern technology, he now thought, is not an instrument or effect of the subject's willful subjection of the world *qua* object, but the far less purposive, less centralized—hence more uncanny (*unheimlich*)—self-organization of entities into mere resource material. Light to read by, news to digest, music to hear, pictures to see, money to spend, and above all enormous amounts of information simply to "process" are now all instantaneously accessible at the flip of a switch. Such "things" are not essentially objects, which is to say targets of representation, but instead "standing-reserve" (*Bestand*), malleable stuff constantly available, ready to be utilized at whim: "Whatever stands by in the sense of standing-reserve no longer stands over against us as object." A jet airliner sitting on the runway, for example, "is ordered to insure the possibility of transportation . . . on call for duty, i.e., ready for takeoff" (322). Similarly, the Rhine vanishes into inconspicuousness when it becomes a mere power supply to the hydroelectric plant on its banks, whereas Hölderlin's hymn "The Rhine" lets the river come into its own and *be* a river (321). Even people are now mere "human resources," for example "the supply of patients for a clinic" (323).

The understanding of being that underlies this unthinking treatment of entities as so much resource material is what Heidegger calls "enframing" (*Ge-stell*) (324). Our contemporary understanding of being sets entities up, enframes them, as stuff to be utilized, exploited, manipulated, processed, and in the end forgotten about—all with the greatest possible efficiency. Contrary to widespread opinion, modern technology, ontologically understood as enframing, is "no mere human doing," for it is as much something happening *to us* as it is something we ourselves undertake or carry out. The way out of what Heidegger calls the "danger" of technology is thus not simply to *undo* it, to break our machines and retreat into naïve fantasies of preindustrial life. One proper response to the blind impulse toward the maximally efficient exploitation of entities is instead simply to meditate on the technological understanding of being *as* an understanding of being, as a "destiny" (*Geschick*), even a "gift" (*Geschenk*) (391 passim). And that means *letting* technical devices—airplanes, radios, computers—*be* the things they are, letting them shine radiantly *as* entities, as opposed to letting them sink into inconspicuousness and oblivion.

Another possible response to the devastation brought on by the technological enframing of entities as standing-reserve can be found in Heidegger's effort to describe humble *non*technological things and so cultivate local *non*technological practices. Heidegger alludes again, though tacitly, to Hölderlin, this time to the ode "Heidelberg," which describes an old stone bridge gracefully spanning the Neckar River. "The highway bridge," by contrast, "is tied into the network of long-distance traffic, paced and calculated for maximum yield" (354). By letting technological things such as superhighways and high-definition television sets *show themselves* as what they are, and by fostering nontechnological practices, Heidegger thinks, we can let the world be an authentic place, a "locale" (*Ort*) in which human beings can genuinely live or "dwell" (*wohnen*) (357), as opposed to a measurable spatial manifold.

Especially in these late essays, written in the 1950s, Heidegger is trying to cultivate a new kind of thinking, a nontechnological form of reflection: not the "calculative" intelligence associated with metaphysics, science, and technology (420, 435), but a "meditative" thinking responsive to the question—which is to say the *mystery*—of being. For centuries philosophers have taken for granted that thinking is essentially active, spontaneous, voluntary. Passivity was supposed to be the hallmark of sense perception and the emotions, which were therefore called the "passions." Even in *Being and Time* Heidegger had glossed his own notion of existential understanding as an active "projection" (*Entwurf*), a "pressing forward (*dringen*) into possibilities."[8]

Measured on a standard of active, calculative effort, the present age is plainly an age of tremendous thought. In terms of medita-

8. *Being and Time*, 184–5.

tive receptivity to the mystery of being, to the gift of our uncanny apprehension of being and our own need for locality and dwelling, however, we are astonishingly thoughtless: "*Most thought-provoking in our thought-provoking time is that we are still not thinking*" (371). Thinking meditatively requires taking up a new relation to language, one in which we no longer simply manipulate words and phrases as transparent instruments of communication, but instead "listen to language in such a way that we let it tell us its saying" (411). Such is the spirit in which Heidegger composed these essays, so strikingly different in tone and intention from the earnest scholarly program of *Being and Time* decades earlier, and such is the spirit in which to read and reflect on them.

Heidegger was a thinker, not a poet—a distinction he softened and problematized, but never denied. His efforts are therefore in the end more like Nietzsche's than like Hölderlin's, in spite of the fact that he insisted that Nietzsche's thinking remained within the orbit of Western metaphysics. Heidegger's later essays are not prose poems, for he is not inviting us, as Hölderlin and Rilke did, to catch a glimpse of the gods who have flown from the modern world. He is instead urging us to become thoughtfully attuned to the understanding of being that has driven our history and that continues to inform both our way of thinking and our way of life. Such a thoughtful attunement, or reattunement, might one day allow us to be receptive to a new world.

Taylor Carman
New York City
June 2008

# EDITOR'S PREFACE

This book offers a selection from the writings of the German thinker Martin Heidegger, born September 26, 1889, in Messkirch, died May 26, 1976, in Freiburg. Its dual purpose is to provide English-speaking students of philosophy and of the arts and sciences with (1) an introduction to Heidegger's thought, and (2) essays particularly thought-provoking for students' own areas of interest. It advances the claim to a "basic" selection only with the proviso that other essays excluded for reasons of length may be as basic for an understanding of Heidegger's thought. Although Martin Heidegger studied plans for the volume during the winter and spring of 1974–75, generously offering suggestions concerning inclusions and exclusions, the plan adopted here cannot be called an "authorized" one.

Eleven selections appear: eight complete essays, two uncut excerpts from larger works, and one abridged piece. With the exception of Reading VI, "Modern Science, Metaphysics, and Mathematics," the sequence of selections is chronological by order of composition.

The major improvements in this second, revised and expanded edition of *Basic Writings* are these: (1) Heidegger's most concise account of his thinking concerning language and propriation (*Ereignis*) has been added (see Reading X, "The Way to Language"); (2) Reading IV, "The Origin of the Work of Art," now appears *complete*, including the Epilogue and the 1956 Addendum; (3) I have checked through each piece, correcting the errors that have come to my attention during the past fifteen years and making more consistent the translation of a number of fundamental words. For example, "clearing" is now used for *Lichtung*, "to propriate" for *sich ereignen*, "propriation" for *Ereignis*. Yet because such changes are expensive to make I have kept them to a minimum, in order to keep the price of the book as low as possible. I have updated the "Suggestions for Further Study" with the help of Robert Bernasconi and Joel Shapiro, but have not really been able to do the same for my General Introduction: the publication of a whole range of Heidegger's Marburg lecture courses in the *Collected Edition* has so enriched and complicated our understanding that I could not easily absorb these new materials into my account. I have tried to deal with some of these new publications in my books, *Intimations of Mortality: Time, Truth, and Finitude in Heidegger's Thinking of Being*, 2nd ed. (University Park, Pa.: Pennsylvania State University Press, 1991) and *Daimon Life: Heidegger and Life-Philosophy* (Bloomington: Indiana University Press, 1992), to which I refer the interested reader.

Two considerations ultimately determined the choice of the selections. First, I tried to offer a glimpse of Heidegger's path of thought from the late 1920s until his death, although restrictions of space forced the exclusion of many signposts along that path. Second, I studied each piece with a view to its autonomy, accessibility, and the special significance of the issues raised in it. Reluctantly, again for reasons of space, I excluded essays on the "history of Being" and on the great thinkers of the Western tradition. Perhaps a second volume will someday be able to offer a selection of Heidegger's attempts to recover and renew the thought of Heraclitus and Parmenides, Plato and Aristotle, Leibniz, Kant, Hegel, and Nietzsche. To friends who have urged the inclusion of these and other materials—and who will be chagrined to find more than one favorite essay missing—I enter the anthologist's plea: even aside from the external pressures that limit his or her freedom, gatherers visit blossoms already most familiar to them, and cannot know or cull the entire garden.

Although I will argue that the later essays refine the project announced and begun even before *Being and Time* (1927) and that therefore the best way to approach Heidegger's career of thought is to read the essays in the order they appear here up to "The End of Philosophy and the Task of Thinking" (1964), editors' arguments are often best ignored. Heidegger himself emphasized that the issues to which these essays respond are what is important. He discouraged biographical or doxographical fixations on "Martin Heidegger" and encouraged questioning on the matters of thinking and for the sake of thinking—*zur Sache des Denkens*. Readers most intrigued by questions in the natural sciences, for example, might well begin with the sixth and seventh selections, "Modern Science, Metaphysics, and Mathematics" and "The Question Concerning Technology," only then going back to the second and first readings, "What Is Metaphysics?" and "Being and Time." Those with a background primarily in the fine arts may want to begin with the fourth essay, "The Origin of the Work of Art," only then entering the territory of the sciences and technology via the eighth, "Building Dwelling Thinking." Those most interested in languages and literatures may want to set out on "The Way to Language," Reading X, first of all. Students intrigued by theory of knowledge may first wish to hear what Heidegger has to say in the third selection, "On the Essence of Truth." Students of history or politics may find the "Letter on Humanism," Reading V, a fruitful beginning. Those who haven't the benefit of a teacher's suggestions and can't think of a place to start might try the ninth selection, "What Calls for Thinking?" In short, the sequence of the readings is not a matter for strict observance; the book is placed at the disposal of all who may find in it food for thought.

Five selections appear in translations prepared especially for this volume: (1) "Being and Time: Introduction," by Joan Stambaugh, in collaboration with J. Glenn Gray and the editor; (2) "What Is Metaphysics?" by the editor; (3) "On the Essence of Truth," by John Sallis; (4) "Letter on Humanism," by Frank A. Capuzzi, in collaboration with J. Glenn Gray and the editor; and (5) "The Way to Language," by the editor.

Footnotes in the readings indicated by arabic numerals are Heidegger's; those marked with an asterisk are by the translator or editor as indicated. All explanatory insertions in Heidegger's texts by translators or the editor appear in square brackets. Quotations on the title pages of the readings are from Heidegger's *From the Experience of Thinking* (1947); the translations are by the late Albert Hofstadter, one of the ablest and most generous of translators.

Permission to reprint copyrighted material was graciously extended by the Henry Regnery Company of Chicago for selections from Martin Heidegger, *What Is A Thing?* (1967), translated by W. B. Barton, Jr., and Vera Deutsch.

Note that the word "man" and the masculine pronouns associated with it, both in Heidegger's essays and in my own remarks, are no more than conveniently brief ways of translating *der Mensch*, "the human being."

My thanks to *Basic Writings*' many friends and helpers over the years, now far too many to list by name, except for my assistants at DePaul University, Anna Vaughn and Ashley Carr, who worked so skillfully on this new edition, and my continuing gratitude to the man who two years before his death served as the book's general editor and tutelary genius—J. Glenn Gray.

D.F.K.
*Chicago*

# GENERAL INTRODUCTION: THE QUESTION OF BEING

by David Farrell Krell

# GENERAL INTRODUCTION: THE QUESTION OF BEING

If it serves its purpose, this entire book will be an introduction to the question of Being in the thought of Martin Heidegger. This "general" introduction to that more demanding one will first try to sketch the prehistory of the question in Heidegger's early years up to its decisive formulation in *Being and Time* (1927). But because only the Introduction to Heidegger's major work appears in these *Basic Writings*, the present introduction, after outlining the prehistory of the question, will offer a brief analysis of *Being and Time*. It will close by trying to show how the later essays advance the project undertaken in that work.

## I

In the summer of 1907 the pastor of Trinity Church in Constance gave a seventeen-year-old high school student a book that was too difficult for him. It was the dissertation of Franz Brentano, *On the Manifold Meaning of Being according to Aristotle* (1862). Martin Heidegger later called that book "the chief help and guide of my first awkward attempts to penetrate into philosophy."[1]

1. Martin Heidegger, "My Way to Phenomenology," in Martin Heidegger, *On Time and Being*, trans. Joan Stambaugh (New York: Harper & Row, 1972), p. 74. See also Heidegger's *Antrittsrede* to the Heidelberg Academy of Sciences, printed in *Jahreshefte* 1957–58, reprinted in Martin Heidegger, *Frühe Schriften* (Frankfurt am Main: V. Klostermann, 1972), pp. ix-xi, and translated by Hans Seigfried in *Man and World*, vol. III, no. 1 (1970), 3–4. In addition to the published sources cited in what follows I am indebted to conversations on various aspects of Heidegger's career with Hannah Arendt, J. Glenn Gray, Friedrich-Wilhelm von Herrmann, and Martin Heidegger.

The young author of that dissertation now being studied by the even younger Freiburg student conceded that his book strove "to solve difficulties experienced scholars have called insoluble."[2] Brentano was trying to unravel the meaning of a word that had long puzzled Aristotle—*to on*, "being." "The question that was raised in earliest times," Aristotle had written, "that we raise today, and that will always be raised and will always be a matter of perplexity [is]: *ti to on*, What is being?"[3]

For his main text Brentano chose a passage in Aristotle's *Metaphysics* (VI, 1, 1026a 33ff.) that reduced the many meanings of "being" to four, and he devoted a chapter to each meaning: (1) being in its essential and inessential senses; (2) being in the sense of the true; (3) being in the sense of potentiality and actuality; and (4) being in the various senses derived from the schema of the categories. Bewildering though this list may be, the text from which it derives was actually one of the least complicated Brentano could have found. Other passages in the *Metaphysics* expanded this list of meanings to include words which in translation read as follows: being as substance, property, on-the-way-to-substance, privation of substantial forms, being that has no existence outside the intellect, being of finished but dependent existences, and being of movement, generation, and corruption. It seemed a bit of an understatement to call "being" a homonym—a word with "manifold meanings." But when the young Heidegger followed Brentano's lead a year later and looked into Aristotle's own works the riddle became even more puzzling. For Aristotle believed that all these equally incomprehensible meanings pointed toward *one* essential sense and insisted that *one* privileged science devote itself to the search for that sense.

> We speak of being in many senses but always with a view to one sense and to one nature. Not simply in the way we use identical expressions but in the way

2. Franz Brentano, *Von der mannigfachen Bedeutung des Seienden nach Aristoteles* (Freiburg-im-Breisgau: Herder, 1862), p. vii. See D. F. Krell, *Intimations of Mortality: Time, Truth, and Finitude in Heidegger's Thinking of Being*, chap. 4.
3. Aristotle, *Metaphysics*, VII, 3, 1028b 2–4.

everything healthy is related to health, inasmuch as it preserves or restores health or is a sign of health. . . . In precisely this way we speak of being in many senses but always with a view to *one* dominant source. . . . And just as there is *one* science of the healthy so it is in all such cases. . . . Obviously therefore it is proper for *one* science to study being insofar as it is being.[4]

Had some Polonius asked the young man what he was reading in his two books on "being" he might well have answered, "Words, words, words." German words from recent times trying to translate Latin words from a bygone age that were trying to translate Greek words from antiquity. But what were the Greek words trying to translate? And whatever it was that for two thousand years had been sinking in the debris of gutted libraries, why be concerned with it now? Why should "being" fascinate a boy who, although studious and devout—the firstborn son of the Messkirch sacristan, one of the boys who rang the bells of the church that gave him his name and who thought he might like to be a university professor some day—preferred swimming and skiing to everything else? Or almost everything else.

It must have been apparent to the young Heidegger that not only did the question of the meaning of "being" elude easy answer, it also withheld its sense as a question. Brentano succeeded in demonstrating that the question of being captivated Aristotle as the single most important question. Heidegger's classical education, emphasizing study of the Greek, Latin, and German languages and literatures, could hardly have failed to demonstrate that Aristotle had almost single-handedly laid the foundations of the sciences. Heidegger knew in some detail Aristotle's contributions to, or creation of, what were later called physics, biology, astronomy, psychology, logic, rhetoric, literary criticism, ethics, and political science. But Aristotle's broadest and deepest question, which demanded an account (*logos*) of the Being of beings (*onta*) and so became known as "ontology," although it incited disputations for the next two thousand years, seemed to have lost all meaning. The question of being? A baffling nexus of fateful signifi-

4. *Metaphysics*, IV, 2, 1003a 33ff.

cance and fatal obscurity. How could even a sense for the question awaken? Whatever the reasons for his early, intense, and never abated passion for it—and we should not expect these or any biographical remarks to solve the enigma—in Heidegger that question evolved with astonishing persistence and in no haste. "The following question concerned me in quite a vague manner: If being [*Seiende*] is predicated with manifold significance, then what is its leading, fundamental signification? What does Being [*Sein*] mean?"[5]

It was furthermore what Nietzsche would have called an untimely question, a thought out of season. Auguste Comte's *Discours sur l'esprit positif* (1844) had determined that human reason was now entering its third and most mature phase of development. Having overcome rank superstition by means of theological fictions and purged theology with distillates of metaphysic, the positive spirit of the modern age now had to abandon the chimerical, arbitrary, vague, and idle questions of ontology or theory of being in favor of the real, certain, precise, and useful undertakings of sciences such as mathematics, astronomy, physics, chemistry, biology, and sociology. Though theory of being might once upon a time have rooted the sciences to a common source, and though Aristotle was surely the taproot of the entire system, the Battle of the Ancients and Moderns had long since uprooted the venerable tree of knowledge and forced its branches to scatter on the winds of positive Progress. While positivism encouraged high-handed neglect or underhanded reduction of such questions as Being, other critical thinkers, as we shall see, attacked from within.

In 1909 Heidegger sought help for his seemingly anachronistic question from a book by Carl Braig, *On Being: An Outline of Ontology* (1896). Braig taught systematic theology at Freiburg University. That same year Heidegger began to study theology under

5. Martin Heidegger, *On Time and Being*, p. 74, for this and the following biographical material.

Professor Braig, stimulated by "the penetrating kind of thinking this teacher concretely demonstrated in every lecture hour" and encouraged by conversations they had during walks after class. Some months later the young theology student learned of a multivolume work that a student of Franz Brentano had published a decade earlier—Edmund Husserl's *Logical Investigations*. Expecting that they too might shed light on the multiple meanings of being, Heidegger borrowed the volumes from the university library. That expectation was disappointed, but Husserl's own project, which his second volume called a "phenomenology," intrigued the young Heidegger. In 1911, after four semesters at the university, he made philosophy his major field of study. Though never losing his interest in theology, Heidegger saw this discipline withdraw from the center of his scholarly work and felt the religion it was to serve becoming less and less centripetal for the life taking shape in him. He read widely in philosophy and in the human and natural sciences, studied the German poets Hölderlin, Rilke, and Georg Trakl, read the novels of Dostoevsky and the works of Søren Kierkegaard, and encountered the newly expanded edition of unpublished notes by Friedrich Nietzsche collected under the title *The Will to Power*. Many of these authors might have discouraged Heidegger's interest in theory of being. Kierkegaard scorned the systematic ontology of Hegel as something between a fairy tale and a swindle; Dostoevsky's heroes eschewed Aristotle and asked instead whether God could be forgiven his complicity in a world where innocents are murdered. Twelve months before Heidegger was born Nietzsche sequestered himself in the Swiss Alps and in *Twilight of the Idols* wrote that the "highest concepts" of Western metaphysics were nothing more than "the last wisps of evaporating reality."[6] Of all the idols vanishing in the twilight, "being" must have been the first to go. In *The Will to Power*

6. *Friedrich Nietzsche Werke*, ed. Karl Schlechta, 6th ed. (Munich: C. Hanser, 1969), II, 958; cf. Martin Heidegger, *An Introduction to Metaphysics*, trans. Ralph Manheim (Garden City, NY: Doubleday-Anchor Books, 1961), p. 29.

Heidegger read that Being was a necessary fiction, an invention of weary folk who cannot endure a world of ceaseless change and eternal Becoming.[7] In Nietzsche's view the history of ontology—which was in fact the history of nihilism—sought a world of definable Being solely in order to rescue man from time. Interest in Being, Nietzsche elsewhere wrote, sprang from revenge against time and its "It-was."[8] Not only was the question of Being anachronistic but its suspicious relationship with time also made its pursuit, at least in traditional metaphysics, a symptom of decadence.

Not only Nietzsche but other sources as well brought together for Heidegger the issues of Being and time. Henri Bergson had been lecturing on time for the past several years in Paris; Edmund Husserl remained particularly intrigued by the phenomenon of our internal consciousness of time. But from Kierkegaard, Dostoevsky, and Nietzsche, Heidegger learned that this question of Being and time, if it were to be pursued at all, would have to be worked out concretely, with attention to historically relevant problems. It would not be enough to shuffle concepts that went back to the age of Aristotle: a new relation to the old language of philosophy would have to be won. The search for a concrete interpretation of the meaning of Being, so mysteriously related to time, so inevitably bound up with language, could not really get under way until Heidegger had completed his formal education and became free to teach. But a sense for the question stirred underground and what had germinated in 1907 would break through the surface twenty years later with *Being and Time*.

Under the direction of Heinrich Rickert, whose neo-Kantian orientation emphasized training in logic, theory of knowledge, and value-theory, and with the help of his teachers in theology, mathe-

7. Friedrich Nietzsche, *The Will to Power*, trans. Walter Kaufmann and R. J. Hollingdale (New York: Random House, 1967), note 585 A; cf. notes 516–17, 531, 570, 572, 579, 581–82, 617, 708.

8. Friedrich Nietzsche, *Thus Spoke Zarathustra*, in *The Portable Nietzsche*, ed. Walter Kaufmann (New York: Viking Press, 1954), pp. 251–53; Schlechta ed., II, 392ff.

matics, and physics, Heidegger prepared a doctoral dissertation entitled *The Doctrine of Judgment in Psychologism: A Critical-Positive Contribution to Logic* (1913). This work vigorously opposed the reduction of logical procedures and norms to psychological processes, a reduction encouraged by the general climate of positivism—the word "positive" in Heidegger's subtitle must not be understood in the Comtean sense!—but resisted by the neo-Kantian schools of Hermann Cohen, Wilhelm Windelband, and Heidegger's own mentors. Often cited in the work were the books of Emil Lask, a former student of Rickert's, influenced too by Edmund Husserl, an energetic opponent of psychologism.[9] However, Heidegger's preoccupation proved to be not psychologism but the *being* of *validity*, especially in the logic of negative assertions and impersonal statements.

It would be a mistake to assume that Heidegger felt perfectly at home with his director's neo-Kantian persuasion. In his first published article (1912) Heidegger had been sharply critical of all the well-known "schools" of modern philosophy since Descartes, which seemed excessively preoccupied with knowledge-theory.[10] Heidegger tentatively supported a brand of "critical realism" that sought a middle path between the "empirical sensationism" of Ernst Mach and the "immanentism" descended from Berkeley's radical idealism and Kant's critical idealism or "phenomenalism." Without wishing to revert to a naïve realism exulting in the self-evidence of the external world, Heidegger rejected epistemology's involvement in problems no longer vital to the conduct of the sciences. No scientific researcher in morphology or microbiology, chemistry or astronomy, doubted the efficacious relation of their work to the outside world; none of them grew apprehensive over the possibility that they were

9. Heidegger's doctoral dissertation now appears in Martin Heidegger, *Frühe Schriften* (cf. note 1, above). Lask is cited on pp. 118–19, Husserl on pp. 5–6 and 68. Note also the remarks on Bertrand Russell, p. 115n.

10. See Martin Heidegger, "The Problem of Reality in Modern Philosophy," trans. Philip J. Bossert, in the *Journal of the British Society for Phenomenology*, vol. IV, no. 1 (1973), 61ff.

working with "mere sensations" or shadows cast by an evil genie. Even in this early piece Heidegger called for "positively progressive work" in philosophy, not in the sense of a positivistic rejection of metaphysics, but in the sense of a reflection that would formulate new problems and stimulate advancement in the natural and historical sciences. He began and concluded his article with a reference to ancient Greek and medieval scholastic philosophy—both dominated by the figure of Aristotle—which had not succumbed to epistemological disputes within the horizonless desert of the subject-object split.

With the outbreak of the war in 1914 Heidegger enlisted in the army, but after two months' service he was discharged for reasons of health. He now began work on his *Habilitationsschrift*, a second dissertation that would allow him to teach in the university as *Privatdozent*. By the spring of 1915 he had largely completed a work entitled *Duns Scotus' Doctrine of Categories and Theory of Meaning*. He dedicated it to Heinrich Rickert. Heidegger later remarked that this writing pointed forward to his preoccupation with Being and language since "doctrine of categories" was a common expression for the Being of beings—it is the last of those meanings Franz Brentano derived from Aristotle—and the "theory of meaning" belonged to *grammatica speculativa*, "the metaphysical reflection on language in its relation to Being."[11] At least as striking in this second dissertation was the tension between Heidegger's development of a problem in pure logic or systematic knowledge-theory and his waxing appreciation of the history and culture of the medieval world. Pseudo-Duns Scotus (that is, as we now know, Thomas of Erfurt) and Heinrich Rickert were not altogether comfortable companions. In the Conclusion written especially for the publication of the work late in 1916 Heidegger did censure the "critical realism" he had

11. Martin Heidegger, "A Dialogue on Language," in Martin Heidegger, *On the Way to Language*, trans. Peter D. Hertz and Joan Stambaugh (New York: Harper & Row, 1971), p. 6; cf. the Foreword to Heidegger's *Frühe Schriften* (where the Habilitation dissertation also appears), p. ix.

endorsed *faute de mieux* five years earlier and did insist that "objectivity has meaning only for a Subject who judges," thus reasserting the priority of a pure logic of concepts in philosophical accounts of judgment.[12] Nevertheless, in the same breath the new *Privatdozent* argued that the proper context for all problems of logic must itself be "translogical" since it is formed by the intersection of philosophy and history. The "genuine optics" of the former was not epistemology but "metaphysics"; the proper issue for metaphysics was not the "Subject" of knowledge-theory but "the living Spirit" of a historical age. At the conclusion of a work committed to the systematic treatment of a problem in logic and theory of knowledge Heidegger wrote, "The epistemological Subject does not express the most meaningful sense of Spirit, much less its full content." It was not the theoretician of knowledge whom Heidegger now wished to confront but the thinker who canvassed and systematized the multiform works of Spirit and who therefore radically transformed philosophy and history for modern thought: Hegel—whom an early devotee had pronounced "the German Aristotle."

A lecture on "The Concept of Time in the Science of History," delivered to the philosophy faculty at Freiburg on July 27, 1915, reflected this same tendency away from pure logic and knowledge-theory toward metaphysics and history. Heidegger alluded to a kind of "metaphysical compulsion" or philosophical "will to power" that properly emboldened philosophers to flee the confinements of pure epistemology in order to pose questions concerning the genuine goals of philosophy and the sciences. His effort in the present instance was to contrast the concept of time in modern physics—from Galileo's free fall experiment to Planck's quantum theory—to that underlying the study of history. The current state of the sciences, historical interpretation (the influence of Wilhelm Dilthey was by now unmistakably active), and the concept of time, all became enduring elements in Heidegger's quest for Being. But they were not

12. *Frühe Schriften*, pp. 347–53, for this and the following.

yet liberated from the epistemological labyrinth of modern subjectivist philosophy: Heidegger's early writings betray the Thesean struggle of his earliest tendencies, toward Greek philosophy and Athenian Aristotle's posing of the question of being, against the Cartesian minotaur.[13]

Once again Heidegger went into the army. Early in 1917 he was stationed in Freiburg with "interior services," working with the military mails; after a full day's work he would repair to the university to conduct his lectures and seminars. Later he was sent to a meteorological station on the western front near Verdun, where he served until the Armistice. While in uniform, in 1917, he married Elfride Petri. Two sons were born to the couple, in 1919 and 1920.

In 1916 Heinrich Rickert accepted the chair of philosophy in Heidelberg vacated by Wilhelm Windelband; Rickert's post in Freiburg went to a Göttingen professor—Edmund Husserl. The author of the *Logical Investigations* (the book that had so impressed Heidegger and convinced him to study philosophy) was by now widely known for his school of phenomenology. Husserl's method of instruction, conducted not so much in a classroom as in what Heidegger calls a "workshop," took the form of "a step-by-step training in phenomenological 'seeing.'"[14] Husserl discouraged the introduction of untested ideas from the philosophical tradition; he rejected appeals to authority or to the great figures in the history of philosophy. Yet Husserl's method of "going to the things themselves," describing phenomena of consciousness as accurately and comprehensively as possible, repeating analyses many times in order to sharpen the analytical focus, began to shed some light on the Aristotelian or

13. In a review of the *Frühe Schriften* John D. Caputo calls attention to the young Heidegger's interest in mathematics, logic, and natural science. "It would be an eye-opening experience for analytic philosophers," he notes, ". . . to see how deeply Heidegger once shared their interests." See "Language, Logic, and Time," in *Research in Phenomenology*, vol. III (1973), 147–55. Yet Heidegger never really abandoned his interest in mathematics and the sciences and remained capable enough in the former to serve on doctoral committees for the mathematics faculty.

14. Martin Heidegger, *On Time and Being*, p. 78.

Greek problem of being, especially *on hōs alēthes*, "being in the sense of the true," or as Heidegger would later say, "the presence of what is present in unconcealment."[15]

However, as Husserl continued to emphasize the development of a system of transcendental phenomenology, first sketched in *Ideas I* (1913), his way and that of his young "assistant" began to diverge. From 1919 on, while preparing his lectures on problems in Husserl's *Logical Investigations* and Aristotle's philosophy, Heidegger began to recognize more clearly and critically the historical antecedents of Husserl's "transcendental subjectivity" and its inheritance of the axiomatic subjectivism of Descartes. Phenomenology's "disinterested observer" paid scant attention to the historical determination of his own goals and methods, so that his manipulations of "acts of consciousness" could hinder rather than promote access to "the things themselves." In contrast, the ancients did not saddle themselves with excessive epistemological equipment in their investigations of being. "What occurs for the phenomenology of acts of consciousness as a self-manifestation of phenomena is thought more originally by Aristotle and in all Greek thinking and existence as *alētheia*, the unconcealedness of what is present, its being revealed, its showing itself."[16] Through parallel studies in Aristotle and Husserlian phenomenology—in the winter of 1921–22 he lectured on "Phenomenological Interpretations (Aristotle)"[17]—Heidegger labored over the question of "the things themselves" in ancient ontology and modern philosophies of knowledge. What was the decisive matter for thinking? "Is it consciousness and its

15. Heidegger locates more precisely the importance of Husserl's work for his own efforts in the sixth of Husserl's *Logical Investigations* (Halle: M. Niemeyer, 1900), where Husserl distinguishes between "sensuous" and "categorial" intuition.

16. Martin Heidegger, *On Time and Being*, p. 79.

17. See Martin Heidegger, *Phänomenologische Interpretationen zu Aristoteles: Einführung in die phänomenologische Forschung* (Frankfurt am Main: V. Klostermann, 1985). For a more detailed view of Heidegger's relation to Husserl see Reading XI; *On the Way to Language*, pp. 5–6, 9, and 269; and Heidegger's Foreword to William J. Richardson, S.J., *Heidegger: Through Phenomenology to Thought* (Hague: M. Nijhoff, 1963), pp. xii-xv.

objectivity or is it the Being of beings in its unconcealedness and concealment?"[18]

As a result of his creative interpretations of Aristotle Heidegger received in 1922 an invitation to take up a professorship at Marburg University. He accepted. Between 1923 and 1928 Heidegger there enjoyed the most stimulating and fruitful years of his entire teaching career. He joined several of his new colleagues, among them the philosopher Nicolai Hartmann and the classicist Paul Friedländer, in a reading group called "Graeca," studying Homer, the tragedians, Pindar, and Thucydides. Most of Heidegger's own lectures and seminars at the university treated topics in the history of philosophy by critically interpreting basic texts such as Descartes's *Meditations*, Kant's *Critique of Pure Reason*, and Hegel's *Logic*; he also offered courses on more general themes in ancient and medieval ontology, including one on the history of the concept of time (Summer, 1925). Particularly influential was his 1924–25 lecture course on Plato's *Sophist*, the Dialogue where the problem of being is central (cf. especially 243d–244a, and see Reading I). Heidegger introduced the problem of being in Plato's *Sophist*, rather typically, by first working through Aristotle's interpretation of *alētheuein* in the sixth book of the *Nicomachean Ethics*, which analyzed the many ways of relating to "truth," that is, ways of letting beings show themselves as they are in their Being.

Not only the younger students who attended Heidegger's lectures but also older colleagues like Hartmann and Paul Natorp—who was instrumental in securing Heidegger's invitation to Marburg—testified to the rigor of his questioning and the startling originality of his insights. Hannah Arendt noted that it was of decisive importance that Heidegger avoided general talk *about* Plato and spent an entire semester closely examining just one of the Dialogues. "Today

18. Martin Heidegger, *On Time and Being*, p. 79.

this sounds quite familiar, because nowadays so many proceed in this way; but no one did so before Heidegger."[19] In this way Plato's theory of Ideas shook off the burden of traditional interpretations that doctrines inevitably accumulate and became a problem for the present. Heidegger had no publications by which he might be recognized: simply by the force of his teaching students throughout Germany came to know of him. Professor Arendt spoke of a "rumor" circulating underground, in unofficial university circles, during the 1920s:

> Thinking has come to life again; the cultural treasures of the past, believed to be dead, are being made to speak, in the course of which it turns out that they propose things altogether different from the familiar, worn-out trivialities they had been presumed to say. There exists a teacher; one can perhaps learn to think.

Another student described Heidegger's impact as a lecturer in this way:

> One can hardly portray Heidegger's arrival in Marburg dramatically enough—not that he tried to make a sensation. His entrance into the lecture hall certainly did betray a sense of self-assurance and a consciousness of his own impact, but what was truly characteristic of his person and his teaching was that he became completely absorbed in his work, and that his work shone forth. With him, lecturing as such became something altogether new: it was no longer a "course of instruction" from a professor who devoted his real energies to research and publication. With Heidegger, book-length monologues lost their usual preeminence. What he gave was more. It was the full concentration of all the powers—powers of genius—in a revolutionary thinker who actually seemed himself to be startled by the intensity of the questions growing more and more radical in him. The passion of thinking was so complete in him that it communicated itself to his listeners, whose fascination nothing could disturb. . . . Who of those who heard him then can ever forget the breathtaking whirlwind of questions he unleashed in the introductory

19. Hannah Arendt, "Martin Heidegger at Eighty," in *The New York Review of Books*, October 21, 1971, p. 51, for this and the following quotation.

> hours of the semester, only to become wholly entangled in the second or third question, so that only in the semester's final hours would dark storm-clouds of statements gather, from which lightning flashed and left us half-dazed?[20]

His students—among them Hans-Georg Gadamer, Jacob Klein, Karl Löwith, Gerhard Krüger, and Walter Bröcker—had more than one reason to be dazed. More than likely they had stayed up half the night discussing German Idealism with Nicolai Hartmann—and on four days of the week Heidegger began his Aristotle lectures at seven o'clock in the morning. They went on empty stomachs and met for outings afterward, at least during the summer semesters, in order to discuss what they had heard. These picnics they dubbed "the Aristotle breakfasts." But weariness and hunger were not the only costs: Gadamer recalls that Heidegger demanded more hard work from them than any other teacher. Yet students and teacher alike thrived. Heidegger's teaching remained from that time on at the very center of his intellectual life: virtually all his written works devolve from lectures and seminar discussions.

## II

One morning during the winter semester of 1925–26 the dean of Marburg's philosophy faculty burst into Heidegger's office.

"Professor Heidegger, you have to publish something, right now. Do you have a manuscript?"

He did.

The faculty had nominated him for the chief philosophical *Lehrstuhl* at Marburg, held previously by Hartmann, but the ministry of culture in Berlin refused the appointment since in the past decade Heidegger had not published a book. Through Edmund Hus-

20. Hans-Georg Gadamer, "Marburger Erinnerungen," in *Alma Mater Philippina* (Marburg am Lahn: Universitätsbund e. V., SS 1973, WS 1973–74, SS 1974), pp. 23–27, 19–24, and 15–19, reprinted in part under the title "Begegnungen mit Martin Heidegger" in the *Frankfurter Allgemeine Zeitung*, September 28, 1974.

serl in Freiburg, Heidegger's manuscript, an unfinished treatise with the title *Sein und Zeit*, *Being and Time*, dedicated to Husserl, found a publisher. Two copies of the page proofs were mailed to the ministry. They were returned marked "Inadequate." When *Being and Time* appeared in February of 1927 the ministry withdrew its disapprobation and granted Heidegger the Marburg chair.

The book thus suffered a premature birth, hectic and deprived of dignity. Of the two major parts projected for *Being and Time* only the first appeared, and even it was incomplete, the third and presumably conclusive division missing. Yet within a few years *Sein und Zeit* won recognition as a truly epoch-making work of twentieth-century European philosophy. To this day it brooks no comparison in terms of influence on Continental science and letters or genuine philosophical achievement. With its appearance the neo-Kantian preoccupation with theory of knowledge and philosophy of values seemed outdated; the customary separation of systematic and historical orientations—against which Heidegger's own earlier work had struggled—no longer held; phenomenology itself received an entirely unexpected reformulation; and the whole history of metaphysics from Plato through Nietzsche came into radical question.[21]

Heidegger began to formulate the question of the meaning of Being as it appears in *Being and Time* during lectures and seminars of 1924, although particular analyses go back to the winter semester of 1919–20. By 1924 he had achieved three decisive insights. First, his training in "phenomenological seeing," with Edmund Husserl instilled an allegiance to "the things themselves," encouraged careful description of phenomena, and implanted the need for a concrete posing of the question. The *logos* of phenomenology would have to "make manifest" the way the things themselves (as phenomena) "showed themselves" to be. (See *Being and Time*, section 7.)

21. Walter Biemel, *Heidegger* (Reinbek bei Hamburg: Rowohlt Taschenbuchverlag, 1973), p. 37.

Second, a renewed study of Aristotle's *Metaphysics* (IX, 10) and *Nicomachean Ethics* (VI, 3ff.), which were main sources for his lecture courses in 1924–25 and 1925–26, revealed the fundamental sense of this "making manifest" in *logos* as disclosing or uncovering and hence determined the basic sense of truth (*alētheia*) to be the *unconcealment* by which all beings show themselves to be. Truth was neither the "correctness" or "correspondence" of assertions with regard to states of affairs nor the "agreement" of subject and object within those assertions; it was rather the *self-showing* that allowed beings to be objects of assertions in the first place. (See "On the Essence of Truth," p. 115ff. and "The Origin of the Work of Art," p. 176) Third, insight into the character of *alētheia* as disclosedness or unconcealment indicated that the leading sense of Being in Aristotle and throughout the Western philosophical tradition was "presence" (*Anwesenheit*). Phenomenology therefore should make manifest what shows itself in unconcealment as what is (at) present. (See *Being and Time*, section 6.) Thus the question of the meaning of Being, raised in a phenomenological manner with a view to the presence of beings in unconcealment, required an investigation into the meaning of time.

But where and with what beings should the investigation begin? This question too Heidegger answered in his Introduction to *Being and Time* (see especially section 2). The question of the meaning of Being could be raised in a phenomenologically concrete manner only by asking about the Being of the question, that is to say, about the way the question presented itself and showed itself to be. Heidegger began in the most curious manner—by thinking about what he was doing. He reflected on this starting point later during the summer semester of 1935.[22] He conceded that an investigation into Being really ought to be able to inquire about the Being of any being—an elephant in the jungles of India or the chemical process of combustion on Mars—any being at all. Yet only one being con-

22. Martin Heidegger, *An Introduction to Metaphysics* (cf. note 6, above), chap. 1.

sistently made itself available each time such a question arose: "the human beings who pose this question." Analysis of the being that raised questions concerning its Being would prepare the way for an inquiry into the meaning of Being in general. But Heidegger resisted the traditional ways of talking about the Being of man in Christian dogma, Cartesian subjectivism, or the disciplines of anthropology and psychology, in order to concentrate on man's character as the questioner. Man questions his own Being and that of other things in the world. He is always—in no matter how vague a way—aware of his being in the world. Heidegger called the Being of this questioner who already has some understanding of Being in general "existence" or *Dasein*. *Being and Time* is the analysis of Dasein, human existence, within the framework of the question of the meaning of Being in general. One of the book's central aims is to resist the inclination nurtured by the metaphysical tradition to interpret the Being of Dasein by means of categories suited not to human beings but to other entities in the universe. All talk of the "composition" of Dasein or of its having been "made" in one fashion or another is conspicuous by its absence; all attempts to interpret Dasein with the same categories used to interpret combustion and elephants are repudiated. Instead, those three decisive insights are put to work. Dasein is the kind of Being that has *logos*—not to be understood derivatively as reason or speech but to be thought as the power to gather and preserve things that are manifest in their Being. This gathering happens already in a fundamental yet unobtrusive way in our everyday dealings, for example, in our use of tools. When we lift a hammer or drive a car we are before we know it enmeshed in a series of meaningful relationships with things. We take up the hammer in order to drive a nail through the shingle into the roof so the rain won't penetrate; we put on the left turn signal well in advance of a turn so that the driver behind can brake and avoid an accident. Such intricate contexts of meaning—which are usually implicit in our activities and become visible only when something goes wrong, when the hammer breaks or the bulb burns out—constitute what Heidegger calls "world." In more general terms, as being-in-the-world, Dasein is the open space where beings reveal themselves in sundry ways, coming out of concealment into their "truth" (*alētheia*) and withdrawing again into obscurity. Dasein is present at the origin of the becoming-present of beings in time. But in what sorts of human activities does the character of Dasein most definitively show itself? How is a phenomenology of *existence* to differ, say, from a sociology or psychology of *man?* Yet another early writing offers insight into the fundamental problem of Dasein; we should take a moment now to refer to it.

Between 1919 and 1921 Heidegger wrote a detailed review of Karl Jaspers's *Psychology of Worldviews*, a work in which Jaspers tried to stake out the boundaries of human psychic life in order to learn "what man is."[23] Jaspers appealed to what he called the "limit situations" that drive the human psyche to extreme kinds of reactions: man recoils against "existential antinomies" or contradictions such as struggle, death, accident, and guilt; his is a frustrated will to the unified, infinite life of Spirit. Heidegger's major complaint about this book was that Jaspers "underestimated and failed to recognize the genuine methodological problem" of his own treatise. So long as he operated with concepts like Spirit, totality, life, and infinity without undertaking a critical examination of the history of such notions, and so long as he applied them to human *Existenz* without giving a preliminary account of the Being of this entity, Jaspers's endeavor remained an arbitrary account of man—albeit an ingenious and suggestive one. Particular analyses of guilt and death greatly impressed Heidegger: it is not difficult to see their influence on some of the most famous sections of *Being and Time*. But the lack of structure, neglect of problems of method, and the ahistorical manner of accepting preconceptions—all these showed Heidegger

23. Jaspers's words, cited in Martin Heidegger, "Anmerkungen zu Karl Jaspers *Psychologie der Weltanschauungen*," in *Karl Jaspers in der Diskussion*, ed. Hans Saner (Munich: R. Piper, 1973), pp. 70–100. I have offered an account of this essay in *Intimations of Mortality*, chap. 1.

the way *not* to go in his own work. It was not sufficient to have a "basic experience" to communicate; the interpretive approach to the question of man's Being would have to be carefully worked out. Because Dasein is itself historical all inquiry concerning it must scrutinize its own history: ontology of Dasein must be *hermeneutical*, that is, aware of its own historical formation and indefatigably attentive to the problem of interpretation. Implied in such awareness of its own interpretive origins is a "destructuring" or dismantling of the transmitted conceptual apparatus, a clearing of the congested arteries of a philosophical tradition that has all the answers but no longer experiences the questions—especially the question of its own provenance and purpose.

> With respect to *what* it experiences, our concrete, factical experience of life has its own tendency to fall into the "objective" meanings of the environment available to experience. . . . With respect to the meaning of its Being, the self can easily be experienced in an objectified sense ("personality" or the "ideal of humanity"). Such a direction for experience comes to the theoretical grasp and to philosophical conception in ever stronger measures as the experienced and known past insinuates itself into the present situation as an objective tradition. As soon as this particular burden of factical life [the past] is seen in terms of tradition . . . , the concrete possibility of bringing phenomena of existence into view and specifying them in genuine conception can manifest itself *only when* the concrete, relevant, and effectively experienced tradition is destructured, precisely in reference to the ways and means by which it specifies self-realizing experience; and *only when*, through the destructuring, the basic motivating experiences that have become effective are dismantled and discussed in terms of their originality. Such destructuring actually remains bound to one's own concrete and fully historical preoccupation with self.[24]

Here and in related passages we hear some of the central motifs of *Sein und Zeit*: the destructuring of the history of ontology, the interpretation of Dasein or existence as "a certain way of Being, a certain meaning of the 'is,'. . . a 'how' of Being," special emphasis

24. Martin Heidegger, in *Karl Jaspers in der Diskussion*, pp. 92–93.

on the historical character of this Being with attention to its factual rootedness in the everyday world and its "manifold relations" with people and things. In this early article Heidegger names that certain "how" of Being *Bekümmerung*, a being preoccupied with itself or taking trouble concerning itself, advancing toward what in *Being and Time* he calls *Sorge* or "care." Care proves to have a *temporal* character. Its explication in *Being and Time* intends to serve as the "transcendental horizon" of the question of Being in general.

Dasein involves itself in all kinds of projects and plans for the future. In a sense it is always ahead of itself. At the same time it must come to terms with certain matters over which it has no control, elements that loom behind it, as it were, appurtenances of the past out of which Dasein is projected or "thrown." Dasein has a history. More, it *is* its own past. Finally, existence gets caught up in issues and affairs of the moment. It lives in the present. Heidegger calls these three constituents of Dasein "existentiality," "facticity," and *Verfallen*—a kind of "ensnarement." Each exhibits a special relation to time: I pursue various possibilities for my future, bear the weight of my own past, and act or drift in the present. Of course at any given moment of my life all three structures are in play. In the second division of *Being and Time* Heidegger shows how time articulates all the structures of human existence displayed in the first division. Not only that. He shows how the temporal analysis allows us to get a grasp on the *whole* of Dasein, conceived as care, from beginning (birth) to end (death). For death is that possibility that invades my present, truncates my future, and monumentalizes my past.

> Death is a possibility of Being that each Dasein must itself take over. With death Dasein stands before itself in its *most proper* potentiality for Being. What is involved in this possibility is nothing less than the being-in-the-world of Dasein as such. Its death is the possibility of being no longer able to be "there." When Dasein stands before itself as this possibility it is *fully* directed toward its very own potentiality for Being. Standing before itself in this way all relations in it to other Daseins are dissolved. This most proper, nonrela-

tional possibility is at the same time the extreme possibility. As potentiality for Being, Dasein cannot surmount the possibility of death. Death is the possibility of the unqualified impossibility of Dasein. Death thus reveals itself as the *most proper, nonrelational, insurmountable possibility.*[25]

Here was an interpretation of the Being of man whose candor not even Nietzsche could doubt, for which Being itself was utterly finite and human fate without reprieve. Its unflinching exposition of the fundamental structures of human being, mood, understanding, and speech, of work, anxiety, concern, and care, of temporality and radical finitude, the intimations of mortality—all these deeply impressed European, Latin American, Indian, and Japanese scholars and writers. Albert Camus described his encounter with Heidegger's analysis of the finitude of Dasein in this way:

> Heidegger coolly considers the human condition and announces that our existence is humiliated. . . . This professor of philosophy writes without trembling and in the most abstract language imaginable that "the finite and limited character of human existence is more primordial than man himself" [cf. *Kant and the Problem of Metaphysics*, section 41]. For him it is no longer necessary to doze; indeed he must remain wakeful unto the very consummation. He persists in this absurd world; he stresses its perishability. He gropes his way amid ruins.[26]

Surely the "finitude of Dasein," Heidegger's attempt to regain the Greek sense of limit and mortality, was not a purely academic or abstract affair. While the son lectured on the problem of death on Friday morning, May 2, 1924, in Freiburg, the father died in Messkirch after a stroke; the son brought one of the first printed copies of *Sein und Zeit* to his mother's sickbed nine days before her death on May 3, 1927. Nevertheless, the author of *Being and Time* himself carefully elaborated the issues of anxiety and death, indeed all the analyses of human being, within the context of the more fun-

25. Martin Heidegger, *Sein und Zeit*, 12th, unaltered ed. (Tübingen: M. Niemeyer, 1972), section 50, p. 250. Throughout these *Basic Writings* the pagination of this German edition of *Being and Time* is cited.

26. Albert Camus, *Le mythe de Sisyphe* (Paris: Gallimard, 1942), pp. 40–41.

damental question of the meaning of Being in general. That the book was considered an "existentialist" manifesto for such a long time testifies to the historic oblivion of the question it raises. Even today readers often find various parts of the analysis of Dasein accessible but miss altogether the sense of the question of Being as such. Understandably so, for precisely this sense is difficult. It cannot be rattled off and put out as information; it remains a problem which here we can only cursorily pose.

Heidegger's analysis of human existence, propaedeutic to the question of the meaning of Being in general yet already projected upon its horizon, establishes the "finitude of Dasein." Gadamer writes:

> What does Being mean? To learn about this question Heidegger proceeded to determine in an ontologically positive way the Being of human existence in itself. He did this instead of understanding it as "merely finite" in contrast to a Being that would be infinitely and perpetually in being.[27]

In *Being and Time* the limit of mortality appears without reference to something unlimited—in open violation of our normal way of conceiving boundaries. There is no Being that can serve as the unwavering horizon against which human being may be measured and found wanting. Perhaps that is the sense also of Heidegger's insistence that Being *needs* mortals and that it is utterly finite (cf. Reading II). Perhaps, too, that is a way of understanding the thrust of Heidegger's research after *Being and Time*: it is not a matter of abandoning finite Dasein in quest of infinite Being but of seeing ever more lucidly the limits within which beings as a whole come to appear. The task for thinking becomes the closure and concealment by which Being withholds itself, the darkness surrounding the source of presence. Pursuit of this task does not take us away from the meaning of Dasein in *Being and Time* but leads us closer to it. True, this treatise stands incomplete. Its second part is missing.

27. Hans-Georg Gadamer, in his Afterword to Martin Heidegger, *Der Ursprung des Kunstwerkes* (Stuttgart: P. Reclam, 1960), p. 105.

More disturbing, Heidegger never published the concluding division of Part One. Projected under the title "Time and Being," this division was to have advanced from the preparatory analysis of everyday existence, through the full determination of the Being of Dasein as temporality, to the question of Temporality and Being in general. Heidegger never brought his investigation full circle. Unlike Parmenides or Hegel, Heidegger could not and did not claim to have conjoined beginning and end in the perfection of circle or system. Even the essay composed in 1961 bearing the title "Time and Being" does not serve as the culminating arc. Nor is that its intention.[28] In the seventh edition of *Sein und Zeit* (1953) Heidegger added a note saying that for the missing third division and second part to be supplied the entire book would have to be rewritten; yet he emphasized that the way taken in the published portion "remains even today a necessary one if the question of Being is to animate our Dasein."

*Being and Time* remains a torso, a fragment of a work. Yet it is Heidegger's magnum opus and provides the impetus for all the later investigations, without exception.

In 1928 Heidegger's alma mater offered him the chair of philosophy vacated by Edmund Husserl, who had retired from teaching. Upon

28. See Martin Heidegger, *On Time and Being*, pp. 1–24; cf. also p. 83. But not all the materials relevant to this problem—the incompleteness of *Being and Time*—can be discussed here. For example, during the latter half of a crucial lecture course entitled *The Basic Problems of Phenomenology*, taught in the summer semester of 1927, i.e., immediately after the appearance of *Being and Time*, Heidegger further explicated his approach to the question of Being. Now he focused on that third stage of the question, the "missing" division of *Sein und Zeit* Part One, called "Time and Being." See *Basic Problems of Phenomenology*, trans. Albert Hofstadter (Bloomington: Indiana University Press, 1982); see also the 1928 course, *Metaphysical Foundations of Logic*, trans. Michael Heim (Bloomington: Indiana University Press, 1984); and, finally, the "first draft" of Division One of *Being and Time*, the 1925 *History of the Concept of Time*, trans. Theodore Kiesiel (Bloomington: Indiana University Press, 1985). I have discussed this difficult matter in *Intimations of Mortality*, chaps. 2–3, and also in chap. 6 of my book, *Of Memory, Reminiscence, and Writing: On the Verge* (Bloomington: Indiana University Press, 1990).

his return to Freiburg Heidegger centered his instruction on Kant and German Idealism. By this time he had completed preparations for a book that would advance the first stage of the "destructuring of the history of ontology," planned as Part Two of *Sein und Zeit*. In *Kant and the Problem of Metaphysics* he confronted the neo-Kantian epistemological interpretation of Kant's first *Critique* with his own perspective of the ontology of Dasein. This confrontation took a particularly dramatic form in April 1929 with the famous "Davos Disputation" between the relatively unknown Heidegger and the widely esteemed neo-Kantian philosopher Ernst Cassirer. While the learned, urbane Cassirer insisted that there were no "essential differences" between their respective positions, Heidegger repeatedly stressed their disagreement. Heidegger was right. Cassirer could never have affirmed the "basic experience" underlying Heidegger's entire project as it was reflected, for example, a few months later in Freiburg in his inaugural lecture, "What Is Metaphysics?" (See Reading II.) Heidegger's reputation as a powerful and original thinker continued to grow—in "official circles" now also.

On April 23, 1933, the combined faculties elected Heidegger rector of the University of Freiburg. Three months earlier Adolf Hitler had been appointed Chancellor of the Weimar Republic; the Nazi party was rapidly consolidating its position in the government. Weary of the political divisiveness, economic crises, and general demoralization that plagued postwar Germany, many German academics—Heidegger among them—supported the Nazi party's call for a German "resurgence." On May 3 and 4 local Freiburg newspapers announced the new rector's "official entrance" into the NSDAP. Suddenly words like *Kampf*, "military service," and "the destiny of the German *Volk*" appeared alongside "science" and "Being" in Heidegger's addresses.[29] On the eve of the Reichstag elections of November 12 Heidegger spoke out in support of Hitlerian

29. See Martin Heidegger, *Die Selbstbehauptung der deutschen Universität*, the "Rektoratsrede" (Breslau: W. G. Korn, 1933), pp. 7, 13–16, 20–21.

policies that had culminated in Germany's withdrawal from the League of Nations—whose birth certificate was the deeply resented Versailles Treaty. Meanwhile the NSDAP-dominated ministry of culture began to pressure university leaders for more politically oriented courses and more ideologically enlightened faculty members to teach them. Even though Heidegger resisted this pressure in some cases, in others he himself willfully applied it. There can be no doubt that he became instrumental in the "synchronization" (*Gleichschaltung*) of the German university with the party-state apparatus. During his tenure as rector he helped to force the university administration, faculty, and student body—not only in Freiburg but throughout Germany—into the National-Socialist mold. At the end of February 1934, because of a series of administrative difficulties and political wrangles in both the party and the university, he resigned the rectorship. By that time he was beginning to recognize the impossibility of the situation and the utter bankruptcy of his hopes for "resurgence." In lectures and seminars he began to criticize, at first cautiously and then more stridently, the Nazi ideology of *Blut und Boden* chauvinism, which preached a racist origin even for poetry.[30] Party adherents bitterly criticized Heidegger in the mid-1930s, and various restrictions were placed on his freedom to publish and to attend conferences. In the summer of 1944 he was declared the most "expendable" member of the university faculty and, along with a recalcitrant ex-dean, sent to the Rhine to dig trenches. Upon his return to Freiburg he was drafted into the People's Militia (*Volkssturm*).

Heidegger's active collaboration with the Nazi party had lasted ten months (from May 1933 to January 1934); a period of passive support and waxing disillusionment followed. His early enthusiastic support of the regime has earned him the virulent enmity of many. The fact that he remained silent after the war about the atrocities

30. One of the most sharply critical texts appears in Martin Heidegger, *The End of Philosophy*, trans. Joan Stambaugh (New York: Harper & Row, 1973), pp. 105ff.

committed against Jews and other peoples in Europe, while at the same time bemoaning the fate of his divided fatherland, has understandably shocked and confused everyone, even those who freely affirm the greatness of his thought. That his early engagement in the Nazi cause was a monstrous error all concede; that his silence is profoundly disturbing all agree; whether that error and the silence sprang from basic and perdurant tendencies of his thought remains a matter of bitter debate.[31]

31. It is of course convenient to decide that Heidegger's involvement in political despotism taints his philosophical work: that is the quickest way to rid the shelves of all sorts of difficult authors from Plato through Hegel and Nietzsche and to make righteous indignation even more satisfying than it usually is. Yet neither will it do to close the eyes and stop up the ears to the dismal matter. This is not the place to discuss it in detail, however, and I will only suggest study of several accounts and reflections. See Hannah Arendt's brief but astute remarks in "Martin Heidegger at Eighty" (cited in note 19, above), at pp. 55–54, n.3. Heidegger's "Rectoral Address" and related materials, translated by Karsten Harries, may be found in Emil Kettering and Günther Neske, eds., *Martin Heidegger and National Socialism: Questions and Answers*, trans. Lisa Harries (New York: Paragon House, 1990). For recent discussion and debate, and the introduction of important new materials concerning the course of Heidegger's involvement, see the *Freiburger Universitätsblätter*, Heft 92 (June 1986), entitled "Martin Heidegger: Ein Philosoph und die Politik," edited by Bernd Martin and Gottfried Schramm, now available in Bernd Martin, ed., *Martin Heidegger und der Nationalsozialismus: Ein Kompendium* (Darmstadt: Wissenschaftliche Buchgesellschaft, 1989). The research of Freiburg historian Hugo Ott, published in a spate of articles in the early 1980s, has now been released in book form under the title *Martin Heidegger: Unterwegs zu seiner Biographie* (Frankfurt: Campus, 1988). For further discussion and debate, new materials, and an excellent bibliography, see the special issue of *The Graduate Faculty Philosophy Journal* (New School for Social Research), vol. 14, no. 2 and vol. 15, no. 1, edited by Marcus Brainard et al., published as a double volume in 1991. Particularly notable philosophical reflections are: Jacques Derrida, *Of Spirit: Heidegger and the Question*, trans. Geoffrey Bennington and Rachel Bowlby (Chicago: University of Chicago Press, 1989); Dominique Janicaud, *L'ombre de cette pensée: Heidegger et la question politique* (Grenoble: Jérôme Millon, 1990), which is in the process of being translated into English; Annemarie Gethmann-Siefert and Otto Pöggeler, eds., *Heidegger und die praktische Philosophie* (Frankfurt: Suhrkamp, 1988); Philippe Lacoue-Labarthe, *Heidegger, Art, and Politics: The Fiction of the Political*, trans. Chris Turner (Oxford, England, and New York: Basil Blackwell, 1990); Reiner Schürmann, *Heidegger on Being and Acting: From Principles to Anarchy* (Bloomington: Indiana University Press, 1986); and Michael E. Zimmerman, *Heidegger's Confrontation with Modernity: Technology, Politics, and Art* (Bloomington: Indiana University Press, 1990). I have tried to speak to some of the philosophical issues

After the war, the French army of occupation, in cooperation with the Freiburg University faculty senate, forbade Heidegger's return to university teaching. They lifted the ban in 1951, a year before his scheduled retirement.

At this point we may try to gain retrospect on Heidegger's teaching activity in Freiburg between 1928 and 1945. We have mentioned that upon his return to Freiburg Heidegger lectured and conducted seminars on Kant and German Idealism. Kant's *Critique of Pure Reason*, Hegel's *Phenomenology of Spirit* and *Science of Logic*, and Schelling's *On the Essence of Human Freedom* were basic texts. His course on Hölderlin during the winter semester of 1934–35 exhibited not only Heidegger's fascination with the poetic word but also his abiding preoccupation with the essence of language as such. In the spring of 1936 he traveled to Rome and lectured on "Hölderlin and the Essence of Poetry." During 1939 he delivered several public lectures on Hölderlin's poem "As on a Holiday. . . ." But if the decade of the 1930s betrays a unity of theme or problem it is that of "the essence of truth." During the years between the winter semester of 1931–32 and the third trimester of 1940 Heidegger offered five courses under this title. Toward the end of 1930 he delivered a public lecture on the same subject to groups in Bremen, Marburg, and Freiburg. Plato's *Republic*, *Theaetetus*, and *Parmenides* often served as the textual basis of these lectures. This decade devoted to *alētheia* bore literary fruit in 1942–43 with the appearance of *Plato's Doctrine of Truth* and *On the Essence of Truth* (for the latter, see Reading III). Toward the close of the 1930s and through the troubled years of the war Heidegger taught five courses on Nietzsche,

---

arising from Heidegger's political debacle in chaps. 4–6 of *Daimon Life: Heidegger and Life-Philosophy* (Bloomington: Indiana University Press, 1992) and in my Introduction to the new two-volume paperback edition of Heidegger's *Nietzsche* (San Francisco: HarperCollins, 1991), "Heidegger Nietzsche Nazism."

who had come to occupy a central position in his view of the destiny of Being in philosophy. These lectures and the treatises based on them make up Heidegger's largest single publication.[32] But his study of "the West's last thinker" compelled a return to the earliest sources of the Western intellectual tradition: in the decade of the 1940s and early 1950s Heidegger lectured on Heraclitus, Parmenides, and once again Aristotle.[33] Throughout his teaching career Heidegger divided his time more or less equally between the Greek and the modern German philosophers. He offered more courses on Aristotle than on anyone else; he lectured on Kant and Hegel almost as often. He discussed Leibniz and Nietzsche as regularly as the early Greek thinkers and Plato. In all cases the questions of Being (*Sein*) as presence and of presence as unconcealment (*alētheia*), effective only as *traces* throughout the history of metaphysics, remained Heidegger's theme. We will return to it after these final biographical remarks.

During the 1950s and 1960s Heidegger wrote and published much, especially on the issue of technology (see Reading VII) and on the phenomenon of language (Reading X). He traveled to Provence in 1958 and 1969 and to Greece in 1962 and 1967. Yet he never strayed far from his Black Forest origins for long. Most of his life was divided between residences in Freiburg or Messkirch (where he had a second study in his brother's home) and sojourns in a ski hut built in Todtnauberg during the Marburg years. Nevertheless, a variety of friendships—with the physicist Werner Heisenberg, the theologian Rudolf Bultmann, the psychologists Ludwig Binswanger, Medard Boss, and Viktor Frankl, the political historian and philosopher Hannah Arendt, the French poet René Char and painter Georges Braque—prevented Heidegger's life from being as pro-

32. Martin Heidegger, *Nietzsche*, 2 vols. (Pfullingen: G. Neske, 1961). Translated in four English volumes (San Francisco: Harper & Row, 1979–87). *Nietzsche* has now appeared in two paperback volumes, cited at the end of note 31.

33. For the first two see Martin Heidegger, *Early Greek Thinking*, trans. D. F. Krell and F. A. Capuzzi (New York: Harper & Row, 1975).

vincial and narrow as it is often portrayed. On the morning of May 26, 1976, Heidegger died at his home in Freiburg. To the very end he worked on projects such as this volume and the much more extensive *Gesamtausgabe* of his writings (begun in 1975). He was possessed of that lucidity Yeats yearned for and achieved—"An old man's eagle mind."

## III

"An understanding of Heidegger's thought," we read in one account of his long career,

> can awaken only when the reader of his works is prepared to understand everything he or she reads as a step toward what is to be thought—as something toward which Heidegger is on the way. Heidegger's thought must be understood *as a way*. It is not a way of many thoughts but one that restricts itself to a single thought. . . . Heidegger has always understood his thinking as going along a way . . . into the neighborhood of Being.[34]

Heidegger ventured onto that path while still a schoolboy and remained true to it.

Yet this linear image of a way into the neighborhood of Being—as though that were somewhere over the rainbow—is annoying. Isn't such dogged persistence a mark of stubbornness or eccentricity; doesn't it ultimately betray a plodding imagination? And isn't the question of Being from first to last an academic one, bloodless and without force, like one of the shades Odysseus awaits in the underworld?

Another student bends the linear image by emphasizing the essential restlessness of Heidegger's passage and the many turns of the path. "Although it always circles about the same thing," he notes,

> Heidegger's thinking does not come to rest. Each time we believe we have finally arrived at the goal and prepare to latch onto it we are thrown into a

34. Otto Pöggeler, *Der Denkweg Martin Heideggers* (Pfullingen: G. Neske, 1963), pp. 8–9.

> new interrogation. Every resting point is shaken. What seemed to be the end and goal becomes a departure for renewed questioning. If Descartes sought an unshakable foundation for philosophizing, Heidegger tries to put precisely this foundation in question.[35]

Heidegger's thought circles about a double theme: the meaning of Being and the propriative event (*Ereignis*) of disclosure. *Sein* and *alētheia* remain the key words, *Sein* meaning coming to presence, and *alētheia* the disclosedness or unconcealment implied in such presence.[36] Of course this double theme has its reverse side. Coming to presence suggests an absence before and after itself, so that withdrawal and departure must always be thought together with *Sein* as presencing; disclosedness or unconcealment suggests a surrounding obscurity, Lethean concealment, so that darkness and oblivion must be thought together with *alētheia*. The propriative event is always simultaneously expropriative (*Enteignis*).

Does this circling about the double theme of presence-absence and unconcealment-concealment remain aware of its own original darkness? Although Heidegger begins by thinking about what he is doing, does he sustain such thinking? In *Being and Time* Heidegger thinks of the being that raises questions. He names it Dasein, the kind of being that is open to Being. His major work is an analysis of existence in terms of its temporal constitution as an approach to the question of the meaning of Being in general. "Nevertheless," Heidegger warns at the end of his book, "our exhibition of the constitution of the Being of Dasein remains only *one way*. Our goal is to work out the question of Being in general. For its part, the *thematic* analysis of existence first needs the light of the idea of Being in general to have been clarified beforehand."[37] The implication is that Heidegger's thought after *Being and Time* pursues the issues of Being (as presence) and truth (as unconcealment) in order to ad-

35. Walter Biemel, *Heidegger*, pp. 8–9.
36. On the double theme or leitmotif of *Sein* and *alētheia*, see Biemel, p. 35.
37. Martin Heidegger, *Sein und Zeit*, p. 436.

vance the question already unfolded with utmost care in *Being and Time*. That is why Heidegger can respond to those who like to speak of a "Heidegger I" and a "Heidegger II" (meaning the author of *Being and Time* and the somehow reformed author of the writings after the *Kehre* or "turning"), that "Heidegger I" is possible only if he is somehow already contained in "Heidegger II."[38] Even in *Being and Time*, as we have seen, Heidegger interprets his thought as a way. At the end of the book he endeavors (as Socrates was fond of saying) to look both fore and aft along it.

> One can never investigate the source and possibility of the "idea" of Being in general. . . without a secure horizon for question and answer. One must seek a *way* of illuminating the fundamental question of ontology and then *go* this way. Whether this is the *sole* or *right* way can be decided only *after one has gone along it*.[39]

However, Heidegger remains his life long on this same way: there is no way he can look back and pass judgment on its rightness—although that does not preclude the possibility of an immanent criticism of *Being and Time*. Heidegger does criticize certain aspects of his thought and language in *Being and Time*—the failure of his analyses of the temporality (*Zeitlichkeit*) of Dasein to cast sufficient light on the Temporality (*Temporalität*) of Being, witnessed perhaps in the failure of the second division to repeat in a detailed fashion the analyses of section 44 on truth from the standpoint of temporality; or the surreptitious predominance there of certain forms of thought and language rooted in the metaphysical tradition, such as the idea of "fundamental ontology" or the readily adopted translation "truth" for *alētheia*. Yet it is wrongheaded to interpret the "turning" as Heidegger's abjuration of *Sein und Zeit*. Nor does it help at all to speak of a "reversal of priorities" from *man* to *Being* in Heidegger's later work or to conceive of the *Kehre* as a stage of

38. See Heidegger's own formulations in his letter to Richardson (see note 17, above), p. xxiii. I have discussed this issue in *Intimations of Mortality*, chap. 6.
39. Martin Heidegger, *Sein und Zeit*, p. 437.

"development" in his thought—a kind of maturing or philosophical growing up. Doubtless, the present collection of essays does not offer enough material from Heidegger's magnum opus to shed real light on the *Kehre* problem. Our remarks here are meant only as a *caveat*. Whether and how Heidegger "develops" need not concern us: better to follow the turning of the matter for thought itself in our own way as best we can.

And if we still insist on images, Heidegger himself offers the aptest one of his own thinking, an image that combines the linearity of the way with the flexure of renewed inquiry. A collection of essays from the 1930s and early 1940s bears the title *Holzwege*, "timber tracks" or "woodpaths."

> "Wood" is an old name for forest. In the wood are paths that mostly wind along until they end quite suddenly in an impenetrable thicket.
>
> They are called "woodpaths."
>
> Each goes its peculiar way, but in the same forest. Often it seems as though one were identical to another. Yet it only seems so.
>
> Woodcutters and foresters are familiar with these paths. They know what it means to be on a woodpath.[40]

To be "on a woodpath" is a popular German expression that means to be on the wrong track or in a cul-de-sac: to be confused and lost. Hence the French translators of Heidegger's *Holzwege* call it *Chemins qui ne mènent nulle part*, "ways that lead nowhere." This is not quite right: woodpaths always lead somewhere—but where they lead cannot be predicted or controlled. They force us to plunge into unknown territory and often to retrace our steps. Surely Heidegger's way is not one of rectilinear progress. He does not aim to cut through the forest of thought in order to reach the other side; nor does he believe it can be circumvented. Nor finally does he commission a land speculator to bulldoze it. *Sein* and *alētheia*, the com-

40. Martin Heidegger, *Holzwege* (Frankfurt am Main: V. Klostermann, 1950), p. 3, the untitled Foreword.

ing to and departing from presence, which is to say, to and from the clearing of unconcealment, occur at each turn of the path. *Holzwege* wend every which way. As important as the double theme of *Sein-alētheia* is for Heidegger's thinking, what remains astonishing is the diversity of issues in his thought—and this makes it much more than a lifeless academic affair. Builders of bridges and high-rise apartments, information technologists, research scientists, painters and poets, farmers and philosophers, each in her or his own way confronts and thinks about beings: from the many inclinations of his solitary way Heidegger wishes to address all these. To build, calculate, investigate, create; to see, hear, say, and cultivate; to think; all are ways men and women involve themselves with beings as a whole. For humans are among the beings that for the time being are. The question of Being is not bloodless after all, but vital.

For what?

For recovery of the chance to ask what is happening with man on this earth the world over, not in terms of headlines but of less frantic and more frightful disclosures.

For maintenance of the critical spirit that can say No and act No (as Nietzsche says) without puncturing the delicate membrane of its Yes.

For nurturing awareness of the possibilities and vulnerabilities implied in these simple words, *am*, *are*, *is*, since Being may be said of all beings and in many senses, though always with a view to one.

For pondering the fact that as we surrender the diverse senses of Being to a sterile uniformity, to a One that can no longer entertain variation and multiplicity, we become immeasurably poorer—and that such poverty makes a difference.

# I

# BEING AND TIME: INTRODUCTION

*We are too late for the gods*
*and too early for Being.*
*Being's poem, just begun, is man.*

Heidegger had the Introduction to *Being and Time* on his desk throughout the period of the book's immediate gestation, 1926–27. At that time he still planned to write an entire second part to the treatise (see the outline on pp. 86–87); thus the Introduction introduces us also to something quite *beyond* the text we possess today as *Being and Time*. Like Hegel's Preface to *The Phenomenology of Spirit*, which came to serve as an introduction to Hegel's entire philosophy, Heidegger's Introduction opens a path to all the later work.

In this text, here printed complete, Heidegger recounts the need to reawaken the question of the meaning of Being. "Being" has long served metaphysics as its most universal and hence undefinable concept. Its meaning is obvious but vacuous. Heidegger argues that interrogation of the meaning of Being requires a *fundamental ontology* whose point of departure is an analysis of *existence*. And not just any sort of existence. Only the being that exists in such a way that its Being is at issue for it, only the being that has an understanding of Being, however vague and amorphous, can raise the question of Being in the first place. Heidegger lets the name *Dasein* (derived historically from *Dass-sein,* the that-it-is of a being) stand for human being or *existence* in the emphatic sense (as *standing out*). In the first division of his treatise he intends to exhibit basic structures of the "average everydayness" of Dasein, i.e., of human being as it is predominantly and customarily. These concretely described structures are then to be grounded in an interpretation of time in the second division. Finally, this grounding should prepare the way for an answer to the question of the meaning of Being in general.

Of course we know that the third division of Part One, "Time and Being," where that response was to unfold, never appeared. (See the General Introduction, above.) Because the third division was in some undisclosed way to "turn" or "reverse" matters from "Being and Time" to "Time and Being," the problem of the incompleteness of *Being and Time* was soon touted as Heidegger's "departure" from that work. In spite of the prevalence of this notion in the secondary literature we may resist any facile opinions concerning Heidegger's *Kehre,* or "turning," by studying carefully the Introduction to *Being and Time* in conjunction with Readings III, V, and XI.

The projected second part of *Being and Time* was to pursue "the task of a destructuring of the history of ontology." If in later years the problem of "the Temporality of Being" called forth Heidegger's most profound meditations, that of the destructuring—which is to be understood literally as a deconstruction or painstaking dismantling—demanded the greatest amount of his time and energy. For the attempt to revitalize traditional formulas and concepts by tracing their history was a task by no means completed in the published part of *Sein und Zeit*. Heidegger's efforts to recover and renew the question of Being, to free it from the encrustations of the metaphysical tradition, remained at the center of his purpose; it was a direct outgrowth of his passion for a concrete way of raising that question, a way founded in "original experiences" of existence. It is significant that this "destructuring" begins with the giants of modern philosophy (specifically, Descartes and Kant) and proceeds *toward* the ancients (specifically, Aristotle).

Finally, the Introduction to *Being and Time* discusses the all-important matter of Heidegger's phenomenological method. Here he responds to the goals and methods promulgated by his teacher Husserl; here he offers a first glimpse of his own ideas of "phenomenon" and "logos." These in turn lay the foundation for the basic issue of truth as disclosure and unconcealment (see Readings III and XI). Heidegger's interpretations of "phenomenon," "logos," and "phenomenology" may therefore be viewed as paving the way for that "turn" presaged in *Being and Time* from the analysis of Dasein to the question of the meaning of Being in general.

Before the Introduction to *Being and Time* Heidegger inserts a brief untitled and unnumbered section. It begins with a quotation from Plato's *Sophist* and then states the purpose of the book. The quotation is noteworthy for at least two reasons. First, it comes immediately after that point in *Sophist* when Theaetetus and the Stranger from Elea realize that the shining forth (*phainesthai*) of "mere appearance" (*to phainomenon*) is completely mysterious to them: their *phenomenology* of appearances will have to become an inquiry into being (*to on*). Second, the quotation comes precisely at the point where the Stranger is confronting an entire *tradition* of stories about being: he will have to destructure that tradition—even at the risk of patricide. The Stranger addresses prior philosophers as follows:

> "'For you have evidently long been aware of this (what you properly mean when you use the expression "being"); but we who once believed we understood it have now become perplexed'" (Plato, *Sophist* 244a). Do we today have an answer to the question of what we properly mean by the word "being"? By no means. And so it is fitting that we raise anew *the question of the meaning of Being*. But are we today perplexed because we cannot understand the expression "Being"? By no means. And so we must first of all awaken an understanding of the meaning of this question. The intention of the following treatise is to work out concretely the question of the meaning of *Being*. Its provisional goal is the interpretation of *time* as the possible horizon of any understanding of Being whatsoever.
>
> The goal we have in view, the investigations implied in such a proposal and demanded by it, as well as the path leading to our goal, require some introductory comment.

# BEING AND TIME

## INTRODUCTION
## THE EXPOSITION OF THE QUESTION OF THE MEANING OF BEING

### CHAPTER ONE
### *The Necessity, Structure, and Priority of the Question of Being*

#### 1. The necessity of an explicit recovery of the question of Being

This question has today been forgotten—although our time considers itself progressive in again affirming "metaphysics." All the same we believe that we are spared the exertion of rekindling a *gigantomachia peri tēs ousias* ["a Battle of Giants concerning Being," Plato, *Sophist* 245e 6–246e 1]. But the question touched upon here is hardly an arbitrary one. It sustained the avid research of Plato and Aristotle but from then on ceased to be heard *as a thematic question of actual investigation*. What these two thinkers gained has been preserved in various distorted and "camouflaged" forms down to Hegel's *Logic*. And what then was wrested from phenomena by the

This translation of the Introduction to *Being and Time* by Joan Stambaugh in collaboration with J. Glenn Gray and the editor appears in this volume for the first time. The whole of *Being and Time* is available in a translation by John Macquarrie and Edward Robinson (New York: Harper & Row, 1962). The German text is Martin Heidegger, *Sein und Zeit*, twelfth, unaltered edition (Tübingen: Max Niemeyer Verlag, 1972), pp. 1–40. *Sein und Zeit* was first published in 1927.

highest exertion of thinking, albeit in fragments and first beginnings, has long since been trivialized.

Not only that. On the foundation of the Greek point of departure for the interpretation of Being a dogmatic attitude has taken shape which not only declares the question of the meaning of Being to be superfluous but sanctions its neglect. It is said that "Being" is the most universal and the emptiest concept. As such it resists every attempt at definition. Nor does this most universal and thus undefinable concept need any definition. Everybody uses it constantly and also already understands what they mean by it. Thus what made ancient philosophizing uneasy and kept it so by virtue of its obscurity has become obvious, clear as day; and this to the point that whoever pursues it is accused of an error of method.

At the beginning of this inquiry the prejudices that implant and nurture ever anew the superfluousness of a questioning of Being cannot be discussed in detail. They are rooted in ancient ontology itself. That ontology in turn can only be interpreted adequately under the guidance of the question of Being which has been clarified and answered beforehand. One must proceed with regard to the soil from which the fundamental ontological concepts grew and with reference to the suitable demonstration of the categories and their completeness. We therefore wish to discuss these prejudices only to the extent that the necessity of a recovery* of the question of the meaning of Being becomes clear. There are three such prejudices.

1. "Being" is the most "universal" concept: *to on esti katholou malista pantōn,*[1] *Illud quod primo cadit sub apprehensione est ens, cuius intellectus includitur in omnibus, quaecumque quis apprehen-*

*The German word *Wiederholung* means literally "repetition." Heidegger uses it not in the sense of a mere reiteration of what preceded, but rather in the sense of fetching something back as a new beginning. Perhaps his use is close to the musical term *recapitulation*, which implies a new beginning incorporating and transforming what preceded. Alternative translations might be "retrieval" or "reprise."—TR./ED.

1. Aristotle, *Metaphysics* III, 4, 1001a 21.

*dit*. "An understanding of Being is always already contained in everything we apprehend in beings."[2] But the "universality" of "Being" is not that of *genus*. "Being" does not delimit the highest region of beings so far as they are conceptually articulated according to genus and species: *oute to on genos* ["Being is not a genus"].[3] The "universality" of Being "*surpasses*" the universality of genus. According to the designation of medieval ontology, "Being" is a *transcendens*. Aristotle himself understood the unity of this transcendental "universal," as opposed to the manifold of the highest generic concepts with material content, as the *unity of analogy*. Despite his dependence upon Plato's ontological position, Aristotle placed the problem of Being on a fundamentally new basis with this discovery. To be sure, he too did not clarify the obscurity of these categorial connections. Medieval ontology discussed this problem in many ways, above all in the Thomist and Scotist schools, without gaining fundamental clarity. And when Hegel finally defines "Being" as the "indeterminate Immediate," and makes this definition the foundation of all the further categorial explications of his *Logic*, he remains within the perspective of ancient ontology—except that he does give up the problem, raised early on by Aristotle, of the unity of Being in contrast to the manifold of "categories" with material content. If one says accordingly that "Being" is the most universal concept, that cannot mean that it is the clearest and that it needs no further discussion. The concept of "Being" is rather the most obscure of all.

2. The concept of "Being" is undefinable. This conclusion was drawn from its highest universality.[4] And correctly so—if *definitio fit per genus proximum et differentiam specificam* [if "definition is

2. Thomas Aquinas, *Summa theologiae* II, 1, Qu. 94, a. 2.

3. Aristotle, *Metaphysics* III, 3, 998b 22.

4. See Pascal, *Pensées et Opuscules* (ed. Brunschvicg), Paris: Hachette, 1912, p. 169: "One cannot undertake to define *being* without falling into this absurdity. For one cannot define a word without beginning in this way: 'It is . . .' This beginning may be expressed or implied. Thus, in order to define *being* one must say, 'It is . . .' and hence employ the word to be defined in its definition."

achieved through the nearest genus and the specific difference"]. Indeed, "Being" cannot be understood as a being. *Enti non additur aliqua natura:* "Being" cannot be defined by attributing beings to it. Being cannot be derived from higher concepts by way of definition and cannot be represented by lower ones. But does it follow from this that "Being" can no longer constitute a problem? By no means. We can conclude only that "Being" is not something like a being. Thus the manner of definition of beings which has its justification within limits—the "definition" of traditional logic which is itself rooted in ancient ontology—cannot be applied to Being. The undefinability of Being does not dispense with the question of its meaning but compels that question.

3. "Being" is the self-evident concept. "Being" is used in all knowing and predicating, in every relation to beings and in every relation to oneself, and the expression is understandable "without further ado." Everybody understands, "The sky *is* blue," "I *am* happy," and similar statements. But this average comprehensibility only demonstrates the incomprehensibility. It shows that an enigma lies *a priori* in every relation and being toward beings as beings. The fact that we live already in an understanding of Being and that the meaning of Being is at the same time shrouded in darkness proves the fundamental necessity of recovering the question of the meaning of "Being."

If what is "self-evident" and this alone—"the covert judgments of common reason" (Kant)—is to become and remain the explicit theme of our analysis (as "the business of philosophers"), then the appeal to self-evidence in the realm of basic philosophical concepts, and indeed with regard to the concept "Being," is a dubious procedure.

But consideration of the prejudices has made it clear at the same time that not only is the *answer* to the question of Being lacking but even the question itself is obscure and without direction. Thus to recover the question of Being means first of all to develop adequately the *formulation* of the question.

## 2. The formal structure of the question of Being

The question of the meaning of Being must be *formulated*. If it is a—or even *the*—fundamental question, such questioning needs the suitable perspicuity. Thus we must briefly discuss what belongs to a question in general in order to be able to make clear that the question of Being is a *distinctive* one.

Every questioning is a seeking. Every seeking takes its direction beforehand from what is sought. Questioning is a knowing search for beings in their thatness and whatness. The knowing search can become an "investigation," as the revealing determination of what the question aims at. As questioning about . . . questioning has *what it asks*. All asking about . . . is in some way an inquiring of. . . . Besides what is asked, what is *interrogated* also belongs to questioning. What is questioned is to be defined and conceptualized in the investigative or specifically theoretical question. As what is really intended, what is to be *ascertained* lies in what is questioned; here questioning arrives at its goal. As an attitude adopted by a being, the questioner, questioning has its own character of Being. Questioning can come about as mere "asking around" or as an explicitly formulated question. What is peculiar to the latter is the fact that questioning becomes lucid in advance with regard to all the above-named constitutive characteristics of the question.

The meaning of Being is the question to be *formulated*. Thus we are confronted with the necessity of explicating the question of Being with regard to the structural moments cited.

As a seeking, questioning needs previous guidance from what it seeks. The meaning of Being must therefore already be available to us in a certain way. We intimated that we are always already involved in an understanding of Being. From this grows the explicit question of the meaning of Being and the tendency toward its concept. We do not *know* what "Being" means. But already when we ask, "What *is* 'Being'?" we stand in an understanding of the "is" without being able to determine conceptually what the "is" means.

We do not even know the horizon upon which we are supposed to grasp and pin down the meaning. *This average and vague understanding of Being is a fact.*

No matter how much this understanding of Being wavers and fades and borders on mere verbal knowledge, this indefiniteness of the understanding of Being that is always already available is itself a positive phenomenon which needs elucidation. However, an investigation of the meaning of Being will not wish to provide this at the outset. The interpretation of the average understanding of Being attains its necessary guideline only with the developed concept of Being. From the clarity of that concept and the appropriate manner of its explicit understanding we shall be able to discern what the obscure or not yet elucidated understanding of Being means, what kinds of obscuration or hindrance of an explicit elucidation of the meaning of Being are possible and necessary.

Furthermore, the average, vague understanding of Being can be permeated by traditional theories and opinions about Being in such a way that these theories, as the sources of the prevailing understanding, remain hidden. What is sought in the question of Being is not something completely unfamiliar, although it is at first totally ungraspable.

What is *asked about* in the question to be elaborated is Being, that which determines beings as beings, that in terms of which beings have always been understood no matter how they are discussed. The Being of beings "is" itself not a being. The first philosophical step in understanding the problem of Being consists in avoiding the *mython tina diēgeisthai,*[5] in not "telling a story," i.e., not determining beings as beings by tracing them back in their origins to another being—as if Being had the character of a possible being. As what is asked about, Being thus requires its own kind of demonstration which is essentially different from discovery of beings. Hence what is to be *ascertained*, the meaning of Being, will require its own conceptualization, which

5. Plato, *Sophist* 242 c.

again is essentially distinct from the concepts in which beings receive their meaningful determination.

Insofar as Being constitutes what is asked about, and insofar as Being means the Being of beings, beings themselves turn out to be what is *interrogated* in the question of Being. Beings are, so to speak, interrogated with regard to their Being. But if they are to exhibit the characteristics of their Being without falsification they must for their part have become accessible in advance as they are in themselves. The question of Being demands that the right access to beings be gained and secured in advance with regard to what it interrogates. But we call many things "in being" [*seiend*], and in different senses. Everything we talk about, mean, and are related to in such and such a way is in being. What and how we ourselves are is also in being. Being is found in thatness and whatness, reality, the being at hand of things [*Vorhandenheit*], subsistence, validity, existence [*Dasein*], and in the "there is" [*es gibt*]. In *which* being is the meaning of Being to be found; from which being is the disclosure of Being to get its start? Is the starting point arbitrary, or does a certain being have priority in the elaboration of the question of Being? Which is this exemplary being and in what sense does it have priority?

If the question of Being is to be explicitly formulated and brought to complete clarity concerning itself, then the elaboration of this question requires, in accord with what has been elucidated up to now, explication of the ways of regarding Being and of understanding and conceptually grasping its meaning, preparation of the possibility of the right choice of the exemplary being, and elaboration of the genuine mode of access to this being. Regarding, understanding and grasping, choosing, and gaining access to, are constitutive attitudes of inquiry and are thus themselves modes of being of a definite being, of *the* being we inquirers ourselves in each case are. Thus to work out the question of Being means to make a being—he who questions—perspicuous in his Being. Asking this question, as a mode of *being* of a being, is itself essentially determined by what is asked about in it—Being. This being which we ourselves in each case are and which includes inquiry among the possibilities of its Being we formulate terminologically as *Dasein*. The explicit and lucid formulation of the question of the meaning of Being requires a prior suitable explication of a being (Dasein) with regard to its Being.*

But does not such an enterprise fall into an obvious circle? To have to determine beings *in their Being* beforehand and then on this foundation first ask the question of Being—what else is that but going around in circles? In working out the question do we not presuppose something that only the answer can provide? Formal objections such as the argument of "circular reasoning," an argument that is always easily raised in the area of investigation of principles, are always sterile when one is weighing concrete ways of investigating. They do not offer anything to the understanding of the issue and they hinder penetration into the field of investigation.

But in fact there is no circle at all in the formulation of our question. Beings can be determined in their Being without the explicit concept of the meaning of Being having to be already available. If this were not so there could not have been as yet any ontological knowledge, and probably no one would deny the factual existence of such knowledge. It is true that "Being" is "presupposed"

*Since the "rationalist school" of Christian Wolff (1679–1754), *Dasein* has been widely used in German philosophy to mean the "existence" (or *Dass-sein*, "that it is"), as opposed to the "essence" (or *Was-sein*, "what it is") of a thing, state of affairs, person, or God. The word connotes especially the existence of living creatures—around 1860 Darwin's "struggle for life" was translated as *Kampf ums Dasein*—and most notably of human beings. Heidegger thus stresses the word's primary nuance: for him Dasein is that kind of existence that is always involved in an understanding of its Being. It must never be confused with the existence of things that lie before us and are on hand or at hand as natural or cultural objects (*Vorhandenheit*, *Zuhandenheit*). In order to stress the special meaning Dasein has for him, Heidegger often hyphenates the word (Da-sein), suggesting "there being," which is to say, the openness to Being characteristic of human existence, which is "there" in the world. (The hyphenated form appears in chapter five of *Being and Time* and in many of the later writings, some of which are included in this volume.) We will follow tradition and let the German word *Dasein* or *Da-sein* stand, translating the former as "existence" or "human being" only when the usage seems to be nonterminological. Finally, in light of Heidegger's interpretation of Being as presence, we note that *Dasein* originally (around 1700) meant nothing more or less than such presence, *Anwesenheit*.—ED.

in all previous ontology, but not as an available *concept*—not as the sort of thing we are seeking. "Presupposing" Being has the character of taking a preliminary look at Being in such a way that on the basis of this look beings that are already given are tentatively articulated in their Being. This guiding look at Being grows out of the average understanding of Being in which we are always already involved *and which ultimately belongs to the essential constitution of Dasein itself.* Such "presupposing" has nothing to do with positing a principle from which a series of propositions is deduced. A "circle in reasoning" cannot possibly lie in the formulation of the question of the meaning of Being, because in answering this question it is not a matter of grounding in deduction but rather of laying bare and exhibiting the ground.

A "circle in reasoning" does not occur in the question of the meaning of Being. Rather, there is a notable "relatedness backward or forward" of what is asked about (Being) to asking as a mode of being of a being. The way what is questioned essentially engages our questioning belongs to the most proper meaning of the question of Being. But this only means that the being that has the character of Dasein has a relation to the question of Being itself, perhaps even a distinctive one. But have we not thereby demonstrated that a particular being has a priority with respect to Being and that the exemplary being that is to function as what is primarily *interrogated* is pregiven? In what we have discussed up to now neither has the priority of Dasein been demonstrated nor has anything been decided about its possible or even necessary function as the primary being to be interrogated. But indeed something like a priority of Dasein has announced itself.

### 3. The ontological priority of the question of Being

Under the guideline of the formal structure of the question as such, the characteristics of the question of Being have made it clear that this question is a unique one, in such a way that its elaboration and indeed solution require a series of fundamental reflections. However, what is distinctive about the question of Being will fully come to light only when that question is sufficiently delineated with regard to its function, intention, and motives.

Up to now the necessity of a recovery of the question was motivated partly by the dignity of its origin but above all by the lack of a definite answer, even by the lack of any adequate formulation. But one can demand to know what purpose this question should serve. Does it remain solely, or *is* it at all, only a matter of free-floating speculation about the most general generalities—*or is it the most basic and at the same time most concrete question?*

Being is always the Being of a being. The totality of beings can, with respect to its various domains, become the field where definite areas of knowledge are exposed and delimited. These areas of knowledge—for example, history, nature, space, life, human being, language, and so on—can in their turn become thematic objects of scientific investigations. Scientific research demarcates and first establishes these areas of knowledge in a rough and ready fashion. The elaboration of the area in its fundamental structures is in a way already accomplished by prescientific experience and interpretation of the domain of Being to which the area of knowledge is itself confined. The resulting "fundamental concepts" comprise the guidelines for the first concrete disclosure of the area. Whether or not the importance of the research always lies in such establishment of concepts, its true progress comes about not so much in collecting results and storing them in "handbooks" as in being forced to ask questions about the basic constitution of each area, these questions being chiefly a reaction to increasing knowledge in each area.

The real "movement" of the sciences takes place in the revision of these basic concepts, a revision which is more or less radical and lucid with regard to itself. A science's level of development is determined by the extent to which it is *capable* of a crisis in its basic concepts. In these immanent crises of the sciences the relation of positive questioning to the matter in question becomes unstable.

Today tendencies to place research on new foundations have cropped up on all sides in the various disciplines.

The discipline which is seemingly the strictest and most securely structured, mathematics, has experienced a "crisis in its foundations." The controversy between formalism and intuitionism centers on obtaining and securing primary access to what should be the proper object of this science. Relativity theory in physics grew out of the tendency to expose nature's own coherence as it is "in itself." As a theory of the conditions of access to nature itself it attempts to preserve the immutability of the laws of motion by defining all relativities; it is thus confronted by the question of the structure of its given area of knowledge, i.e., by the problem of matter. In biology the tendency has awakened to get behind the definitions mechanism and vitalism have given to "organism" and "life" and to define anew the kind of Being of living beings as such. In the historical and humanistic disciplines the drive toward historical actuality itself has been strengthened by the transmission and portrayal of tradition: the history of literature is to become the history of critical problems. Theology is searching for a more original interpretation of man's being toward God, prescribed by the meaning of faith and remaining within it. Theology is slowly beginning to understand again Luther's insight that its system of dogma rests on a "foundation" that does not stem from a questioning in which faith is primary and whose conceptual apparatus is not only insufficient for the range of problems in theology but rather covers them up and distorts them.

Fundamental concepts are determinations in which the area of knowledge underlying all the thematic objects of a science attain an understanding that precedes and guides all positive investigation. Accordingly these concepts first receive their genuine evidence and "grounding" in a correspondingly preliminary research into the area of knowledge itself. But since each of these areas arises from the domain of beings themselves, this preliminary research that creates the fundamental concepts amounts to nothing else than interpreting these beings in terms of the basic constitution of their Being. This kind of investigation must precede the positive sciences—and it *can* do so. The work of Plato and Aristotle is proof of this. Laying the foundations of the sciences in this way is different in principle from "logic" limping along behind, investigating here and there the status of a science in terms of its "method." Such laying of foundations is productive logic in the sense that it leaps ahead, so to speak, into a definite realm of Being, discloses it for the first time in its constitutive Being, and makes the acquired structures available to the positive sciences as lucid directives for inquiry. Thus, for example, what is philosophically primary is not a theory of concept-formation in historiology, nor the theory of historical knowledge, nor even the theory of history as the object of historiology; what is primary is rather the interpretation of properly historical beings with regard to their historicity. Similarly, the positive result of Kant's *Critique of Pure Reason* consists in its beginning to work out what belongs to any nature whatsoever, and not in a "theory" of knowledge. His transcendental logic is an *a priori* logic of the realm of Being called nature.

But such inquiry—ontology taken in its broadest sense without reference to specific ontological directions and tendencies—itself still needs a guideline. It is true that ontological inquiry is more original than the ontic inquiry of the positive sciences. But it remains naive and opaque if its investigations into the Being of beings leave the meaning of Being in general undiscussed. And precisely the ontological task of a genealogy of the different possible ways of Being (which is not to be construed deductively) requires a preliminary understanding of "what we properly mean by this expression 'Being.'"

The question of Being thus aims at an *a priori* condition of the possibility not only of the sciences which investigate beings of such and such a type—and are thereby already involved in an understanding of Being; but it aims also at the condition of the possibility of the ontologies which precede the ontic sciences and found them.

*All ontology, no matter how rich and tightly knit a system of categories it has at its disposal, remains fundamentally blind and perverts its most proper intent if it has not previously clarified the meaning of Being sufficiently and grasped this clarification as its fundamental task.*

Ontological research itself, correctly understood, gives the question of Being its ontological priority over and above merely resuming an honored tradition and making progress on a problem until now opaque. But this scholarly, scientific priority is not the only one.

## 4. The ontic* priority of the question of Being

Science in general can be defined as the totality of fundamentally coherent true propositions. This definition is not complete, nor does it get at the meaning of science. As ways in which man behaves, sciences have this being's (man's) kind of Being. We are defining this being terminologically as *Dasein*. Scientific research is neither the sole nor the primary kind of possible Being of this being. Moreover, Dasein itself is distinctly different from other beings. We must make this distinct difference visible in a preliminary way. Here the discussion must anticipate subsequent analyses, which only later will become properly demonstrative.

Dasein is a being that does not simply occur among other beings. Rather it is ontically distinguished by the fact that in its Being this being is concerned *about* its very Being. Thus it is constitutive of

*Throughout *Being and Time* Heidegger contrasts the "ontic" to the "ontological." As we have seen, "ontological" refers to the Being of beings (*onta*) or to any account (*logos*) of the same; hence it refers to a particular discipline (traditionally belonging to metaphysics) or to the content or method of this discipline. On the contrary, "ontic" refers to any manner of dealing with beings that does *not* raise the ontological question. Most disciplines and sciences remain "ontic" in their treatment of beings. What it means to speak of the "ontic priority" of the question of the meaning of Being—a paradox that should give us pause—the present section elucidates. Compare the parallel but not identical opposition of "existentiell" and "existential" in this same section, below.—Ed.

the Being of Dasein to have, in its very Being, a relation of Being to this Being. And this in turn means that Dasein understands itself in its Being in some way and with some explicitness. It is proper to this being that it be disclosed to itself with and through its Being. *Understanding of Being is itself a determination of the Being of Dasein.* The ontic distinction of Dasein lies in the fact that it *is* ontological.

To be ontological does not yet mean to develop ontology. Thus if we reserve the term ontology for the explicit, theoretical question of the meaning of beings, the intended ontological character of Dasein is to be designated as pre-ontological. That does not signify being simply ontical, but rather being in the manner of an understanding of Being.

We shall call the very Being to which Dasein can relate in one way or another, and somehow always does relate, existence [*Existenz.*] And because the essential definition of this being cannot be accomplished by ascribing to it a "what" that specifies its material content, because its essence lies rather in the fact that it has always to be its Being as its own, the term Dasein, as a pure expression of Being, has been chosen to designate this being.

Dasein always understands itself in terms of its existence, in terms of its possibility to be itself or not to be itself. Dasein has either chosen these possibilities itself, stumbled upon them, or already grown up in them. Existence is decided only by each Dasein itself in the manner of seizing upon or neglecting such possibilities. We come to terms with the question of existence always only through existence itself. We shall call *this* kind of understanding of itself *existentiell* understanding. The question of existence is an ontic "affair" of Dasein. For this the theoretical perspicuity of the ontological structure of existence is not necessary. The question of structure aims at the analysis of what constitutes existence. We shall call the coherence of these structures *existentiality.* Its analysis does not have the character of an existentiell understanding but rather an *existential* one. The task of an existential analysis of Dasein is prescribed with regard to its possibility and necessity in the ontic constitution of Dasein.*

ut since existence defines Dasein, the ontological analysis of this being always requires a previous glimpse of existentiality. However, we understand existentiality as the constitution-of-Being of the being that exists. But the idea of Being already lies in the idea of such a constitution of Being. And thus the possibility of carrying out the analysis of Dasein depends upon the prior elaboration of the question of the meaning of Being in general.

Sciences and disciplines are ways of being of Dasein in which Dasein relates also to beings that it need not itself be. But *being in a world* belongs essentially to Dasein. Thus the understanding of Being that belongs to Dasein just as originally implies the understanding of something like "world" and the understanding of the Being of beings accessible within the world. Ontologies that have beings unlike Dasein as their theme are accordingly founded and motivated in the ontic structure of Dasein itself. This structure includes in itself the determination of a pre-ontological understanding of Being.

Thus *fundamental ontology,* from which alone all other ontologies can originate, must be sought in the *existential analysis of Dasein.*

Dasein accordingly takes priority in several ways over all other beings. The first priority is an *ontic* one: this being is defined in its Being by existence. The second priority is an *ontological* one: on the basis of its determination as existence Dasein is in itself "ontological." But just as originally Dasein possesses—in a manner constitutive of its understanding of existence—an understanding of the Being of all beings unlike itself. Dasein therefore has its third priority as the ontic-ontological condition of the possibility of all

*Heidegger coins the term *existentiell* (here translated as "existentiell") to designate the way Dasein in any given case actually exists by realizing or ignoring its various possibilities—in other words, by living its life. *One* of those possibilities is to inquire into the *structure* of its life and possibilities; the kind of understanding thereby gained Heidegger calls *existenzial* (here translated as "existential"). The nexus of such structures he call *Existentialität* (here translated as "existentiality").—Ed.

ontologies. Dasein has proven to be what, before all other beings, is ontologically the primary being to be interrogated.

However, the roots of the existential analysis, for their part, are ultimately *existentiell*—they are *ontic*. Only when philosophical research and inquiry themselves are grasped in an existentiell way—as a possibility of being of each existing Dasein—does it become possible at all to disclose the existentiality of existence and therewith to get hold of a sufficiently grounded set of ontological problems. But with this the ontic priority of the question of Being as well has become clear.

The ontic-ontological priority of Dasein was already seen early on, without Dasein itself being grasped in its genuine ontological structure or even becoming a problem with such an aim. Aristotle says, *hē psychē ta onta pōs estin*.[6] The soul (of man) is in a certain way beings. The "soul" which constitutes the Being of man discovers in its ways to be—*aisthēsis* and *noēsis*—all beings with regard to their thatness and whatness, that is to say, always also in their Being. Thomas Aquinas discussed this statement—which refers back to Parmenides' ontological thesis—in a manner characteristic of him. Thomas is engaged in the task of deriving the "transcendentals," i.e., the characteristics of Being that lie beyond every possible generic determination of a being in its material content, every *modus specialis entis*, and that are necessary attributes of every "something," whatever it might be. For him the *verum* too is to be demonstrated as being such a *transcendens*. This is to be accomplished by appealing to a being which in conformity with its kind of Being is suited to "come together" with any being whatsoever. This distinctive being, the *ens quod natum est convenire cum omni ente* ["the being whose nature it is to meet with all other beings"], is the soul (*anima*).[7] The priority of Dasein over and above

6. *De anima*, III, 8, 431b 21; cf. ibid., III, 5, 430a 14ff. [The Teubner edition which Heidegger cites removes the *panta* from this famous phrase, which in most English editions reads, "The soul is in a certain way *all* beings."—ED.]

7. *Quaestiones de veritate*, Qu. 1, a. 1 c; cf. the occasionally stricter exposition, which deviates from what was cited, of a "deduction" of the transcendentals in the brief work *De natura generis*.

all other beings which emerges here without being ontologically clarified obviously has nothing in common with a vapid subjectivizing of the totality of beings.

The demonstration of the ontic-ontological distinctiveness of the question of Being is grounded in the preliminary indication of the ontic-ontological priority of Dasein. But the analysis of the structure of the question of Being as such (section 2) came up against the distinctive function of this being within the formulation of that very question. Dasein revealed itself to be that being which must first be elaborated in a sufficiently ontological manner if the inquiry is to become a lucid one. But now it has become evident that the ontological analysis of Dasein in general constitutes fundamental ontology, that Dasein consequently functions as the being that is to be *interrogated* fundamentally in advance with respect to its Being.

If the interpretation of the meaning of Being is to become a task, Dasein is not only the primary being to be interrogated; in addition to this it is the being that always already in its Being is related to *what is sought* in this question. But then the question of Being is nothing else than the radicalization of an essential tendency of Being that belongs to Dasein itself, namely, of the pre-ontological understanding of Being.

## Chapter Two
## *The Double Task in Working Out the Question of Being*
## *The Method of the Investigation and Its Outline*

### 5. The ontological analysis of Dasein as exposure of the horizon for an interpretation of the meaning of Being in general

In designating the tasks that lie in "formulating" the question of Being, we showed that not only must we pinpoint the particular being that is to function as the primary object of interrogation but also that an explicit appropriation and securing of correct access to this being is required. We discussed which being it is that takes over the major role within the question of Being. But how should this being, Dasein, become accessible and, so to speak, be envisaged in a perceptive interpretation?

The ontic-ontological priority that has been demonstrated for Dasein could lead to the mistaken opinion that this being would have to be what is primarily given also ontically-ontologically, not only in the sense that such a being could be grasped immediately but also that the prior givenness of its manner of being would be just as "immediate." True, Dasein is ontically not only what is near or even nearest—we ourselves *are* it in each case. Nevertheless, or precisely for this reason, it is ontologically what is farthest away. True, it belongs to its most proper Being to have an understanding of this Being and to sustain a certain interpretation of it. But this does not at all mean that the most readily available pre-ontological interpretation of its own Being could be adopted as an adequate guideline, as though this understanding of Being perforce stemmed from a thematic ontological reflection on the most proper constitution of its Being. Rather, in accordance with the manner of being belonging to it, Dasein tends to understand its own Being in terms of that being to which it is essentially, continually, and most closely related—the "world." In Dasein itself and therewith in its own understanding of Being, as we shall show, the way the world is understood is ontologically reflected back upon the interpretation of Dasein.

The ontic-ontological priority of Dasein is therefore the reason why the specific constitution of the Being of Dasein—understood in the sense of the "categorial" structure that belongs to it—remains hidden from it. Dasein is ontically "closest" to itself, while ontologically farthest away; but pre-ontologically it is surely not foreign to itself.

For the time being we have only indicated that an interpretation of this being is confronted with peculiar difficulties rooted in the mode of being of the thematic object and the way it is thematized. These difficulties do not result from some shortcoming of our powers of knowledge or lack of a suitable way of conceiving—a lack seemingly easy to remedy.

Not only does an understanding of Being belong to Dasein, but this understanding also develops or decays according to the actual manner of being of Dasein at any given time; for this reason it has a wealth of interpretations at its disposal. Philosophical psychology, anthropology, ethics, "politics," poetry, biography, and the discipline of history pursue in different ways and to varying extents the behavior, faculties, powers, possibilities, and vicissitudes of Dasein. But the question remains whether these interpretations were carried out in as original an existential manner as their existentiell originality perhaps merited. The two do not necessarily go together, but they also do not exclude one another. Existentiell interpretation can require existential analysis, provided philosophical knowledge is understood in its possibility and necessity. Only when the fundamental structures of Dasein are adequately worked out with explicit orientation toward the problem of Being will the previous results of the interpretation of Dasein receive their existential justification.

Hence the first concern in the question of Being must be an analysis of Dasein. But then the problem of gaining and securing the access that leads to Dasein becomes really crucial. Expressed negatively, no arbitrary idea of Being and reality, no matter how "self-evident" it is, may be brought to bear on this being in a dogmatically constructed way; no "categories" prescribed by such ideas may be forced upon Dasein without ontological deliberation. The manner of access and interpretation must instead be chosen in such a way that this being can show itself to itself on its own terms. Furthermore, this manner should show that being as it is *at first and for the most part*—in its average *everydayness*. Not arbitrary and accidental structures but essential ones are to be demonstrated in this everydayness, structures that remain determinative in every mode of being of factual Dasein. By looking at the fundamental constitution of the everydayness of Dasein we shall bring out in a preparatory way the Being of this being.

The analysis of Dasein thus understood is wholly oriented toward the guiding task of working out the question of Being. Its limits are thereby determined. It cannot hope to provide a complete ontology of Dasein, which of course must be supplied if something like a "philosophical" anthropology is to rest on a philosophically adequate basis. With a view to a possible anthropology or its ontological foundation, the following interpretation will provide only a few "parts," although not unessential ones. However, the analysis of Dasein is not only incomplete but at first also *preliminary*. It only brings out the Being of this being, without interpreting its meaning. Its aim is rather to expose the horizon for the most original interpretation of Being. Once we have reached that horizon the preparatory analysis of Dasein requires recovery on a higher, properly ontological basis.

The meaning of the Being of that being we call Dasein proves to be *temporality* [*Zeitlichkeit*]. In order to demonstrate this we must recover our interpretation of those structures of Dasein that shall have been indicated in a preliminary way—this time as modes of temporality. While it is true that with this interpretation of Dasein as temporality the answer to the guiding question about the meaning of Being in general is not given as such, the soil from which we may reap it will nevertheless be prepared.

We intimated that a pre-ontological Being belongs to Dasein as its ontic constitution. Dasein *is* in such a way that, by being, it understands something like Being. Remembering this connection, we must show that *time* is that from which Dasein tacitly understands and interprets something like Being at all. Time must be brought to light and genuinely grasped as the horizon of every understanding and interpretation of Being. For this to become clear we need an *original explication of time as the horizon of the understanding of Being, in terms of temporality as the Being of Dasein which understands Being.* This task as a whole requires that the concept of time thus gained be distinguished from the common understanding of it. The latter has become explicit in an interpretation of time which reflects the traditional concept that has persisted since Aristotle and beyond Bergson. We must thereby make clear that and in what way this concept of time and the common understanding of time in general originate from temporality. In this way the common concept of time receives again its rightful autonomy—contrary to Bergson's thesis that time understood in the common way is really space.

For a long while, "time" has served as the ontological—or rather ontic—criterion for naively distinguishing the different regions of beings. "Temporal" beings (natural processes and historical events) are separated from "atemporal" beings (spatial and numerical relationships). We are accustomed to distinguishing the "timeless" meaning of propositions from the "temporal" course of propositional statements. Further, a "gap" between "temporal" being and "supratemporal" eternal being is found, and the attempt made to bridge the gap. "Temporal" here means as much as being "in time," an obscure enough definition to be sure. The fact remains that time in the sense of "being in time" serves as a criterion for separating the regions of Being. How time comes to have this distinctive ontological function, and even with what right precisely something like time serves as such a criterion, and most of all whether in this naive ontological application of time its genuinely possible ontological relevance is expressed, has neither been asked nor investigated up to now. "Time," especially on the horizon of the common understanding of it, has chanced to acquire this "obvious" ontological function "of itself," as it were, and has retained it to the present day.

In contrast we must show, on the basis of the question of the meaning of Being which shall have been worked out, *that—and in what way—the central range of problems of all ontology is rooted in the phenomenon of time correctly viewed and correctly explained.*

If Being is to be conceived in terms of time, and if the various modes and derivatives of Being, in their modifications and derivations, are in fact to become intelligible through consideration of time, then Being itself—and not only beings that are "in time"—is made visible in its "temporal" ["*zeitlich*"] character. But then "temporal" can no longer mean only "being in time." The "atemporal" and the "supratemporal" are also "temporal" with respect to their Being; this not only by way of privation when compared to "temporal" beings which are "in time," but in a *positive* way which, of course, must first be clarified. Because the expression "temporal" belongs to both prephilosophical and philosophical usage, and because that expression will be used in a different sense in the following investigations, we shall call the original determination of the meaning of Being and its characters and modes which devolve from time its *Temporal* [temporale] determination. The fundamental ontological task of the interpretation of Being as such thus includes the elaboration of the *Temporality of Being* [Temporalität des Seins.] In the exposition of the problem of Temporality the concrete answer to the question of the meaning of Being is first given.

Because Being is comprehensible only on the basis of the consideration of time, the answer to the question of Being cannot lie in an isolated and blind proposition. The answer is not grasped by repeating what is stated propositionally, especially when it is transmitted as a free-floating result, so that we merely take notice of a standpoint which perhaps deviates from the way the matter has been previously treated. Whether the answer is "novel" is of no importance and remains extrinsic. What is positive about the answer must lie in the fact that it is *old* enough to enable us to learn to comprehend possibilities prepared by the "ancients." In conformity to its most proper sense, the answer provides a directive for concrete ontological research, that is, a directive to begin its investigative inquiry within the horizon exhibited—and that is all it provides.

If the answer to the question of Being thus becomes the guiding directive for research, then it is sufficiently given only if the specific mode of being of previous ontology—the vicissitudes of its questioning, its findings, and its failures—becomes visible as necessary to the very character of Dasein.

### 6. The task of a destructuring of the history of ontology*

All research—especially when it moves in the sphere of the central question of Being—is an ontic possibility of Dasein. The Being of Dasein finds its meaning in temporality. But temporality is at the same time the condition of the possibility of historicity as a temporal mode of being of Dasein itself, regardless of whether and how it is a being "in time." As a determination, historicity is prior to what is called history (world-historical occurrences). Historicity means the constitution of Being of the "occurrence" of Dasein as such; it is the ground for the fact that something like the discipline of "world history" is at all possible and historically belongs to world history. In its factual Being Dasein always is as and "what" it already was. Whether explicitly or not, it *is* its past. It is its own past not only in such a way that its past, as it were, pushes itself along "behind" it, and that it possesses what is past as a property that is still at hand and occasionally has an effect on it. Dasein "is" its past in the manner of *its* Being which, roughly expressed, actually "occurs" out of its future. In its manner of being at any given time, and accordingly also with the understanding of Being that belongs to it, Dasein grows into a customary interpretation of itself and grows up in that interpretation. It understands itself in terms of this interpretation at first, and within a certain range, constantly. This understanding discloses the possibilities of its Being and regulates them. Its own past—and that always means that of its "generation"—does not *follow after* Dasein but rather always goes already ahead of it.

*Heidegger's word *Destruktion* does not mean "destruction" in the usual sense—which the German word *Zerstörung* expresses. The word *destructuring* should serve to keep the negative connotations at a distance and to bring out the neutral, ultimately constructive, sense of the original.—TR./ED.

This elemental historicity of Dasein can remain concealed from it. But it can also be discovered in a certain way and be properly cultivated. Dasein can discover, preserve, and explicitly pursue tradition. The discovery of tradition and the disclosure of what it "transmits," and how it does this, can be undertaken as an independent task. In this way Dasein advances to the mode of being of historical inquiry and research. But the discipline of history—more precisely, the historicality underlying it—as the manner of being of inquiring Dasein, is possible only because Dasein is determined by historicity in the ground of its Being. If historicity remains concealed from Dasein, and so long as it does so, the possibility of historical inquiry and discovery of history is denied it. If the discipline of history is lacking, that is no evidence *against* the historicity of Dasein; rather it is evidence for this constitution-of-Being in a deficient mode. Only because it is "historic" in the first place can an age lack the discipline of history.

On the other hand, if Dasein has seized upon its inherent possibility not only of making its existence perspicuous but also of inquiring into the meaning of existentiality itself, that is to say, of provisionally inquiring into the meaning of Being in general; and if insight into the essential historicity of Dasein has opened up in such inquiry; then it is inevitable that inquiry into Being, which was designated with regard to its ontic-ontological necessity, is itself characterized by historicity. The elaboration of the question of Being must therefore receive its directive to inquire into its own history from the most proper ontological sense of the inquiry itself, as a historical one; that means to become historical in order to come to the positive appropriation of the past, to come into full possession of its most proper possibilities of inquiry. The question of the meaning of Being is led to understand itself as historical in accordance with its own way of proceeding, i.e., as the provisional explication of Dasein in its temporality and historicity.

However, the preparatory interpretation of the fundamental structures of Dasein with regard to its usual and average way of being—in which it is also first of all historical—will make the following clear: Dasein not only has the inclination to be ensnared in the world in which it is and to interpret itself in terms of that world by its reflected light; at the same time Dasein is also ensnared in a tradition which it more or less explicitly grasps. This tradition deprives Dasein of its own leadership in questioning and choosing. This is especially true of *that* understanding (and its possible development) which is rooted in the most proper Being of Dasein—the ontological understanding.

The tradition that hereby gains dominance makes what it "transmits" so little accessible that at first and for the most part it covers it over instead. What has been handed down it hands over to obviousness; it bars access to those original "wellsprings" out of which the traditional categories and concepts were in part genuinely drawn. The tradition even makes us forget such a provenance altogether. Indeed it makes us wholly incapable of even understanding that such a return is necessary. The tradition uproots the historicity of Dasein to such a degree that it only takes an interest in the manifold forms of possible types, directions, and standpoints of philosophizing in the most remote and strangest cultures, and with this interest tries to veil its own lack of foundation. Consequently, in spite of all historical interest and zeal for a philologically "viable" interpretation, Dasein no longer understands the most elementary conditions which alone make a positive return to the past possible—in the sense of its productive appropriation.

At the outset (section 1) we showed that the question of the meaning of Being was not only unresolved, not only inadequately formulated, but in spite of all interest in "metaphysics" has even been forgotten. Greek ontology and its history, which through many twists and turns still determine the conceptual character of philosophy today, are proof of the fact that Dasein understands itself and Being in general in terms of the "world." The ontology that thus arises is ensnared by the tradition, which allows it to sink to the level of the obvious and become mere material for reworking (as it was for Hegel).

Greek ontology thus uprooted becomes a fixed body of doctrine in the Middle Ages. But its systematics is not at all a mere joining together of traditional elements into a single structure. Within the limits of its dogmatic adoption of the fundamental Greek interpretations of Being, this systematics contains a great deal of unpretentious work which does make advances. In its *scholastic* mold, Greek ontology makes the essential transition via the *Disputationes metaphysicae* of Suarez into the "metaphysics" and transcendental philosophy of the modern period; it still determines the foundations and goals of Hegel's *Logic*. Insofar as certain distinctive domains of Being become visible in the course of this history and henceforth chiefly dominate the range of problems (Descartes's *ego cogito*, subject, the "I," reason, spirit, person), the beings just cited remain unquestioned with respect to the Being and structure of their being, this corresponding to the thorough neglect of the question of Being. But the categorial content of traditional ontology is transferred to these beings with corresponding formalizations and purely negative restrictions, or else dialectic is called upon to help with an ontological interpretation of the substantiality of the subject.

If the question of Being is to achieve clarity regarding its own history, a loosening of the sclerotic tradition and a dissolving of the concealments produced by it are necessary. We understand this task as the *destructuring* of the traditional content of ancient ontology, which is to be carried out along the *guidelines of the question of Being*. This destructuring is based on the original experiences in which the first and subsequently guiding determinations of Being were gained.

This demonstration of the provenance of the fundamental ontological concepts, as the investigation that displays their "birth certificate," has nothing to do with a pernicious relativizing of ontological standpoints. The destructuring has just as little the *negative* sense of disburdening ourselves of the ontological tradition. On the contrary, it should stake out the positive possibilities of the tradition, and that always means to fix its *boundaries*. These are factually given with the specific formulation of the question and the prescribed demarcation of the possible field of investigation. The destructuring is not related negatively to the past: its criticism concerns "today" and the dominant way we treat the history of ontology, whether it be conceived as the history of opinions, ideas, or problems. However, the destructuring does not wish to bury the past in nullity; it has a *positive* intent. Its negative function remains tacit and indirect.

The destructuring of the history of ontology essentially belongs to the formulation of the question of Being and is possible solely within such a formulation. Within the scope of this treatise, which has as its goal a fundamental elaboration of the question of Being, the destructuring can be carried out only with regard to the fundamentally decisive stages of that history.

In accord with the positive tendency of the destructuring, the question must first be asked whether and to what extent in the course of the history of ontology in general the interpretation of Being has been thematically connected with the phenomenon of time. We must also ask whether the range of problems concerning Temporality that necessarily belongs here was fundamentally worked out, or could have been. Kant is the first and only one who traversed a stretch of the path toward investigating the dimension of Temporality—or allowed himself to be driven there by the compelling force of the phenomena themselves. Only when the problem of Temporality is pinned down can we succeed in casting light on the obscurity of his doctrine of the schematism. Furthermore, in this way we can also show *why* this area had to remain closed to Kant in its proper dimensions and in its central ontological function. Kant himself knew that he was venturing forth into an obscure area: "This schematism of our understanding as regards appearances and their mere form is an art hidden in the depths of the human soul, the true devices of which are hardly ever to be divined from Nature and laid uncovered before our eyes."[1] What it is that Kant shrinks back from here, as it were, must be brought to light

1. Kant, *Critique of Pure Reason*, B 180–81.

thematically and in principle if the expression "Being" is to have a demonstrable meaning. Ultimately the phenomena to be explicated in the following analysis under the rubric "Temporality" are precisely those that determine the *most covert* judgments of "common reason," analysis of which Kant calls the "business of philosophers."

In pursuing the task of destructuring on the guideline of the problem of Temporality the following treatise will attempt to interpret the chapter on the schematism and the Kantian doctrine of time developed there. At the same time we must show why Kant could never gain insight into the problem of Temporality. Two things prevented this insight. On the one hand, the neglect of the question of Being in general, and in connection with this, the lack of a thematic ontology of Dasein—in Kantian terms, the lack of a prior ontological analysis of the subjectivity of the subject. Instead, Kant dogmatically adopted Descartes's position—notwithstanding all his essential advances. Despite his taking this phenomenon back into the subject, Kant's analysis of time remained oriented toward the traditional, common understanding of it. It is this that finally prevented Kant from working out the phenomenon of a "transcendental determination of time" in its own structure and function. As a consequence of this double effect of the tradition, the decisive *connection* between *time* and the "*I think*" remained shrouded in complete obscurity. It did not even become a problem.

By taking over Descartes's ontological position Kant neglects something essential: an ontology of Dasein. In terms of Descartes's innermost tendency this omission is a decisive one. With the *cogito sum* Descartes claims to prepare a new and secure foundation for philosophy. But what he leaves undetermined in this "radical" beginning is the manner of being of the *res cogitans*, more precisely, *the meaning of the Being of the* "*sum.*" Working out the tacit ontological foundations of the *cogito sum* will constitute the second stage of our destructuring of, and path back into, the history of ontology. The interpretation will demonstrate not only that Descartes had to neglect the question of Being altogether but also why he held the opinion that the absolute "certainty" of the *cogito* exempted him from the question of the meaning of the Being of this being.

However, with Descartes it is not just a matter of neglect and thus of a complete ontological indeterminateness of the *res cogitans sive mens sive animus* ["the thinking thing, whether it be mind or soul"]. Descartes carries out the fundamental reflections of his *Meditations* by applying medieval ontology to this being which he posits as the *fundamentum inconcussum* ["unshakable foundation"]. The *res cogitans* is ontologically determined as *ens*, and for medieval ontology the meaning of the Being of the *ens* is established in the understanding of it as *ens creatum*. As the *ens infinitum* God is the *ens increatum*. But createdness, in the broadest sense of something's being produced, is an essential structural moment of the ancient concept of Being. The ostensibly new beginning of philosophizing betrays the imposition of a fatal prejudice. On the basis of this prejudice later times neglect a thematic ontological analysis of "the mind" ["*Gemüt*"] which would be guided by the question of Being; likewise they neglect a critical confrontation with the inherited ancient ontology.

Everyone familiar with the medieval period sees that Descartes is "dependent" upon medieval scholasticism and uses its terminology. But with this "discovery" nothing is gained philosophically as long as it remains obscure to what a profound extent medieval ontology influences the way posterity determines or fails to determine ontologically the *res cogitans*. The full extent of this influence cannot be estimated until the meaning and limitations of ancient ontology have been shown by our orientation toward the question of Being. In other words, the destructuring sees itself assigned the task of interpreting the very basis of ancient ontology in light of the problem of Temporality. Here it becomes evident that the ancient interpretation of the Being of beings is oriented toward the "world" or "nature" in the broadest sense and that it indeed gains its understanding of Being from "time." The outward evidence of this—but of course *only* outward—is the determination of the meaning of

Being as *parousia* or *ousia*, which means ontologically and temporally "presence" ["*Anwesenheit*"]. Beings are grasped in their Being as "presence"; that is to say, they are understood with regard to a definite mode of time, the *present*.

The problem of Greek ontology must, like that of any other, take its guideline from Dasein itself. In the ordinary and also the philosophical "definition," Dasein, that is, the Being of man, is delineated as *zōon logon echon*, that creature whose Being is essentially determined by its being able to speak. *Legein* (see section 7 B) is the guideline for arriving at the structures of Being of the beings we encounter in discourse and discussion. That is why the ancient ontology developed by Plato becomes "dialectic." The possibility of a more radical conception of the problem of Being grows with the continuing development of the ontological guideline itself, i.e., of the "hermeneutics" of the *logos*. "Dialectic," which was a genuine philosophic embarrassment, becomes superfluous. Aristotle has "no understanding of it" for *this* reason, that he places it on a more radical foundation and transcends it. *Legein* itself, or *noein*—the simple apprehension of something at hand in its pure being at hand [*Vorhandenheit*], which Parmenides already used as a guide for interpreting Being—has the Temporal structure of a pure "making present" of something. Beings, which show themselves in and for this making present and which are understood as beings proper, are accordingly interpreted with regard to the present; that is to say, they are conceived as presence (*ousia*).

However, this Greek interpretation of Being comes about without any explicit knowledge of the guideline functioning in it, without taking cognizance of or understanding the fundamental ontological function of time, without insight into the ground of the possibility of this function. On the contrary, time itself is taken to be one being among others. The attempt is made to grasp time itself in the structure of its Being on the horizon of an understanding of Being which is oriented toward time in an inexplicit and naive way.

Within the framework of the following fundamental elaboration of the question of Being we cannot offer a detailed Temporal interpretation of the foundations of ancient ontology—especially of its scientifically highest and purest stage, i.e., in Aristotle. Instead, we offer an interpretation of Aristotle's treatise on time,[2] which can be chosen as the way of discerning the basis and limits of the ancient science of Being.

Aristotle's treatise on time is the first detailed interpretation of this phenomenon that has come down to us. It essentially determined all the following interpretations, including that of Bergson. From our analysis of Aristotle's concept of time it becomes retrospectively clear that the Kantian conception moves within the structures developed by Aristotle. This means that Kant's fundamental ontological orientation—despite all the differences implicit in a new inquiry—remains the Greek one.

The question of Being attains true concreteness only when we carry out the destructuring of the ontological tradition. By so doing we can thoroughly demonstrate the inescapability of the question of the meaning of Being and so demonstrate the meaning of our talk about a "recovery" of the question.

In this field where "the matter itself is deeply veiled,"[3] any investigation will avoid overestimating its results. For such inquiry is constantly forced to face the possibility of disclosing a still more original and more universal horizon from which it could draw the answer to the question "What does 'Being' mean?" We can discuss such possibilities seriously and with a positive result only if the question of Being has been reawakened and we have reached the point where we can come to terms with it in a controlled fashion.

## 7. The phenomenological method of the investigation

With the preliminary characterization of the thematic object of the investigation (the Being of beings, or the meaning of Being in general) its method would appear to be already prescribed. The task of

2. Aristotle, *Physics*, IV, 10–14; 217b 29–224a 17.
3. Kant, *Critique of Pure Reason*, B 121.

ontology is to set in relief the Being of beings and to explicate Being. And the method of ontology remains questionable in the highest degree as long as we wish merely to consult historically transmitted ontologies or similar efforts. Since the term "ontology" is used in a formally broad sense for this investigation, the approach of clarifying its method by pursuing the history of that method is automatically precluded.

In using the term "ontology" we do not specify any definite philosophical discipline standing in relation to others. It should not at all be our task to satisfy the demands of any established discipline. On the contrary, such a discipline can be developed only from the compelling necessity of definite questions and procedures demanded by the "things themselves."

With the guiding question of the meaning of Being the investigation arrives at the fundamental question of philosophy in general. The treatment of this question is *phenomenological*. With this term the treatise dictates for itself neither a "standpoint" nor a "direction," because phenomenology is neither of these and can never be as long as it understands itself. The expression "phenomenology" signifies primarily a *concept of method*. It does not characterize the "what" of the objects of philosophical research in terms of their content but the "how" of such research. The more genuinely effective a concept of method is and the more comprehensively it determines the fundamental conduct of a science, the more originally is it rooted in confrontation with the things themselves and the farther away it moves from what we call a technical device—of which there are many in the theoretical disciplines.

The term "phenomenology" expresses a maxim that can be formulated: "To the things themselves!" It is opposed to all free-floating constructions and accidental findings; it is also opposed to taking over concepts only seemingly demonstrated; and likewise to pseudo-questions which often are spread abroad as "problems" for generations. But one might object that this maxim is, after all, abundantly self-evident and, moreover, an expression of the principle of all scientific knowledge. It is not clear why this commonplace should be explicitly put in the title of our research. In fact we are dealing with "something self-evident" which we want to get closer to, insofar as that is important for clarification of the procedure in our treatise. We shall explicate only the preliminary concept of phenomenology.

The expression has two components, phenomenon and logos. These go back to the Greek terms *phainomenon* and *logos*. Viewed extrinsically, the word "phenomenology" is formed like the terms theology, biology, sociology, translated as the science of God, of life, of the community. Accordingly, phenomenology would be the *science of phenomena*. The preliminary concept of phenomenology is to be exhibited by characterizing what is meant by the two components, phenomenon and logos, and by establishing the meaning of the *combined* word. The history of the word itself, which originated presumably with the Wolffian school, is not important here.

A. *The concept of phenomenon*

The Greek expression *phainomenon*, from which the term "phenomenon" derives, comes from the verb *phainesthai*, meaning "to show itself." Thus *phainomenon* means what shows itself, the self-showing, the manifest. *Phainesthai* itself is a "middle voice" construction of *phainō*, to bring into daylight, to place in brightness. *Phainō* belongs to the root *pha-*, like *phōs*, light or brightness, i.e., that within which something can become manifest, visible in itself. Thus the meaning of the expression "phenomenon" is *established* as *what shows itself in itself*, what is manifest. The *phainomena*, "phenomena," are thus the totality of what lies in the light of day or can be brought to light. Sometimes the Greeks simply identified this with *ta onta* (beings). Beings can show themselves from themselves in various ways, depending on the mode of access to them. The possibility even exists that they can show themselves as they are *not* in themselves. In this self-showing beings "look like. . . ." Such self-showing we call *seeming* [Scheinen]. And so the expression *phainomenon*, phenomenon, means in Greek: what looks like something, what "seems," "semblance." *Phainomenon agathon* means a good that looks like—but "in reality" is not what it gives itself out to be. It is extremely important for further understanding of the concept of phenomenon to see how what is named in both meanings of *phainomenon* ("phenomenon" as self-showing and "phenomenon" as semblance) are structurally connected. Only because something claims to show itself in accordance with its meaning at all, that is, claims to be a phenomenon, *can* it show itself *as* something it is *not*, or *can* it "only look like. . . ." The original meaning (phenomenon, what is manifest) already contains and is the basis of *phainomenon* ("semblance"). We attribute to the term "phenomenon" the positive and original meaning of *phainomenon* terminologically, and separate the phenomenon of semblance from it as a privative modification. But what *both* terms express has at first nothing at all to do with what is called "appearance" or even "mere appearance."

One speaks of "appearances or symptoms of illness." What is meant by this are occurrences in the body that show themselves and in this self-showing as such "indicate" something that does *not* show itself. When such occurrences emerge, their self-showing coincides with the being at hand [*Vorhandensein*] of disturbances that do not show themselves. Appearance, as the appearance "of something," thus precisely does *not* mean that something shows itself; rather, it means that something makes itself known which does not show itself. It makes itself known through something that does show itself. Appearing is a *not showing itself*. But this "not" must by no means be confused with the privative not which determines the structure of semblance. What does *not* show itself, in the manner of what appears, can also never seem. All indications, presentations, symptoms, and symbols have the designated formal, fundamental structure of appearing, although they do differ among themselves.

Although "appearing" is never a self-showing in the sense of phenomenon, appearing is possible only *on the basis* of a *self-showing* of something. But this, the self-showing that makes appearing possible, is not appearing itself. Appearing is a *making itself known* through something that shows itself. If we then say that with the word "appearance" we are pointing to something in which something appears without itself being an appearance, then the concept of phenomenon is not thereby delimited but *presupposed*. However, this presupposition remains hidden because the expression "to appear" in this definition of "appearance" is used in an equivocal sense. That in which something "appears" means that in which something makes itself known, that is, does not show itself; in the expression "without itself being an 'appearance'" appearance means the *self-showing*. But this self-showing essentially belongs to the "wherein" in which something makes itself known. Accordingly, phenomena are *never* appearances, but every appearance is dependent upon phenomena. If we define phenomenon with the help of a concept of "appearance" that is still unclear, then everything is turned upside down, and a "critique" of phenomenology on this basis is surely a bizarre enterprise.

The expression "appearance" itself in turn can have a double meaning. First, *appearing* in the sense of making itself known as something that does not show itself and, second, in the sense of what does the making known—what in its self-showing indicates something that does not show itself. Finally, one can use appearing as the term for the genuine meaning of phenomenon as self-showing. If one designates these three different states of affairs as "appearance," confusion is inevitable.

However, this confusion is considerably increased by the fact that "appearance" can take on still another meaning. If one understands what does the making known—what in its self-showing indicates the nonmanifest—as what comes to the fore in the nonmanifest itself, and radiates from it in such a way that what is nonmanifest is thought of as what is essentially *never* manifest—if one understands the matter in this way, then appearance is tantamount to a bringing to the fore, or to what is brought to the fore. However, the latter does not constitute the proper Being of what actually conducts something to the fore. Hence appearance has the sense of "mere appearance." That which makes known, itself brought to the fore, indeed shows itself; but it does so in such a way that, as the emanation of what it makes known, it precisely and continually veils what it is in itself. But then again this not-showing which veils is not semblance. Kant uses the term "appearance" in this twofold way. On the one hand, appearances are for him the "objects of empirical intuition," what shows itself in intuition. This self-showing (phenomenon in the genuine, original sense) is, on the other hand, "appearance" as the emanation of something that makes itself known but *conceals* itself in the appearance.

Since a phenomenon is constitutive for "appearance" in the sense of making itself known through a self-showing, and since this phenomenon can turn into semblance in a privative way, appearance can also turn into mere semblance. Under a certain kind of light someone can look as if he were flushed. The redness that shows itself can be taken as making known the presence of fever; this in turn would indicate a disturbance in the organism.

*Phenomenon*—the self-showing in itself—means a distinctive way something can be encountered. On the other hand, *appearance* means a referential relation in beings themselves such that what does the *referring* (the making known) can fulfill its possible function only if it shows itself in itself—only if it is a "phenomenon." Both appearance and semblance are themselves grounded in the phenomenon, albeit in different ways. The confusing multiplicity of "phenomena" designated by the terms *phenomenon*, *semblance*, *appearance*, *mere appearance*, can be unraveled only if the concept of phenomenon is understood from the very beginning as the self-showing in itself.

But if in the way we grasp the concept of phenomenon we leave undetermined which beings are to be addressed as phenomena, and if we leave altogether open whether the self-showing is actually a particular being or a characteristic of the Being of beings, then we are dealing solely with the *formal* concept of phenomenon. If by the self-showing we understand those beings that are accessible, for example, in Kant's sense of empirical intuition, the formal concept of phenomenon can be applied legitimately. In this use phenomenon has the meaning of the *common* concept of phenomenon. But this common one is not the phenomenological concept of phenomenon. On the horizon of the Kantian problem what is understood phenomenologically by the term phenomenon (disregarding other differences) can be illustrated when we say that what already shows itself in appearances prior to and always accompanying what we commonly understand as phenomena, though unthematically, can be brought thematically to self-showing. This self-showing as such in itself ("the forms of intuition") are the phenomena of phenomenology. For clearly space and time must be able to show themselves in this way. They must be able to become phenomena if Kant claims to make a valid transcendental statement when he says that space is the *a priori* "wherein" of an order.

Now, if the phenomenological concept of phenomenon is to be understood at all (regardless of how the self-showing may be more closely determined), we must inevitably presuppose insight into the sense of the formal concept of phenomenon and the legitimate application of phenomenon in its ordinary meaning. However, before getting hold of the preliminary concept of phenomenology we must define the meaning of *logos*, in order to make clear in which sense phenomenology can be "a science of" phenomena at all.

### B. *The concept of logos*

The concept of *logos* has many meanings in Plato and Aristotle, indeed in such a way that these meanings diverge, without a basic meaning positively taking the lead. This is in fact only an illusion which lasts so long as an interpretation is not able to grasp adequately the basic meaning in its primary content. If we say that the basic meaning of *logos* is speech, this literal translation becomes valid only when we define what speech itself means. The later history of the word *logos*, and especially the manifold and arbitrary interpretations of subsequent philosophy, conceal constantly the proper meaning of speech—which is manifest enough. *Logos* is "translated," and that always means interpreted, as reason, judgment, concept, definition, ground, relation. But how can "speech" be so susceptible of modification that *logos* means all the things mentioned, indeed in scholarly usage? Even if *logos* is understood in the sense of a statement, and statement as "judgment," this apparently correct translation can still miss the fundamental meaning—especially if judgment is understood in the sense of some contemporary "theory of judgment." *Logos* does not mean judgment, in any case not primarily, if by judgment we understand "connecting two things" or "taking a position" either by endorsing or rejecting.

Rather, *logos* as speech really means *dēloun*, to make manifest "what is being talked about" in speech. Aristotle explicates this function of speech more precisely as *apophainesthai*.[4] *Logos* lets something be seen (*phainesthai*), namely what is being talked about, and indeed *for* the speaker (who serves as the medium) or for those who speak with each other. Speech "lets us see," from itself, *apo* . . ., what is being talked about. In speech (*apophansis*), insofar as it is genuine, *what* is said should be derived *from* what is being talked about. In this way spoken communication, in what it says, makes manifest what it is talking about and thus makes it accessible to another. Such is the structure of *logos* as *apophansis*. Not every "speech" suits *this* mode of making manifest, in the sense of letting something be seen by indicating it. For example, requesting (*euchē*) also makes something manifest, but in a different way.

When fully concrete, speech (letting something be seen) has the character of speaking or vocalization in words. *Logos* is *phonē*, in-

4. See *De interpretatione*, chaps. 1–6. See further, *Metaphysics* VII, 4 and *Nicomachean Ethics*, Bk. VI.

deed *phonē meta phantasias*—vocalization in which something always is sighted.

Only *because* the function of *logos* as *apophansis* lies in letting something be seen by indicating it can *logos* have the structure of *synthesis*. Here synthesis does not mean to connect and conjoin representations, to manipulate psychical occurrences, which then gives rise to the "problem" of how these connections, as internal, correspond to what is external and physical. The *syn* [of *synthesis*] here has a purely apophantical meaning: to let something be seen in its *togetherness* with something, to let something be seen *as* something.

Furthermore, because *logos* lets something be seen, it can *therefore* be true or false. But everything depends on staying clear of any concept of truth construed in the sense of "correspondence" or "accordance" [*Übereinstimmung*]. This idea is by no means the primary one in the concept of *alētheia*. The "being true" of *logos* as *alētheuein* means: to take beings that are being talked *about* in *legein* as *apophainesthai* out of their concealment; to let them be seen as something unconcealed (*alēthes*); to *discover* them. Similarly "being false," *pseudesthai*, is tantamount to deceiving in the sense of covering up: putting something in front of something else (by way of letting it be seen) and thereby proffering it *as* something it is *not*.

But because "truth" has this meaning, and because *logos* is a specific mode of letting something be seen, *logos* simply may *not* be acclaimed as the primary "place" of truth. If one defines truth as what "properly" pertains to judgment, which is quite customary today, and if one invokes Aristotle in support of this thesis, such invocation is without justification and the Greek concept of truth thoroughly misunderstood. In the Greek sense what is "true"—indeed more originally true than the *logos* we have been discussing—is *aisthēsis*, the straightforward sensuous apprehending of something. To the extent that an *aisthēsis* aims at its *idia* [what is its own]—the beings genuinely accessible only *through* it and *for* it, for example, *looking* at colors—apprehending is always true. This means that looking always discovers colors, hearing always discovers tones. What is in the purest and most original sense "true"—that is, what only discovers in such a way that it can never cover up anything—is pure *noein*, straightforwardly observant apprehension of the simplest determinations of the Being of beings as such. This *noein* can never cover up, can never be false; at worst it can be a nonapprehending, *agnoein*, not sufficing for straightforward, appropriate access.

What no longer takes the form of a pure letting be seen, but rather in its indicating always has recourse to something else and so always lets something be seen *as* something, acquires a structure of synthesis and therewith the possibility of covering up. However, "truth of judgment" is only the opposite of this covering up; it is a *multiply-founded* phenomenon of truth. Realism and idealism alike thoroughly miss the meaning of the Greek concept of truth from which alone the possibility of something like a "theory of Ideas" can be understood at all as philosophical *knowledge*. And because the function of *logos* lies in letting something be seen straightforwardly, in *letting* beings be *apprehended*, *logos* can mean *reason*. Moreover, because *logos* is used in the sense not only of *legein* but also of *legomenon*—what is pointed to as such; and because the latter is nothing other than the *hypokeimenon*—what always already is at hand at the *basis* of every discourse and discussion in progress; for these reasons *logos* qua *legomenon* means ground, *ratio*. Finally, because *logos* as *legomenon* can also mean what is addressed, as something that has become visible in its relation to something else, in its "relatedness," *logos* acquires the meaning of a *relationship with* and a *relating to* something.

This interpretation of "apophantic speech" may suffice to clarify the primary function of *logos*.

### C. *The preliminary concept of phenomenology*

When we bring to mind concretely what has been exhibited in the interpretation of "phenomenon" and "logos" we are struck by an inner relation between what is meant by these terms. The expression "phenomenology" can be formulated in Greek as *legein ta phainomena*. But *legein* means *apophainesthai*. Hence phenomenology means: *apophainesthai ta phainomena*—to let what shows itself be seen from itself, just as it shows itself from itself. That is the formal meaning of the type of research that calls itself "phenomenology." But this expresses nothing other than the maxim formulated above: "To the things themselves!"

Accordingly, the term "phenomenology" differs in meaning from such expressions as "theology" and the like. Such titles designate the objects of the respective disciplines in terms of their content. "Phenomenology" neither designates the object of its researches nor is it a title that describes their content. The word only tells us something about the *how* of the demonstration and treatment of *what* this discipline considers. Science "of" the phenomena means that it grasps its objects in *such* a way that everything about them to be discussed must be directly indicated and directly demonstrated. The basically tautological expression "descriptive phenomenology" has the same sense. Here description does not mean a procedure like that of, say, botanical morphology. The term rather has the sense of a prohibition, insisting that we avoid all nondemonstrative determinations. The character of description itself, the specific sense of the *logos*, can be established only from the "compelling nature" ["*Sachheit*"] of what is "described," i.e., of what is to be brought to scientific determinateness in the way phenomena are encountered. The meaning of the formal and common concepts of the phenomenon formally justifies our calling every way of indicating beings as they show themselves in themselves "phenomenology."

Now, what must be taken into account if the formal concept of the phenomenon is to be deformalized to the phenomenological one, and how does this differ from the common concept? What is it that phenomenology is to "let be seen"? What is it that is to be called "phenomenon" in a distinctive sense? What is it that by its very essence becomes the *necessary* theme when we indicate something *explicitly?* Manifestly it is something that does *not* show itself at first and for the most part, something that is *concealed*, in contrast to what at first and for the most part does show itself. But at the same time it is something that essentially belongs to what at first and for the most part shows itself, indeed in such a way that it constitutes its meaning and ground.

But what remains *concealed* in an exceptional sense, or what falls back and is *covered up* again, or shows itself only in a *distorted* way, is not this or that being but rather, as we have shown in our foregoing observations, the *Being* of beings. It can be covered up to such a degree that it is forgotten and the question about it and its meaning is in default. Thus what demands to become a phenomenon in a distinctive sense, in terms of its most proper content, phenomenology has taken into its "grasp" thematically as its object.

Phenomenology is the way of access to, and the demonstrative manner of determination of, what is to become the theme of ontology. *Ontology is possible only as phenomenology.* The phenomenological concept of phenomenon, as self-showing, means the Being of beings—its meaning, modifications, and derivatives. This self-showing is nothing arbitrary, nor is it something like an appearing. The Being of beings can least of all be something "behind which" something else stands, something that "does not appear."

Essentially, nothing else stands "behind" the phenomena of phenomenology. Nevertheless, what is to become a phenomenon can be concealed. And precisely because phenomena are at first and for the most part *not* given phenomenology is needed. Being covered up is the counterconcept to "phenomenon."

There are various ways phenomena can be covered up. In the first place, a phenomenon can be covered up in the sense that it has not yet been *discovered* at all. There is neither knowledge nor lack of knowledge about it. In the second place, a phenomenon can be *buried over.* This means it was once discovered but then got covered up again. This covering up can be total, but more commonly, what was once discovered may still be visible, though only as semblance. However, where there is semblance there is "Being." This kind of covering up, "distortion," is the most frequent and the most dangerous kind because here the possibilities of being deceived and misled are especially pertinacious. Within a "system" the structures and concepts of Being that are available but concealed with respect to their autochthony may perhaps claim their rights. On the basis of their integrated structure in a system they present themselves as something "clear" which is in no need of further justification and which therefore can serve as a point of departure for a process of deduction.

The covering up itself, whether it be understood in the sense of concealment, being buried over, or distortion, has in turn a twofold possibility. There are accidental coverings and necessary ones, the latter grounded in the substantive nature of the discovered. It is possible for every phenomenological concept and proposition drawn from genuine origins to degenerate when communicated as a statement. It gets circulated in a vacuous fashion, loses its autochthony, and becomes a free-floating thesis. Even in the concrete work of phenomenology lurks possible inflexibility and the inability to grip what was originally "grasped." And the difficulty of this research consists precisely in making it self-critical in a positive sense.

The way of encountering Being and the structures of Being in the mode of phenomenon must first be *won* from the objects of phenomenology. Thus the *point of departure* of the analysis, the *access to* the phenomenon, and *passage through* the prevalent coverings must secure their own method. The idea of an "originary" and "intuitive" grasp and explication of phenomena must be opposed to the naïveté of an accidental, "immediate," and unreflective "beholding."

On the basis of the preliminary concept of phenomenology just delimited, the terms "phenomenal" and "phenomenological" can now be given fixed meanings. What is given and is explicable in the way we encounter the phenomenon is called "phenomenal." In this sense we speak of phenomenal structures. Everything that belongs to the manner of indication and explication, and constitutes the conceptual tools this research requires, is called "phenomenological."

Because phenomenon in the phenomenological understanding is always just what constitutes Being, and furthermore because Being is always the Being of beings, we must first of all bring beings themselves forward in the right way if we are to have any prospect of exposing Being. These beings must likewise show themselves in the way of access that genuinely belongs to them. Thus the common concept of phenomenon becomes phenomenologically relevant. The preliminary task of a "phenomenological" securing of that being which is to serve as our example, as the point of departure for the analysis proper, is always already prescribed by the goal of this analysis.

As far as content goes, phenomenology is the science of the Being of beings—ontology. In our elucidation of the tasks of ontology the necessity arose of a fundamental ontology which would have as its theme that being which is ontologically and ontically distinctive, namely, Dasein. This must be done in such a way that our ontology confronts the cardinal problem, the question of the meaning of Being in general. From the investigation itself we shall see that the methodological meaning of phenomenological description is *interpretation*. The *logos* of the phenomenology of Dasein has the character of *hermēneuein*, through which the proper meaning of Being and the basic structures of the very Being of Dasein are *made known* to the understanding of Being that belongs to Dasein itself. Phenomenology of Dasein is *hermeneutics* in the original signification of that word, which designates the work of interpretation. But since discovery of the meaning of Being and of the basic structures of Dasein in general exhibits the horizon for every further ontological research into beings unlike Dasein, the present hermeneutic is at the same time "hermeneutics" in the sense that it works out the conditions of the possibility of every ontological investigation. Finally, since Dasein has ontological priority over all other beings—as a being that has the possibility of existence [*Existenz*]—hermeneutics, as the interpretation of the Being of Dasein, receives a specific third and, philosophically understood, *primary* meaning of an analysis of the existentiality of existence. To the extent that this hermeneutic elaborates the historicity of Dasein ontologically as the ontic condition of the possibility of the discipline of history, it contains the roots of what can be called "hermeneutics" only in a derivative sense: the methodology of the historical humanistic disciplines.

As the fundamental theme of philosophy, Being is no sort of genus of beings; yet it pertains to every being. Its "universality" must be sought in a higher sphere. Being and its structure transcend every being and every determination of beings there might be. *Being is the* ***transcendens*** *pure and simple.* The transcendence of the Being of Dasein is a distinctive one since in it lies the possibility and necessity of the most radical *individuation*. Every disclosure of Being as the *transcendens* is *transcendental* knowledge. *Phenomenological truth* (*disclosedness of Being*) is ***veritas transcendentalis***.

Ontology and phenomenology are not two different disciplines that among others belong to philosophy. Both terms characterize philosophy itself, its object and procedure. Philosophy is universal, phenomenological ontology, taking its departure from the hermeneutic of Dasein, which as an analysis of *existence* has fastened the end of the guideline of all philosophical inquiry at the point from which it *arises* and to which it *returns*.

The following investigations would not have been possible without the foundation laid by Edmund Husserl; with his *Logical Investigations* phenomenology achieved a breakthrough. Our elucidations of the preliminary concept of phenomenology show that what is essential to it does not consist in its *actuality* as a philosophical "movement." Higher than actuality stands *possibility*.

We can understand phenomenology solely by seizing upon it as a possibility.[5]

With regard to the awkwardness and "inelegance" of expression in the following analyses we may remark that it is one thing to report narratively about *beings* and another to grasp beings in their *Being*. For the latter task not only most of the words are lacking but above all the "grammar." If we may allude to earlier and in their own right altogether incomparable researches on the analysis of Being, then we should compare the ontological sections in Plato's *Parmenides* or the fourth chapter of the seventh book of Aristotle's *Metaphysics* with a narrative passage from Thucydides. Then we would see the stunning character of the formulations by which their philosophers challenged the Greeks. Since our powers are essentially inferior, and also since the area of Being to be disclosed ontologically is far more difficult than that presented to the Greeks, the complexity of our concept-formation and the severity of our expression will increase.

## 8. The outline of the treatise

The question of the meaning of Being is the most universal and the emptiest. But at the same time the possibility inheres of its keenest particularization in every individual Dasein. If we are to gain the fundamental concept of "Being" and the prescription of the ontologically requisite conceptuality in all its necessary variations, we need a concrete guideline. The "special character" of the investigation does not belie the universality of the concept of Being. For we may advance to Being by way of a special interpretation of a particular being, Dasein, in which the horizon for an understanding

5. If the following investigation takes any steps forward in disclosing "the things themselves" the author must above all thank E. Husserl, who by providing his own incisive personal guidance and by very generously sharing his unpublished investigations familiarized the author during his student years in Freiburg with the most diverse areas of phenomenological research.

and a possible interpretation of Being is to be won. But this being is in itself "historic," so that its most proper ontological illumination necessarily becomes a "historical" interpretation.

The elaboration of the question of Being is a two-pronged task; our treatise therefore has two parts.

*Part One:* The interpretation of Dasein with a view to temporality and the explication of time as the transcendental horizon of the question of Being.

*Part Two:* Basic features of a phenomenological destructuring of the history of ontology on the guideline of the problem of Temporality.

The first part consists of three divisions:

1. The preparatory fundamental analysis of Dasein.
2. Dasein and temporality.
3. Time and Being.

The second part likewise has three divisions:

1. Kant's doctrine of the schematism and of time, as the preliminary stage of a problem of Temporality.*
2. The ontological foundation of Descartes's *cogito sum* and the incorporation of medieval ontology in the problem of the *res cogitans*.
3. Aristotle's treatise on time as a way of discerning the phenomenal basis and limits of ancient ontology.†

*See Martin Heidegger, *Kant und das Problem der Metaphysik*, first published in 1929, fourth, expanded edition (Frankfurt am Main: V. Klostermann, 1973), pp. xii-xviii. A new translation of *Kant and the Problem of Metaphysics* by Richard Taft is now available (Bloomington: Indiana University Press, 1990).—Ed.

†See Martin Heidegger, *The Basic Problems of Phenomenology*, section 19, for Heidegger's remarkable destructuring of the Aristotelian treatise on time.—Ed.

# II

# WHAT IS METAPHYSICS?

*The world's darkening never reaches to the light of Being.*

On July 24, 1929, Heidegger delivered his inaugural lecture to the Freiburg University faculties. That lecture, translated here complete but without the Afterword and Introduction appended to it in 1943 and 1949, stressed several key issues in his then recently published *Being and Time* and also pointed forward to his later studies.

In *Being and Time* Heidegger undertook a concrete description of the being that questions Being. At various watersheds in his history Western man has advanced descriptions of himself in sundry forms: he has always expressed opinions about who he is. Until recently Occidental man has consistently described himself as the rational animal, the living creature that thinks and has knowledge. Heidegger sought the fundamental source of this knowledge—indeed knowledge of any kind—and found it in the disclosedness (*Erschlossenheit*) of Dasein as being-in-the-world. Knowledge and rational opinions are certainly one kind of disclosure, but Heidegger located a more generalized and primary sort: Dasein is disclosed as such in the *moods* in which man finds himself, such as joy, boredom, excitement, or anxiety. Why at any particular time we find ourselves in "good" or "bad" moods we do not know. "And Dasein cannot know anything of the sort because the possibilities of disclosure that belong to knowledge do not extend far enough for the original disclosure of moods . . ." (*Being and Time,* section 29).

Now Heidegger searched for a particular mood that would disclose something essential about man's existence as a whole. Partly thanks to his reading of Kierkegaard he found it in *anxiety,* which is not fear of this or that but a malaise at once less identifiable and more oppressive. "That in the face of which one has anxiety," Heidegger emphasized in *Being and Time,* "is being-in-the-world as such" (section 40). In anxiety I realize that I have been "thrown" into the world and that my life and death—my Being as such—is an issue I must face. In anxiety, "Dasein finds itself *face to face* with the nothing of the possible impossibility of its own existence" (section 53). Nothing in particular makes me anxious. . . . Heidegger became interested in the equivocal sense of this "nothing" both in our everyday speech and in the language of metaphysics. "The nothing that anxiety brings before us unveils the nullity that determines Dasein in its *ground*—which is its being thrown into death" (section 62).

However, it is important to note that for Heidegger anxiety and the revelation of the nothing are not symptoms of pathological man. Dasein is the place for the disclosure of Being as such and in general, a matter which therefore must somehow be bound to the nothing. But Being—at least one kind of being—has always been the business of ontology or metaphysics. What is metaphysics? Metaphysics is interpretation of beings and forgetfulness of Being and that means neglect of the essence of the *Nihil*.

Thus Heidegger's preoccupation with the nothing becomes an important theme that bridges his early and later work and serves to characterize his unique approach to philosophy. "The nothing" comes to be a name for the source not only of all that is dark and riddlesome in existence—which seems to rise from nowhere and to return to it—but also of the openness of Being as such and the brilliance surrounding whatever comes to light. Because metaphysics has in one way or another sought to banish or ignore this unaccountable source of man and Being, Heidegger follows Nietzsche in identifying the history of metaphysics as nihilism. Nihilism does not result from excessive preoccupation with the nothing. On the contrary, only by asking the question of the nothing can nihilism be countered. Such asking is not quickly satisfied. In the "Letter on Humanism" (Reading V) it emerges once again in the context of further discussion of *Being and Time* vis-à-vis the metaphysical tradition.

# WHAT IS METAPHYSICS?

"What is metaphysics?" The question awakens expectations of a discussion about metaphysics. This we will forgo. Instead we will take up a particular metaphysical question. In this way it seems we will let ourselves be transposed directly into metaphysics. Only in this way will we provide metaphysics the proper occasion to introduce itself.

Our plan begins with the unfolding of a metaphysical inquiry, then tries to elaborate the question, and concludes by answering it.

## The Unfolding of a Metaphysical Inquiry

From the point of view of sound common sense, philosophy is in Hegel's words "the inverted world." Hence the peculiar nature of our undertaking requires a preliminary sketch. This will take shape about a twofold character of metaphysical interrogation.

First, every metaphysical question always encompasses the whole range of metaphysical problems. Each question is itself always the whole. Therefore, second, every metaphysical question can be asked only in such a way that the questioner as such is present together with the question, that is, is placed in question. From this

---

This translation of *Was ist Metaphysik?* appears for the first time in this book. The German text was first published by the Bonn firm of Friedrich Cohen; it is included in Martin Heidegger, *Wegmarken* (Frankfurt am Main: Vittorio Klostermann Verlag, 1967), pp. 1–19. I am indebted to two previous translations: (1) the 1937 French translation by Henry Corbin in Martin Heidegger, *Questions I* (Paris: Gallimard, 1968), pp. 47–72, and (2) an earlier English translation by R. F. C. Hull and Alan Crick in *Existence and Being*, edited by Werner Brock (Chicago: Henry Regnery, 1949), pp. 325–349.

we conclude that metaphysical inquiry must be posed as a whole and from the essential position of the existence [*Dasein*] that questions. We are questioning, here and now, for ourselves. Our existence—in the community of researchers, teachers, and students—is determined by science. What happens to us, essentially, in the grounds of our existence, when science becomes our passion?

The scientific fields are quite diverse. The ways they treat their objects of inquiry differ fundamentally. Today only the technical organization of universities and faculties consolidates this burgeoning multiplicity of disciplines; the practical establishment of goals by each discipline provides the only meaningful source of unity. Nonetheless, the rootedness of the sciences in their essential ground has atrophied.

Yet when we follow their most proper intention, in all the sciences we relate ourselves to beings themselves. Precisely from the point of view of the sciences or disciplines no field takes precedence over another, neither nature over history nor vice versa. No particular way of treating objects of inquiry dominates the others. Mathematical knowledge is no more rigorous than philological-historical knowledge. It merely has the character of "exactness," which does not coincide with rigor. To demand exactness in the study of history is to violate the idea of the specific rigor of the humanities. The relation to the world that pervades all the sciences as such lets them—each according to its particular content and mode of being—seek beings themselves in order to make them objects of investigation and to determine their grounds. According to the idea behind them, in the sciences we approach what is essential in all things. This distinctive relation to the world in which we turn toward beings themselves is supported and guided by a freely chosen attitude of human existence. To be sure, man's prescientific and extrascientific activities also are related to beings. But science is exceptional in that, in a way peculiar to it, it gives the matter itself explicitly and solely the first and last word. In such impartiality of inquiring, determining, and grounding, a peculiarly delineated submission to beings themselves obtains, in order that they may reveal themselves. This position of service in research and theory evolves in such a way as to become the ground of the possibility of a proper though limited leadership in the whole of human existence. The special relation science sustains to the world and the attitude of man that guides it can of course be fully grasped only when we see and comprehend what happens in the relation to the world so attained. Man—one being among others—"pursues science." In this "pursuit" nothing less transpires than the irruption by one being called "man" into the whole of beings, indeed in such a way that in and through this irruption beings break open and show what they are and how they are. The irruption that breaks open, in its way, helps beings above all to themselves.

This trinity—relation to the world, attitude, and irruption—in its radical unity brings a luminous simplicity and aptness of Dasein to scientific existence. If we are to take explicit possession of the Dasein illuminated in this way for ourselves, then we must say:

That to which the relation to the world refers are beings themselves—and nothing besides.
That from which every attitude takes its guidance are beings themselves—and nothing further.
That with which the scientific confrontation in the irruption occurs are beings themselves—and beyond that nothing.

But what is remarkable is that, precisely in the way scientific man secures to himself what is most properly his, he speaks of something different. What should be examined are beings only, and besides that—nothing; beings alone, and further—nothing; solely beings, and beyond that—nothing.

What about this nothing? Is it an accident that we talk this way so automatically? Is it only a manner of speaking—and nothing besides?

However, what trouble do we take concerning this nothing? The nothing is rejected precisely by science, given up as a nullity. But when we give up the nothing in such a way do we not concede it? Can we, however, speak of concession when we concede nothing? But perhaps our confused talk already degenerates into an empty squabble over words. Against it science must now reassert its seriousness and soberness of mind, insisting that it is concerned solely with beings. The nothing—what else can it be for science but an outrage and a phantasm? If science is right, then only one thing is sure: science wishes to know nothing of the nothing. Ultimately this is the scientifically rigorous conception of the nothing. We know it, the nothing, in that we wish to know nothing about it.

Science wants to know nothing of the nothing. But even so it is certain that when science tries to express its proper essence it calls upon the nothing for help. It has recourse to what it rejects. What incongruous state of affairs reveals itself here?

With this reflection on our contemporary existence as one determined by science we find ourselves enmeshed in a controversy. In the course of this controversy a question has already evolved. It only requires explicit formulation: How is it with the nothing?

## The Elaboration of the Question

The elaboration of the question of the nothing must bring us to the point where an answer becomes possible or the impossibility of any answer becomes clear. The nothing is conceded. With a studied indifference science abandons it as what "there is not."

All the same, we shall try to ask about the nothing. What is the nothing? Our very first approach to this question has something unusual about it. In our asking we posit the nothing in advance as something that "is" such and such; we posit it as a being. But that is exactly what it is distinguished from. Interrogating the nothing—asking what and how it, the nothing, is—turns what is interrogated into its opposite. The question deprives itself of its own object.

Accordingly, every answer to this question is also impossible from the start. For it necessarily assumes the form: the nothing "is" this or that. With regard to the nothing, question and answer alike are inherently absurd.

But it is not science's rejection that first of all teaches us this. The commonly cited ground rule of all thinking, the proposition that contradiction is to be avoided, universal "logic" itself, lays low this question. For thinking, which is always essentially thinking about something, must act in a way contrary to its own essence when it thinks of the nothing.

Since it remains wholly impossible for us to make the nothing into an object, have we not already come to the end of our inquiry into the nothing—assuming that in this question "logic" is of supreme importance, that the intellect is the means, and thought the way, to conceive the nothing originally and to decide about its possible exposure?

But are we allowed to tamper with the rule of "logic"? Is not intellect the taskmaster in this question of the nothing? Only with its help can we at all define the nothing and pose it as a problem—which, it is true, only devours itself. For the nothing is the negation of the totality of beings; it is nonbeing pure and simple. But with that we bring the nothing under the higher determination of the negative, viewing it as the negated. However, according to the reigning and never-challenged doctrine of "logic," negation is a specific act of the intellect. How then can we in our question of the nothing, indeed in the question of its questionability, wish to brush the intellect aside? Are we altogether sure about what we are presupposing in this matter? Do not the "not," negatedness, and thereby negation too represent the higher determination under which the nothing falls as a particular kind of negated matter? Is the nothing given only because the "not," i.e., negation, is given? Or is it the other way around? Are negation and the "not" given only because the nothing is given? That has not been decided; it has not even been raised expressly as a question. We assert that the nothing is more original than the "not" and negation.

If this thesis is right, then the possibility of negation as an act of the intellect, and thereby the intellect itself, are somehow dependent upon the nothing. Then how can the intellect hope to decide about the nothing? Does the ostensible absurdity of question and answer with respect to the nothing in the end rest solely in a blind conceit of the far-ranging intellect?

But if we do not let ourselves be misled by the formal impossibility of the question of the nothing; if we pose the question in spite of this; then we must at least satisfy what remains the basic demand for the possible advancing of every question. If the nothing itself is to be questioned as we have been questioning it, then it must be given beforehand. We must be able to encounter it.

Where shall we seek the nothing? Where will we find the nothing? In order to find something must we not already know in general that it is there? Indeed! At first and for the most part man can seek only when he has anticipated the being at hand of what he is looking for. Now the nothing is what we are seeking. Is there ultimately such a thing as a search without that anticipation, a search to which pure discovery belongs?

Whatever we may make of it, we do know the nothing, if only as a word we rattle off every day. For this common nothing that glides so inconspicuously through our chatter, blanched with the anemic pallor of the obvious, we can without hesitation furnish even a "definition":

The nothing is the complete negation of the totality of beings.

Does not this characterization of the nothing ultimately provide an indication of the direction from which alone the nothing can come to meet us?

The totality of beings must be given in advance so as to be able to fall prey straightway to negation—in which the nothing itself would then be manifest.

But even if we ignore the questionableness of the relation between negation and the nothing, how should we who are essentially finite make the whole of beings totally penetrable in itself and also for us? We can of course think the whole of beings in an "idea," then negate what we have imagined in our thought, and thus "think" it negated. In this way we do attain the formal concept of the imagined nothing but never the nothing itself. But the nothing is nothing, and if the nothing represents total indistinguishability no distinction can obtain between the imagined and the "proper" nothing. And the "proper" nothing itself—is not this the camouflaged but absurd concept of a nothing that is? For the last time now the objections of the intellect would call a halt to our search, whose legitimacy, however, can be demonstrated only on the basis of a fundamental experience of the nothing.

As surely as we can never comprehend absolutely the whole of beings in themselves we certainly do find ourselves stationed in the midst of beings that are revealed somehow as a whole. In the end an essential distinction prevails between comprehending the whole of beings in themselves and finding oneself in the midst of beings as a whole. The former is impossible in principle. The latter happens all the time in our existence. It does seem as though we cling to this or that particular being, precisely in our everyday preoccupations, as though we were completely abandoned to this or that region of beings. No matter how fragmented our everyday existence may appear to be, however, it always deals with beings in a unity of the "whole," if only in a shadowy way. Even and precisely when we are not actually busy with things or ourselves, this "as a whole" overcomes us—for example in genuine boredom. Boredom is still distant when it is only this book or that play, that business or this idleness, that drags on. It irrupts when "one is bored." Profound boredom, drifting here and there in the abysses of our existence like a muffling fog, removes all things and human beings and oneself along with them into a remarkable indifference. This boredom reveals beings as a whole.

Another possibility of such revelation is concealed in our joy in the presence of the Dasein—and not simply of the person—of a human being whom we love.

Such being attuned, in which we "are" one way or another and which determines us through and through, lets us find ourselves among beings as a whole. The founding mode of attunement [*die Befindlichkeit der Stimmung*] not only reveals beings as a whole in various ways, but this revealing—far from being merely incidental—is also the basic occurrence of our Da-sein.

What we call a "feeling" is neither a transitory epiphenomenon of our thinking and willing behavior nor simply an impulse that provokes such behavior nor merely a present condition we have to put up with somehow or other.

But just when moods of this sort bring us face to face with beings as a whole they conceal from us the nothing we are seeking. Now we come to share even less in the opinion that the negation of beings as a whole that are revealed to us in mood places us before the nothing. Such a thing could happen only in a correspondingly original mood which in the most proper sense of unveiling reveals the nothing.

Does such an attunement, in which man is brought before the nothing itself, occur in human existence?

This can and does occur, although rarely enough and only for a moment, in the fundamental mood of anxiety. By this anxiety we do not mean the quite common anxiousness, ultimately reducible to fearfulness, which all too readily comes over us. Anxiety is basically different from fear. We become afraid in the face of this or that particular being that threatens us in this or that particular respect. Fear in the face of something is also in each case a fear for something in particular. Because fear possesses this trait of being "fear in the face of" and "fear for," he who fears and is afraid is captive to the mood in which he finds himself. Striving to rescue himself from this particular thing, he becomes unsure of everything else and completely "loses his head."

Anxiety does not let such confusion arise. Much to the contrary, a peculiar calm pervades it. Anxiety is indeed anxiety in the face of . . ., but not in the face of this or that thing. Anxiety in the face of . . . is always anxiety for . . ., but not for this or that. The indeterminateness of that in the face of which and for which we become anxious is no mere lack of determination but rather the essential impossibility of determining it. In a familiar phrase this indeterminateness comes to the fore.

In anxiety, we say, "one feels ill at ease [*es ist einem unheimlich*]." What is "it" that makes "one" feel ill at ease? We cannot say what it is before which one feels ill at ease. As a whole it is so for one. All things and we ourselves sink into indifference. This, however, not in the sense of mere disappearance. Rather, in this very receding things turn toward us. The receding of beings as a whole that closes in on us in anxiety oppresses us. We can get no hold on things. In the slipping away of beings only this "no hold on things" comes over us and remains.

Anxiety reveals the nothing.

We "hover" in anxiety. More precisely, anxiety leaves us hanging because it induces the slipping away of beings as a whole. This implies that we ourselves—we humans who are in being—in the midst of beings slip away from ourselves. At bottom therefore it is not as though "you" or "I" feel ill at ease; rather, it is this way for some "one." In the altogether unsettling experience of this hovering where there is nothing to hold onto, pure Da-sein is all that is still there.

Anxiety robs us of speech. Because beings as a whole slip away, so that just the nothing crowds round, in the face of anxiety all utterance of the "is" falls silent. That in the malaise of anxiety we often try to shatter the vacant stillness with compulsive talk only proves the presence of the nothing. That anxiety reveals the nothing man himself immediately demonstrates when anxiety has dissolved. In the lucid vision sustained by fresh remembrance we must say that that in the face of which and for which we were anxious was "properly"—nothing. Indeed: the nothing itself—as such—was there.

With the fundamental mood of anxiety we have arrived at that occurrence in human existence in which the nothing is revealed and from which it must be interrogated.

How is it with the nothing?

## The Response to the Question

We have already won the answer which for our purposes is at least at first the only essential one when we take heed that the question of the nothing remains actually posed. This requires that we actively complete the transformation of man into his Da-sein that every instance of anxiety occasions in us, in order to get a grip on the nothing revealed there as it makes itself known. At the same time this demands that we expressly hold at a distance those designations of the nothing that do not result from its claims.

The nothing reveals itself in anxiety—but not as a being. Just as little is it given as an object. Anxiety is no kind of grasping of the nothing. All the same, the nothing reveals itself in and through anxiety, although, to repeat, not in such a way that the nothing becomes manifest in our malaise quite apart from beings as a whole. Rather, we said that in anxiety the nothing is encountered at one with beings as a whole. What does this "at one with" mean?

In anxiety beings as a whole become superfluous. In what sense does this happen? Beings are not annihilated by anxiety, so that nothing is left. How could they be, when anxiety finds itself precisely in utter impotence with regard to beings as a whole? Rather, the nothing makes itself known with beings and in beings expressly as a slipping away of the whole.

No kind of annihilation of the whole of beings as such takes place in anxiety; just as little do we produce a negation of beings as a whole in order to attain the nothing for the first time. Apart from the consideration that the expressive function of a negating assertion remains foreign to anxiety as such, we also come always too late with such a negation that should produce the nothing. The nothing rises to meet us already before that. We said it is encountered "at one with" beings that are slipping away as a whole.

In anxiety there occurs a shrinking back before . . . that is surely not any sort of flight but rather a kind of bewildered calm. This "back before" takes its departure from the nothing. The nothing itself does not attract; it is essentially repelling. But this repulsion is itself as such a parting gesture toward beings that are submerging as a whole. This wholly repelling gesture toward beings that are in retreat as a whole, which is the action of the nothing that oppresses Dasein in anxiety, is the essence of the nothing: nihilation. It is neither an annihilation of beings nor does it spring from a negation. Nihilation will not submit to calculation in terms of annihilation and negation. The nothing itself nihilates.

Nihilation is not some fortuitous incident. Rather, as the repelling gesture toward the retreating whole of beings, it discloses these beings in their full but heretofore concealed strangeness as what is radically other—with respect to the nothing.

In the clear night of the nothing of anxiety the original openness of beings as such arises: that they are beings—and not nothing. But this "and not nothing" we add in our talk is not some kind of appended clarification. Rather, it makes possible in advance the revelation of beings in general. The essence of the originally nihilating nothing lies in this, that it brings Da-sein for the first time before beings as such.

Only on the ground of the original revelation of the nothing can human existence approach and penetrate beings. But since existence in its essence relates itself to beings—those which it is not and that which it is—it emerges as such existence in each case from the nothing already revealed.

Da-sein means: being held out into the nothing.

Holding itself out into the nothing, Dasein is in each case already beyond beings as a whole. This being beyond beings we call "transcendence." If in the ground of its essence Dasein were not transcending, which now means, if it were not in advance holding itself out into the nothing, then it could never be related to beings nor even to itself.

Without the original revelation of the nothing, no selfhood and no freedom.

With that the answer to the question of the nothing is gained. The nothing is neither an object nor any being at all. The nothing comes forward neither for itself nor next to beings, to which it would, as it were, adhere. For human existence, the nothing makes possible the openedness of beings as such. The nothing does not merely serve as the counterconcept of beings; rather, it originally belongs to their essential unfolding as such. In the Being of beings the nihilation of the nothing occurs.

But now a suspicion we have been suppressing too long must finally find expression. If Dasein can relate itself to beings only by holding itself out into the nothing and can exist only thus; and if the nothing is originally disclosed only in anxiety; then must we not hover in this anxiety constantly in order to be able to exist at all? And have we not ourselves confessed that this original anxiety is rare? But above all else, we all do exist and relate ourselves to beings which we may or may not be—without this anxiety. Is this not an arbitrary invention and the nothing attributed to it a flight of fancy?

Yet what does it mean that this original anxiety occurs only in rare moments? Nothing else than that the nothing is at first and for the most part distorted with respect to its originality. How, then? In this way: we usually lose ourselves altogether among beings in a certain way. The more we turn toward beings in our preoccupations the less we let beings as a whole slip away as such and the more we turn away from the nothing. Just as surely do we hasten into the public superficies of existence.

And yet this constant if ambiguous turning away from the nothing accords, within certain limits, with the most proper significance of the nothing. In its nihilation the nothing directs us precisely toward beings. The nothing nihilates incessantly without our really knowing of this occurrence in the manner of our everyday knowledge.

What testifies to the constant and widespread though distorted revelation of the nothing in our existence more compellingly than negation? But negation does not conjure the "not" out of itself as a means for making distinctions and oppositions in whatever is given, inserting itself, as it were, in between what is given. How could negation produce the not from itself when it can make denials only when something deniable is already granted to it? But how could the deniable and what is to be denied be viewed as something susceptible to the not unless all thinking as such has caught sight of the not already? But the not can become manifest only when its origin, the nihilation of the nothing in general, and therewith the nothing itself, is disengaged from concealment. The not does not originate through negation; rather, negation is grounded in the not that springs from the nihilation of the nothing. But negation is also only one way of nihilating, that is, only one sort of behavior that has been grounded beforehand in the nihilation of the nothing.

In this way the above thesis in its main features has been proven: the nothing is the origin of negation, not vice versa. If the power of the intellect in the field of inquiry into the nothing and into Being is thus shattered, then the destiny of the reign of "logic" in philosophy is thereby decided. The idea of "logic" itself disintegrates in the turbulence of a more original questioning.

No matter how much or in how many ways negation, expressed or implied, permeates all thought, it is by no means the sole authoritative witness of the revelation of the nothing belonging essentially to Dasein. For negation cannot claim to be either the sole or the leading nihilative behavior in which Dasein remains shaken by the nihilation of the nothing. Unyielding antagonism and stinging rebuke have a more abysmal source than the measured negation of thought. Galling failure and merciless prohibition require some deeper answer. Bitter privation is more burdensome.

These possibilities of nihilative behavior—forces in which Dasein bears its thrownness without mastering it—are not types of mere negation. That does not prevent them, however, from speaking out in the "no" and in negation. Indeed here for the first time the barrenness and range of negation betray themselves. The saturation of existence by nihilative behavior testifies to the constant though doubtlessly obscured manifestation of the nothing that only anxiety originally reveals. But this implies that the original anxiety in existence is usually repressed. Anxiety is there. It is only sleeping. Its breath quivers perpetually through Dasein, only slightly in those who are jittery, imperceptibly in the "Oh, yes" and the "Oh, no" of men of affairs; but most readily in the reserved, and most assuredly in those who are basically daring. But those daring ones are sustained by that on which they expend themselves—in order thus to preserve the ultimate grandeur of existence.

The anxiety of those who are daring cannot be opposed to joy or even to the comfortable enjoyment of tranquilized bustle. It stands—outside all such opposition—in secret alliance with the cheerfulness and gentleness of creative longing.

Original anxiety can awaken in existence at any moment. It needs no unusual event to rouse it. Its sway is as thoroughgoing as its possible occasionings are trivial. It is always ready, though it only seldom springs, and we are snatched away and left hanging.

Being held out into the nothing—as Dasein is—on the ground of concealed anxiety makes man a lieutenant of the nothing. We are so finite that we cannot even bring ourselves originally before the nothing through our own decision and will. So profoundly does finitude entrench itself in existence that our most proper and deepest limitation refuses to yield to our freedom.

Being held out into the nothing—as Dasein is—on the ground of concealed anxiety is its surpassing of beings as a whole. It is transcendence.

Our inquiry concerning the nothing is to bring us face to face with metaphysics itself. The name "metaphysics" derives from the Greek *meta ta physika*. This peculiar title was later interpreted as characterizing the inquiry, the *meta* or *trans* extending out "over" beings as such.

Metaphysics is inquiry beyond or over beings, which aims to recover them as such and as a whole for our grasp.

In the question concerning the nothing such an inquiry beyond or over beings, beings as a whole, takes place. It proves thereby to be a "metaphysical" question. At the outset we ascribed a twofold character to such questions: first, each metaphysical question always encompasses the whole of metaphysics; second, every metaphysical question implicates the interrogating Dasein in each case in the question.

To what extent does the question concerning the nothing permeate and embrace the whole of metaphysics?

For a long time metaphysics has expressed the nothing in a proposition clearly susceptible of more than one meaning: *ex nihilo nihil fit*—from nothing, nothing comes to be. Although in discussions of the proposition the nothing itself never really becomes a problem, the respective views of the nothing nevertheless express the guiding fundamental conception of beings. Ancient metaphysics conceives the nothing in the sense of nonbeing, that is, unformed matter, matter which cannot take form as an in-formed being that would offer an outward appearance or aspect (*eidos*). To be in being is to be a self-forming form that exhibits itself as such in an image (as a spectacle). The origins, legitimacy, and limits of this conception of Being are as little discussed as the nothing itself. On the other hand, Christian dogma denies the truth of the proposition *ex nihilo nihil fit* and thereby bestows on the nothing a transformed significance, the sense of the complete absence of beings apart from God: *ex nihilo fit—ens creatum* [From nothing comes—created being]. Now the nothing becomes the counterconcept to being proper, the *summum ens*, God as *ens increatum*. Here too the interpretation of the nothing designates the basic conception of beings. But the metaphysical discussion of beings stays on the same level as the question of the nothing. The questions of Being and of the nothing as such are not posed. Therefore no one is bothered by the difficulty that if God creates out of nothing precisely He must be able to relate Himself to the nothing. But if God is God he cannot know the nothing, assuming that the "Absolute" excludes all nothingness.

This cursory historical review shows the nothing as the counter-concept to being proper, that is, as its negation. But if the nothing becomes any problem at all, then this opposition does not merely undergo a somewhat more significant determination; rather, it awakens for the first time the proper formulation of the metaphysical question concerning the Being of beings. The nothing does not remain the indeterminate opposite of beings but reveals itself as belonging to the Being of beings.

"Pure Being and pure Nothing are therefore the same." This proposition of Hegel's (*Science of Logic*, vol. I, *Werke* III, 74) is correct. Being and the nothing do belong together, not because both—from the point of view of the Hegelian concept of thought—agree in their indeterminateness and immediacy, but rather because Being itself is essentially finite and reveals itself only in the transcendence of Dasein which is held out into the nothing.

Assuming that the question of Being as such is the encompassing question of metaphysics, then the question of the nothing proves to be such that it embraces the whole of metaphysics. But the question of the nothing pervades the whole of metaphysics since at the same time it forces us to face the problem of the origin of negation, that is, ultimately, to face up to the decision concerning the legitimacy of the rule of "logic" in metaphysics.

The old proposition *ex nihilo nihil fit* is therefore found to contain another sense, one appropriate to the problem of Being itself, that runs: *ex nihilo omne ens qua ens fit* [From the nothing all beings as beings come to be]. Only in the nothing of Dasein do beings as a whole, in accord with their most proper possibility—that is, in a finite way—come to themselves. To what extent then has the question of the nothing, if it is a metaphysical question, implicated our questioning Dasein? We have characterized our existence, experienced here and now, as essentially determined by science. If our existence so defined is posed in the question of the nothing, then it must have become questionable through this question.

Scientific existence possesses its simplicity and aptness in that it relates to beings themselves in a distinctive way and only to them. Science would like to dismiss the nothing with a lordly wave of the hand. But in our inquiry concerning the nothing it has by now become manifest that scientific existence is possible only if in advance it holds itself out into the nothing. It understands itself for what it is only when it does not give up the nothing. The presumed soberness of mind and superiority of science become laughable when it does not take the nothing seriously. Only because the nothing is manifest can science make beings themselves objects of investigation. Only if science exists on the basis of metaphysics can it advance further in its essential task, which is not to amass and classify bits of knowledge but to disclose in ever-renewed fashion the entire region of truth in nature and history.

Only because the nothing is manifest in the ground of Dasein can the total strangeness of beings overwhelm us. Only when the strangeness of beings oppresses us does it arouse and evoke wonder. Only on the ground of wonder—the revelation of the nothing—does the "why?" loom before us. Only because the "why" is possible as such can we in a definite way inquire into grounds, and ground them. Only because we can inquire and ground is the destiny of our existence placed in the hands of the researcher.

The question of the nothing puts us, the questioners, in question. It is a metaphysical question.

Human existence can relate to beings only if it holds itself out into the nothing. Going beyond beings occurs in the essence of Dasein. But this going beyond is metaphysics itself. This implies that metaphysics belongs to the "nature of man." It is neither a division of academic philosophy nor a field of arbitrary notions. Metaphysics is the basic occurrence of Dasein. It is Dasein itself. Because the truth of metaphysics dwells in this groundless ground it stands in closest proximity to the constantly lurking possibility of deepest error. For this reason no amount of scientific rigor attains to the seriousness of metaphysics. Philosophy can never be measured by the standard of the idea of science.

If the question of the nothing unfolded here has actually questioned us, then we have not simply brought metaphysics before us in an extrinsic manner. Nor have we merely been "transposed" to it. We cannot be transposed there at all, because insofar as we exist we are always there already. *Physei gar, ō phile, enesti tis philosophia tēi tou andros dianoiai* ["For by nature, my friend, man's mind dwells in philosophy"] (Plato, *Phaedrus*, 279a). So long as man exists, philosophizing of some sort occurs. Philosophy—what we call philosophy—is metaphysics' getting under way, in which philosophy comes to itself and to its explicit tasks. Philosophy gets under way only by a peculiar insertion of our own existence into the fundamental possibilities of Dasein as a whole. For this insertion it is of decisive importance, first, that we allow space for beings as a whole; second, that we release ourselves into the nothing, which is to say, that we liberate ourselves from those idols everyone has and to which they are wont to go cringing; and finally, that we let the sweep of our suspense take its full course, so that it swings back into the basic question of metaphysics which the nothing itself compels: Why are there beings at all, and why not rather nothing?

# III

# ON THE ESSENCE OF TRUTH

*The splendor of the simple.*

According to Franz Brentano "being" in the sense of "the true" was the second of being's manifold meanings in Aristotle. The difficulty proved to be that "true" also meant many different things. Judgments, propositions, and mathematical formulas could be true or false but so could something we perceive or remember, dream or imagine; "things" (*pragmata*) might be true or false in a sense and so might people.

Now, all these senses of "true" are analogous: while distinct in meaning they all tend toward "one sense and one dominant source." Brentano tried to get at that one basic sense by offering a familiar example from geometry: at a certain point in the demonstration of a geometrical theorem one can only ask, "*Is* this, or is it *not?*" meaning, "Is this true, or false?" Hence the meaning of "the true" turns out to be the that-it-is of something, its Being—which of course has manifold senses!

Even as a youth Heidegger was intrigued by the intimate relationship of Being and truth. Brentano stated but did not solve this problem which stimulated Heidegger's thought through the years. "On the Essence of Truth" (discussed in the General Introduction, pp. 30ff.) stems from the decade of the 1930s but points back to *Being and Time* and forward to virtually all the later works. Section 44 of *Being and Time* is entitled "Dasein, Disclosedness, and Truth." It is divided into three subsections which treat (a) the traditional concept of truth and its ontological foundations, (b) the original phenomenon of truth and the derivative character of the traditional concept, and (c) truth's mode of being, and its presupposition. The traditional concept derives from Aristotle, and Brentano is surely right in conjoining truth and Being—if only as a problem. Aristotle's discussion of truth as *homoiōsis,* a kind of "likening" between things and the soul's experience of them, transmitted through various Jewish and Arabian philosophers, influences medieval scholastic philosophy; the latter's formulations survive in modern and even contemporary philosophy of knowledge. Aquinas speaks of *adaequatio intellectūs et rei,* the correspondence of intellect and thing, Kant of "the agreement of knowledge with its object," while some contemporary logical positivists define truth as "empirical verifiability"—the conformity of assertion to matter of fact. Heidegger wishes to know what is "tacitly posited" in the idea of truth as correspondence and what sort of Being the agreement between knowledge and its object or conformity of proposition and fact must have. The upshot is that a discovery (*Entdecken*) of beings that lets them be seen is always presupposed in all correspondence or adequation of judgment and state of affairs. Hence section 44 refers back to the meaning of *apophansis* in section 7 (cf. p. 78ff., above): the original meaning of truth appears in the word "phenomenology" as a "taking beings out of concealment, letting them be seen in their unconcealment (uncoveredness)." But discovery of beings is grounded in the disclosedness (*Erschlossenheit*) of world and Dasein. Disclosedness or unconcealment (*Unverborgenheit*) is therefore the most original meaning of truth.

However, disclosedness never goes unchallenged. Dasein discovers beings but also covers them over: aware of its possibilities, Dasein is nevertheless "thrown" into the world and "ensnared" by it. Hence Dasein is "equally in truth and in untruth." Open to beings and to its own being possible, Dasein nonetheless relinquishes this openness in exchange for the security of whatever "they" say is true. It lets truth slip into the same oblivion as Being and finds its "truths" as so many scintillating beings there before it, polished yet manipulable. The most dazzlingly finished become "eternal truths." Presupposed in such truths of faith or science or even the university of life, however, is a kind of opening or openness by virtue of which something can and does show itself and let itself be seen. This opening resists depiction. Indeed the attempt to speak of it becomes embroiled in the most complicated abstrusities in order to let this quite simple thing—which is no thing at all—show itself and become manifest.

To *let* unconcealment *show itself:* this is perhaps the most succinct formulation of the task of Heidegger's thinking. At the heart of the task stands the question of *freedom* (see sections 3 and 4 of the present essay), a freedom that refers us back to the discussion of Dasein as *transcendence* (in section 7 of *Being and Time,* pp. 85–86). However, "freedom" and "transcendence" no longer mean what traditional morals and metaphysics take them to mean. Both refer to the mystery of the openness or "clearing" (*Lichtung*) of Being, "the clearing that shelters." Finally, the task requires that we think historically. The word *Wesen* ("essence") in the title of the essay is to be thought historically as an "essential unfolding" (see Reading VII).

*A note on the text*. Heidegger indicates that the first paragraph (he surely means the first *two* paragraphs) of the final section of the truth essay ("9. Note") was appended to the second edition of the essay in 1949. In fact, it is clear that the entire note offers a retrospect on the essay. It tells us that the title "essence of truth" was to be reversed in a sequel on "the truth of essence," a phrase employed in sections 7 and 8 (pp. 132 and 135) of the essay. Heidegger was unable to carry out this reversal, for reasons that become clearer in Reading V. An indication of the growing importance of the *history of Being* in Heidegger's thinking is his adoption of the archaic spelling *Seyn* (here rendered as "Beyng," a form that disappears from English after the sixteenth century) as a name for the ontological difference—the difference between Being and beings—that dominates any given epoch in the history of Being.

# ON THE ESSENCE OF TRUTH

Our topic is the *essence* of truth. The question regarding the essence of truth is not concerned with whether truth is a truth of practical experience or of economic calculation, the truth of a technical consideration or of political sagacity, or, in particular, a truth of scientific research or of artistic composition, or even the truth of thoughtful reflection or of cultic belief. The question of essence disregards all this and attends to the one thing that in general distinguishes every "truth" as truth.

Yet with this question concerning essence do we not soar too high into the void of generality that deprives all thinking of breath? Does not the extravagance of such questioning bring to light the groundlessness of all philosophy? A radical thinking that turns to what is actual must surely from the first insist bluntly on establishing the actual truth that today gives us a measure and a stand against the confusion of opinions and reckonings. In the face of this actual need, what use is the question concerning the essence of truth, this "abstract" question that disregards everything actual? Is not the question of essence the most unessential and superfluous that could be asked?

No one can evade the evident certainty of these considerations. None can lightly neglect their compelling seriousness. But what is

---

This translation of *Vom Wesen der Wahrheit* by John Sallis was prepared especially for this volume. It appears here complete. A previous English translation by R. F. C. Hull and Alan Crick was published in Martin Heidegger, *Existence and Being*, edited by Werner Brock (Chicago: Henry Regnery, 1949), pp. 292–324. The German text is contained in Martin Heidegger, *Wegmarken* (Frankfurt am Main: Vittorio Klostermann Verlag, 1967), pp. 73–97. This translation is based on the fourth edition of the essay, published by Klostermann in 1961.

it that speaks in these considerations? "Sound" common sense. It harps on the demand for palpable utility and inveighs against knowledge of the essence of beings, which essential knowledge has long been called "philosophy."*

Common sense has its own necessity; it asserts its rights with the weapon peculiarly suitable to it, namely, appeal to the "obviousness" of its claims and considerations. However, philosophy can never refute common sense, for the latter is deaf to the language of philosophy. Nor may it even wish to do so, since common sense is blind to what philosophy sets before its essential vision.

Moreover, we ourselves remain within the sensibleness of common sense to the extent that we suppose ourselves to be secure in those multiform "truths" of practical experience and action, of research, composition, and belief. We ourselves intensify that resistance which the "obvious" has to every demand made by what is questionable.

Therefore even if some questioning concerning truth is necessary, what we then demand is an answer to the question as to where we stand today. We want to know what our situation is today. We call for the goal that should be posited for man in and for his history. We want the actual "truth." Well then—truth!

But in calling for the actual "truth" we must already know what truth as such means. Or do we know this only by "feeling" and "in a general way"? But is not such vague "knowing" and our indifference regarding it more desolate than sheer ignorance of the essence of truth?

## 1. The Usual Concept of Truth

What do we ordinarily understand by "truth?" This elevated yet at the same time worn and almost dulled word "truth" means what

*Throughout the translation *das Seiende* is rendered as "being" or "beings," *ein Seiendes* as "a being," *Sein* as "Being," *das Seiende im Ganzen* as either "being as a whole" or "beings as a whole" depending on the context.—Tr.

makes a true thing true. What is a true thing? We say, for example, "It is a true joy to cooperate in the accomplishment of this task." We mean that it is purely and actually a joy. The true is the actual. Accordingly, we speak of true gold in distinction from false. False gold is not actually what it appears to be. It is merely a "semblance" and thus is not actual. What is not actual is taken to be the opposite of the actual. But what merely seems to be gold is nevertheless something actual. Accordingly, we say more precisely: actual gold is genuine gold. Yet both are "actual," the circulating counterfeit no less than the genuine gold. What is true about genuine gold thus cannot be demonstrated merely by its actuality. The question recurs: what do "genuine" and "true" mean here? Genuine gold is that actual gold the actuality of which is in accordance [*in der Übereinstimmung steht*] with what, always and in advance, we "properly" mean by "gold." Conversely, wherever we suspect false gold, we say: "Here something is not in accord" [*stimmt nicht*]. On the other hand, we say of whatever is "as it should be": "It is in accord." The *matter* is in accord [*Die Sache stimmt*].

However, we call true not only an actual joy, genuine gold, and all beings of such kind, but also and above all we call true or false our statements about beings, which can themselves be genuine or not with regard to their kind, which can be thus or otherwise in their actuality. A statement is true if what it means and says is in accordance with the matter about which the statement is made. Here too we say, "It is in accord." Now, though, it is not the matter that is in accord but rather the *proposition*.

The true, whether it be a matter or a proposition, is what accords, the accordant [*das Stimmende*]. Being true and truth here signify accord, and that in a double sense: on the one hand, the consonance [*Einstimmigkeit*] of a matter with what is supposed in advance regarding it and, on the other hand, the accordance of what is meant in the statement with the matter.

This dual character of the accord is brought to light by the traditional definition of truth: *veritas est adaequatio rei et intellectūs.*

This can be taken to mean: truth is the correspondence [*Angleichung*] of the matter to knowledge. But it can also be taken as saying: truth is the correspondence of knowledge to the matter. Admittedly, the above definition is usually stated only in the formula *veritas est adaequatio intellectūs ad rem* [truth is the adequation of intellect to thing]. Yet truth so conceived, propositional truth, is possible only on the basis of material truth [*Sachwahrheit*], of *adaequatio rei ad intellectum* [adequation of thing to intellect]. Both concepts of the essence of *veritas* have continually in view a conforming to . . . [*Sichrichten nach* . . .], and hence think truth as *correctness* [Richtigkeit].

Nonetheless, the one is not the mere inversion of the other. On the contrary, in each case *intellectus* and *res* are thought differently. In order to recognize this we must trace the usual formula for the ordinary concept of truth back to its most recent (i.e., the medieval) origin. *Veritas* as *adaequatio rei ad intellectum* does not imply the later transcendental conception of Kant—possible only on the basis of the subjectivity of man's essence—that "objects conform to our knowledge." Rather, it implies the Christian theological belief that, with respect to what it is and whether it is, a matter, as created (*ens creatum*), *is* only insofar as it corresponds to the idea preconceived in the *intellectus divinus*, i.e., in the mind of God, and thus measures up to the idea (is correct) and in this sense is "true." The *intellectus humanus* too is an *ens creatum*. As a capacity bestowed upon man by God, it must satisfy its *idea*. But the understanding measures up to the idea only by accomplishing in its propositions the correspondence of what is thought to the matter, which in its turn must be in conformity with the *idea*. If all beings are "created," the possibility of the truth of human knowledge is grounded in the fact that matter and proposition measure up to the idea in the same way and therefore are fitted to each other on the basis of the unity of the divine plan of creation. *Veritas* as *adaequatio rei* (*creandae*) *ad intellectum* (*divinum*) guarantees *veritas* as *adaequatio intellectūs* (*humani*) *ad rem* (*creatam*). Throughout, *veritas* essentially implies *convenientia*, the coming of beings themselves, as created, into agreement with the Creator, an "accord" with regard to the way they are determined in the order of creation.

But this order, detached from the notion of creation, can also be represented in a general and indefinite way as a world-order. The theologically conceived order of creation is replaced by the capacity of all objects to be planned by means of a worldly reason [*Weltvernunft*] which supplies the law for itself and thus also claims that its procedure is immediately intelligible (what is considered "logical"). That the essence of propositional truth consists in the correctness of statements needs no further special proof. Even where an effort is made—with a conspicuous lack of success—to explain how correctness is to occur, it is already presupposed as being the essence of truth. Likewise, material truth always signifies the consonance of something at hand with the "rational" concept of its essence. The impression arises that this definition of the essence of truth is independent of the interpretation of the essence of the Being of all beings, which always includes a corresponding interpretation of the essence of man as the bearer and executor of *intellectus*. Thus the formula for the essence of truth (*veritas est adaequatio intellectūs et rei*) comes to have its general validity as something immediately evident to everyone. Under the domination of the obviousness that this concept of truth seems to have but that is hardly attended to as regards its essential grounds, it is considered equally obvious that truth has an opposite, and that there is untruth. The untruth of the proposition (incorrectness) is the nonaccordance of the statement with the matter. The untruth of the matter (nongenuineness) signifies nonagreement of a being with its essence. In each case untruth is conceived as a nonaccord. The latter falls outside the essence of truth. Therefore when it is a question of comprehending the pure essence of truth, untruth, as such an opposite of truth, can be put aside.

But then is there any further need at all for a special unveiling of the essence of truth? Is not the pure essence of truth already adequately represented in the generally accepted concept, which is upset by no theory and is secured by its obviousness? Moreover, if we take the tracing back of propositional truth to material truth to be what in the first instance it shows itself to be, namely a theological explanation, and if we then keep the philosophical definition completely pure of all admixture of theology and limit the concept of truth to propositional truth, then we encounter an old—though not the oldest—tradition of thinking, according to which truth is the accordance (*homoiōsis*) of a statement (*logos*) with a matter (*pragma*). What is it about statements that here remains still worthy of question—granted that we know what is meant by accordance of a statement with the matter? Do we know that?

## 2. The Inner Possibility of Accordance

We speak of accordance in various senses. We say, for example, considering two five-mark coins lying on the table: they are in accordance with one another. They come into accord in the oneness of their outward appearance. Hence they have the latter in common, and thus they are in this regard alike. Furthermore, we speak of accordance whenever, for example, we state regarding one of the five-mark coins: this coin is round. Here the statement is in accordance with the thing. Now the relation obtains, not between thing and thing, but rather between a statement and a thing. But wherein are the thing and the statement supposed to be in accordance, considering that the relata are manifestly different in their outward appearance? The coin is made of metal. The statement is not material at all. The coin is round. The statement has nothing at all spatial about it. With the coin something can be purchased. The statement about it is never a means of payment. But in spite of all their dissimilarity the above statement, as true, is in accordance with the coin. And according to the usual concept of truth this accord is supposed to be a correspondence. How can what is completely dissimilar, the statement, correspond to the coin? It would have to become the coin and in this way relinquish itself entirely. The statement never succeeds in doing that. The moment it did, it would no longer be able as a statement to be in accordance with the thing. In the correspondence the statement must remain—indeed even first become—what it is. In what does its essence, so thoroughly different from every thing, consist? How is the statement able to correspond to something else, the thing, precisely by persisting in its own essence?

Correspondence here cannot signify a thing-like approximation between dissimilar kinds of things. The essence of the correspondence is determined rather by the kind of relation that obtains between the statement and the thing. As long as this "relation" remains undetermined and is not grounded in its essence, all dispute over the possibility and impossibility, over the nature and degree, of the correspondence loses its way in a void. But the statement regarding the coin relates "itself" to this thing in that it presents [*vor-stellt*] it and says of the presented how, according to the particular perspective that guides it, it is disposed. What is stated by the presentative statement is said of the presented thing in just such manner *as* that thing, as presented, is. The "such-as" has to do with the presenting and its presented. Disregarding all "psychological" preconceptions as well as those of any "theory of consciousness," to present here means to let the thing stand opposed as object. As thus placed, what stands opposed must traverse an open field of opposedness [*Entgegen*] and nevertheless must maintain its stand as a thing and show itself as something withstanding [*ein Ständiges*]. This appearing of the thing in traversing a field of opposedness takes place within an open region, the openness of which is not first created by the presenting but rather is only entered into and taken over as a domain of relatedness. The relation of the presentative statement to the thing is the accomplishment of that *bearing* [Verhältnis] which originally and always comes to prevail as a comportment [*Verhalten*]. But all comportment is distinguished by the fact that, standing in the open region, it adheres to something opened up *as such.** What is thus opened up, solely in this strict sense, was experienced early in Western thinking as "what is present" and for a long time has been named "being."

Comportment stands open to beings. Every open relatedness is a comportment. Man's open stance varies depending on the kind of beings and the way of comportment. All working and achieving, all action and calculation, keep within an open region within which beings, with regard to what they are and how they are, can properly take their stand and become capable of being said. This can occur only if beings present themselves along with the presentative statement so that the latter subordinates itself to the directive that it speak of beings *such-as* they are. In following such a directive the statement conforms to beings. Speech that directs itself accordingly is correct (true). What is thus said is the correct (the true).

A statement is invested with its correctness by the openness of comportment; for only through the latter can what is opened up really become the standard for the presentative correspondence. Open comportment must let itself be assigned this standard. This means that it must take over a pregiven standard for all presenting. This belongs to the openness of comportment. But if the correctness (truth) of statements becomes possible only through this openness of comportment, then what first makes correctness possible must with more original right be taken as the essence of truth.

Thus the traditional assignment of truth exclusively to statements as the sole essential locus of truth falls away. Truth does not originally reside in the proposition. But at the same time the question arises as to the ground of the inner possibility of the open comportment that pregives a standard, which possibility alone lends to

*The text reads, "ein Offenbares *als ein solches*." In ordinary German *offenbar* means "evident," "manifest." However, the context that it has here through its link with "open region" (*das Offene*), "open stance" (*Offenständigkeit*), and "openness" (*Offenheit*) already suggests the richer sense that the word has for Heidegger: that of something's being so opened up as to reveal itself, to be manifest (as, for example, a flower in bloom), in contrast to something's being so closed or sealed up within itself that it conceals itself.—Tr.

propositional correctness the appearance of fulfilling the essence of truth at all.

### 3. The Ground of the Possibility of Correctness

Whence does the presentative statement receive the directive to conform to the object and to accord by way of correctness? Why is this accord involved in determining the essence of truth? How can something like the accomplishment of a pregiven directedness occur? And how can the initiation into an accord occur? Only if this pregiving has already entered freely into an open region for something opened up which prevails there and which binds every presenting. To free oneself for a binding directedness is possible only by *being free* for what is opened up in an open region. Such being free points to the heretofore uncomprehended essence of freedom. The openness of comportment as the inner condition of the possibility of correctness is grounded in freedom. *The essence of truth is freedom.*

But does not this proposition regarding the essence of correctness substitute one obvious item for another? In order to be able to carry out any act, and therefore one of presentative stating and even of according or not according with a "truth," the actor must of course be free. However, the proposition in question does not really mean that an unconstrained act belongs to the execution of the statement, to its pronouncement and reception; rather, the proposition says that freedom is the *essence* of truth itself. In this connection "essence" is understood as the ground of the inner possibility of what is initially and generally admitted as known. Nevertheless, in the concept of freedom we do not think truth, and certainly not at all its essence. The proposition that the essence of truth (correctness of statements) is freedom must consequently seem strange.

To place the essence of truth in freedom—does not this mean to submit truth to human caprice? Can truth be any more radically undermined than by being surrendered to the arbitrariness of this "wavering reed"? What forced itself upon sound judgment again and again in the previous discussion now all the more clearly comes to light: truth is here driven back to the subjectivity of the human subject. Even if an objectivity is also accessible to this subject, such objectivity along with subjectivity, still remains something human and at man's disposal.

Certainly deceit and dissimulation, lies and deception, illusion and semblance—in short, all kinds of untruth—are ascribed to man. But of course untruth is also the opposite of truth. For this reason, as the nonessence of truth, it is appropriately excluded from the sphere of the question concerning the pure essence of truth. This human origin of untruth indeed only serves to confirm by contrast the essence of truth "in itself" as holding sway "beyond" man. Metaphysics regards such truth as the imperishable and eternal, which can never be founded on the transitoriness and fragility that belong to man's essence. How then can the essence of truth still have its subsistence and its ground in human freedom?

Resistance to the proposition that the essence of truth is freedom is based on preconceptions, the most obstinate of which is that freedom is a property of man. The essence of freedom neither needs nor allows any further questioning. Everyone knows what man is.

## 4. The Essence of Freedom

However, indication of the essential connection between truth as correctness and freedom uproots those preconceptions—granted of course that we are prepared for a transformation of thinking. Consideration of the essential connection between truth and freedom leads us to pursue the question of the essence of man in a regard that assures us an experience of a concealed essential ground of man (of Dasein), and in such a manner that the experience transposes us in advance into the originally essential domain of truth. But here it becomes evident also that freedom is the ground of the inner possibility of correctness only because it receives its own essence from the more original essence of uniquely essential truth. Freedom was first determined as freedom for what is opened up in an open region. How is this essence of freedom to be thought? That which is opened up, that to which a presentative statement as correct corresponds, are beings opened up in an open comportment. Freedom for what is opened up in an open region lets beings be the beings they are. Freedom now reveals itself as letting beings be.

Ordinarily we speak of letting be whenever, for example, we forgo some enterprise that has been planned. "We let something be" means we do not touch it again, we have nothing more to do with it. To let something be has here the negative sense of letting it alone, of renouncing it, of indifference and even neglect.

However, the phrase required now—to let beings be—does not refer to neglect and indifference but rather the opposite. To let be is to engage oneself with beings. On the other hand, to be sure, this is not to be understood only as the mere management, preservation, tending, and planning of the beings in each case encountered or sought out. To let be—that is, to let beings be as the beings which they are—means to engage oneself with the open region and its openness into which every being comes to stand, bringing that openness, as it were, along with itself. Western thinking in its beginning conceived this open region as *ta alēthea*, the unconcealed. If we translate *alētheia* as "unconcealment" rather than "truth," this translation is not merely more literal; it contains the directive to rethink the ordinary concept of truth in the sense of the correctness of statements and to think it back to that still uncomprehended disclosedness and disclosure of beings. To engage oneself with the disclosedness of beings is not to lose oneself in them; rather, such engagement withdraws in the face of beings in order that they might reveal themselves with respect to what and how they are, and in order that presentative correspondence might take its standard from them. As this letting-be it exposes itself to beings as such and transposes all comportment into the open region. Letting-be, i.e., freedom, is intrinsically exposing, ek-sistent.* Considered in regard to the essence of truth, the essence of freedom manifests itself as exposure to the disclosedness of beings.

Freedom is not merely what common sense is content to let pass under this name: the caprice, turning up occasionally in our choosing, of inclining in this or that direction. Freedom is not mere absence of constraint with respect to what we can or cannot do. Nor is it on the other hand mere readiness for what is required and necessary (and so somehow a being). Prior to all this ("negative" and "positive" freedom), freedom is engagement in the disclosure of beings as such. Disclosedness itself is conserved in ek-sistent engagement, through which the openness of the open region, i.e., the "there" ["*Da*"], is what it is.

In Da-sein the essential ground, long ungrounded, on the basis of which man is able to ek-sist, is preserved for him. Here "existence" does not mean *existentia* in the sense of occurring or being at hand. Nor on the other hand does it mean, in an "existentiell" fashion, man's moral endeavor on behalf of his "self," based on his psychophysical constitution. Ek-sistence, rooted in truth as freedom, is exposure to the disclosedness of beings as such. Still uncomprehended, indeed, not even in need of an essential grounding, the ek-sistence of historical man begins at that moment when the first thinker takes a questioning stand with regard to the unconcealment of beings by asking: what are beings? In this question unconcealment is experienced for the first time. Being as a whole reveals itself as *physis*, "nature," which here does not yet mean a particular sphere of beings but rather beings as such as a whole, specifically in the sense of upsurgent presence [*aufgehendes Anwesen*]. History begins only when beings themselves are expressly drawn up into their unconcealment and conserved in it, only when this conservation is conceived on the basis of questioning regarding

*This variant of the word *Existenz* indicates the ecstatic character of freedom, its standing outside itself.—Tr.

beings as such. The primordial disclosure of being as a whole, the question concerning beings as such, and the beginning of Western history are the same; they occur together in a "time" which, itself unmeasurable, first opens up the open region for every measure.

But if ek-sistent Da-sein, which lets beings be, sets man free for his "freedom" by first offering to his choice something possible (a being) and by imposing on him something necessary (a being), human caprice does not then have freedom at its disposal. Man does not "possess" freedom as a property. At best, the converse holds: freedom, ek-sistent, disclosive Da-sein, possesses man—so originally that only *it* secures for humanity that distinctive relatedness to being as a whole as such which first founds all history. Only ek-sistent man is historical. "Nature" has no history.

Freedom, understood as letting beings be, is the fulfillment and consummation of the essence of truth in the sense of the disclosure of beings. "Truth" is not a feature of correct propositions that are asserted of an "object" by a human "subject" and then "are valid" somewhere, in what sphere we know not; rather, truth is disclosure of beings through which an openness essentially unfolds [*west*]. All human comportment and bearing are exposed in its open region. Therefore man *is* in the manner of ek-sistence.

Because every mode of human comportment is in its own way open and plies itself to that toward which it comports itself, the restraint of letting-be, i.e., freedom, must have granted it its endowment of that inner directive for correspondence of presentation to beings. That man ek-sists now means that for historical humanity the history of its essential possibilities is conserved in the disclosure of beings as a whole. The rare and the simple decisions of history arise from the way the original essence of truth essentially unfolds.

However, because truth is in essence freedom, historical man can, in letting beings be, also *not* let beings be the beings which they are and as they are. Then beings are covered up and distorted. Semblance comes to power. In it the nonessence of truth comes to the fore. However, because ek-sistent freedom as the essence of truth is not a property of man; because on the contrary man ek-sists and so becomes capable of history only as the property of this freedom; the nonessence of truth cannot first arise subsequently from mere human incapacity and negligence. Rather, untruth must derive from the essence of truth. Only because truth and untruth are, *in essence, not* irrelevant to one another, but rather belong together, is it possible for a true proposition to enter into pointed opposition to the corresponding untrue proposition. The question concerning the essence of truth thus first reaches the original domain of what is at issue when, on the basis of a prior glimpse of the full essence of truth, it has included a consideration of untruth in its unveiling of that essence. Discussion of the nonessence of truth is not the subsequent filling of a gap but rather the decisive step toward an adequate posing of the *question* concerning the essence of truth. Yet how are we to comprehend the nonessence in the essence of truth? If the essence of truth is not exhausted by the correctness of statements, then neither can untruth be equated with the incorrectness of judgments.

## 5. The Essence of Truth

The essence of truth reveals itself as freedom. The latter is ek-sistent, disclosive letting beings be. Every mode of open comportment flourishes in letting beings be and in each case is a comportment to this or that being. As engagement in the disclosure of being as a whole as such, freedom has already attuned all comportment to being as a whole. However, being attuned (attunement)* can never be understood as "experience" and "feeling," because it is thereby simply deprived of its essence. For here it is

*The text reads, "*Die Gestimmtheit* (*Stimmung*). . . ." *Stimmung* refers not only to the kind of attunement that a musical instrument receives by being tuned but also to the kind of attunement that constitutes a mood or a disposition of Dasein. The important etymological connection between *Stimmung* and the various formations based on *stimmen* (to accord) is not retained in the translation.—Tr.

interpreted on the basis of something ("life" and "soul") that can maintain the semblance of the title of essence only as long as it bears in itself the distortion and misinterpretation of being attuned. Being attuned, i.e., ek-sistent exposedness to beings as a whole, can be "experienced" and "felt" only because the "man who experiences," without being aware of the essence of the attunement, is always engaged in being attuned in a way that discloses beings as a whole. Every mode of historical man's comportment—whether accentuated or not, whether understood or not—is attuned, and by this attunement is drawn up into beings as a whole. The openedness of being as a whole does not coincide with the sum of all immediately familiar beings. On the contrary: where beings are not very familiar to man and are scarcely and only roughly known by science, the openedness of beings as a whole can prevail more essentially than it can where the familiar and well-known has become boundless, and nothing is any longer able to withstand the business of knowing, since technical mastery over things bears itself without limit. Precisely in the leveling and planing of this omniscience, this mere knowing, the openedness of beings gets flattened out into the apparent nothingness of what is no longer even a matter of indifference, but rather is simply forgotten.

Letting beings be, which is an attuning, a bringing into accord, prevails throughout and anticipates all the open comportment that flourishes in it. Man's comportment is brought into definite accord throughout by the openedness of being as a whole. However, from the point of view of everyday calculations and preoccupations this "as a whole" appears to be incalculable and incomprehensible. It cannot be understood on the basis of the beings opened up in any given case, whether they belong to nature or to history. Although it ceaselessly brings everything into definite accord, still it remains indefinite, indeterminable; it then coincides for the most part with what is most fleeting and most unconsidered. However, what brings into accord is not nothing, but rather a concealing of beings as a whole. Precisely because letting be always lets beings be in a particular comportment that relates to them and thus discloses them, it conceals beings as a whole. Letting-be is intrinsically at the same time a concealing. In the ek-sistent freedom of Da-sein a concealing of being as a whole propriates [*ereignet sich*]. Here there *is* concealment.

## 6. Untruth as Concealing

Concealment deprives *alētheia* of disclosure yet does not render it *sterēsis* (privation); rather, concealment preserves what is most proper to *alētheia* as its own. Considered with respect to truth as disclosedness, concealment is then undisclosedness and accordingly the untruth that is most proper to the essence of truth. The concealment of beings as a whole does not first show up subsequently as a consequence of the fact that knowledge of beings is always fragmentary. The concealment of beings as a whole, untruth proper, is older than every openedness of this or that being. It is also older than letting-be itself, which in disclosing already holds concealed and comports itself toward concealing. What conserves letting-be in this relatedness to concealing? Nothing less than the concealing of what is concealed as a whole, of beings as such, i.e., the mystery; not a particular mystery regarding this or that, but rather the one mystery—that, in general, mystery (the concealing of what is concealed) as such holds sway throughout man's Da-sein.

In letting beings as a whole be, which discloses and at the same time conceals, it happens that concealing appears as what is first of all concealed. Insofar as it ek-sists, Da-sein conserves the first and broadest undisclosedness, untruth proper. The proper nonessence of truth is the mystery. Here nonessence does not yet have the sense of inferiority to essence in the sense of what is general (*koinon*, *genos*), its *possibilitas* and the ground of its possibility. Nonessence is here what in such a sense would be a pre-essential essence. But "nonessence" means at first and for the most part the deformation of that already inferior essence. Indeed, in each of these significations the nonessence remains always in its own way essential to the essence and never becomes unessential in the sense of irrelevant. But to speak of nonessence and untruth in this manner goes very much against the grain of ordinary opinion and looks like a dragging up of forcibly contrived *paradoxa*. Because it is difficult to eliminate this impression, such a way of speaking, paradoxical only for ordinary *doxa* (opinion), is to be renounced. But surely for those who know about such matters the "non-" of the primordial nonessence of truth, as untruth, points to the still unexperienced domain of the truth of Being (not merely of beings).

As letting beings be, freedom is intrinsically the resolutely open bearing that does not close up in itself.* All comportment is grounded in this bearing and receives from it directedness toward beings and disclosure of them. Nevertheless, this bearing toward concealing conceals itself in the process, letting a forgottenness of the mystery take precedence and disappearing in it. Certainly man takes his bearings [*verhält sich*] constantly in his comportment toward beings; but for the most part he acquiesces in this or that being and its particular openedness. Man clings to what is readily available and controllable even where ultimate matters are concerned. And if he sets out to extend, change, newly assimilate, or secure the openedness of the beings pertaining to the most various domains of his activity and interest, then he still takes his directives from the sphere of readily available intentions and needs.

However, to reside in what is readily available is intrinsically not to let the concealing of what is concealed hold sway. Certainly, among readily familiar things there are also some that are puzzling,

*"Resolutely open bearing" seeks to translate *das entschlossene Verhältnis*. *Entschlossen* is usually rendered as "resolute," but such a translation fails to retain the word's structural relation to *verschlossen*, "closed" or "shut up." Significantly, this connection is what makes it possible for Heidegger to transform the sense of the word: he takes the prefix as a privation rather than as indicating establishment of the condition designated by the word to which it is affixed. Thus, as the text here makes quite clear, *entschlossen* signifies just the opposite of that kind of "resolve" in which one makes up one's mind in such fashion as to close off all other possibilities: it is rather a kind of keeping *un-closed*.—Tr.

unexplained, undecided, questionable. But these self-certain questions are merely transitional, intermediate points in our movement within the readily familiar and thus not essential. Wherever the concealment of beings as a whole is conceded only as a limit that occasionally announces itself, concealing as a fundamental occurrence has sunk into forgottenness.

But the forgotten mystery of Dasein is not eliminated by the forgottenness; rather, the forgottenness bestows on the apparent disappearance of what is forgotten a peculiar presence [*Gegenwart*]. By disavowing itself in and for forgottenness, the mystery leaves historical man in the sphere of what is readily available to him, leaves him to his own resources. Thus left, humanity replenishes its "world" on the basis of the latest needs and aims, and fills out that world by means of proposing and planning. From these man then takes his standards, forgetting being as a whole. He persists in them and continually supplies himself with new standards, yet without considering either the ground for taking up standards or the essence of what gives the standard. In spite of his advance to new standards and goals, man goes wrong as regards the essential genuineness of his standards. He is all the more mistaken the more exclusively he takes himself, as subject, to be the standard for all beings. The inordinate forgetfulness of humanity persists in securing itself by means of what is readily available and always accessible. This persistence has its unwitting support in that *bearing* by which Dasein not only ek-sists but also at the same time *in-sists*, i.e., holds fast to what is offered by beings, as if they were open of and in themselves.

*As ek-sistent, Dasein is insistent.* Even in insistent existence the mystery holds sway, but as the forgotten and hence "unessential" essence of truth.

## 7. Untruth as Errancy

As insistent, man is turned toward the most readily available beings. But he insists only by being already ek-sistent, since, after all, he takes beings as his standard. However, in taking its standard, humanity is turned away from the mystery. The insistent turning toward what is readily available and the ek-sistent turning away from the mystery belong together. They are one and the same. Yet turning toward and away from is based on a turning to and fro proper to Dasein. Man's flight from the mystery toward what is readily available, onward from one current thing to the next, passing the mystery by—this is *erring.**

Man errs. Man does not merely stray into errancy. He is always astray in errancy, because as ek-sistent he in-sists and so already is caught in errancy. The errancy through which man strays is not something which, as it were, extends alongside man like a ditch into which he occasionally stumbles; rather, errancy belongs to the inner constitution of the Da-sein into which historical man is admitted. Errancy is the free space for that turning in which insistent ek-sistence adroitly forgets and mistakes itself constantly anew. The concealing of the concealed being as a whole holds sway in that disclosure of specific beings, which, as forgottenness of concealment, becomes errancy.

Errancy is the essential counter-essence to the primordial essence of truth. Errancy opens itself up as the open region for every opposite to essential truth. Errancy is the open site for and ground of *error.* Error is not merely an isolated mistake but the realm (the domain) of the history of those entanglements in which all kinds of erring get interwoven.

In conformity with its openness and its relatedness to beings as a whole, every mode of comportment has its mode of erring. Error extends from the most ordinary wasting of time, making a mistake, and miscalculating, to going astray and venturing too far in one's essential attitudes and decisions. However, what is ordinarily and even according to the teachings of philosophy recognized as error,

*"To err" may translate *irren* only if it is understood in its root sense derived from the Latin *errare*, "to wander from the right way," and only secondarily in the sense "to fall into error."—Tr.

incorrectness of judgments and falsity of knowledge, is only one mode of erring and, moreover, the most superficial one. The errancy in which any given segment of historical humanity must proceed for its course to be errant is essentially connected with the openness of Dasein. By leading him astray, errancy dominates man through and through. But, as leading astray, errancy at the same time contributes to a possibility that man is capable of drawing up from his ek-sistence—the possibility that, by experiencing errancy itself and by not mistaking the mystery of Da-sein, he *not* let himself be led astray.

Because man's in-sistent ek-sistence proceeds in errancy, and because errancy as leading astray always oppresses in some manner or other and is formidable on the basis of this oppression of the mystery, specifically as something forgotten, in the ek-sistence of his Dasein man is *especially* subjected to the rule of the mystery and the oppression of errancy. He is in the *needful condition of being constrained* by the one and the other. The full essence of truth, including its most proper nonessence, keeps Dasein in need by this perpetual turning to and fro. Dasein is a turning into need. From man's Dasein and from it alone arises the disclosure of necessity and, as a result, the possibility of being transposed into what is inevitable.

The disclosure of beings as such is simultaneously and intrinsically the concealing of being as a whole. In the simultaneity of disclosure and concealing, errancy holds sway. Errancy and the concealing of what is concealed belong to the primordial essence of truth. Freedom, conceived on the basis of the in-sistent ek-sistence of Dasein, is the essence of truth (in the sense of the correctness of presenting) only because freedom itself originates from the primordial essence of truth, the rule of the mystery in errancy. Letting beings be takes its course in open comportment. However, letting beings as such be as a whole occurs in a way befitting its essence only when from time to time it gets taken up in its primordial essence. Then resolute openness toward the mystery [*Ent-schlossen-*

***heit zum Geheimnis***] is under way into errancy as such. Then the question of the essence of truth gets asked more originally. Then the ground of the intertwining of the essence of truth with the truth of essence reveals itself. The glimpse into the mystery out of errancy is a question—in the sense of that unique question of what being as such is as a whole. This questioning thinks the question of the *Being* of beings, a question that is essentially misleading and thus in its manifold meaning is still not mastered. The thinking of Being, from which such questioning primordially originates, has since Plato been understood as "philosophy," and later received the title "metaphysics."

## 8. Philosophy and the Question of Truth

In the thinking of Being the liberation of man for ek-sistence, the liberation that grounds history, is put into words. These are not merely the "expression" of an opinion but always already the ably conserved articulation of the truth of being as a whole. How many have ears for these words matters not. Who those are that can hear them determines man's standpoint in history. However, in the same period in which the beginning of philosophy takes place, the *marked* domination of common sense (sophistry) also begins.

Sophistry appeals to the unquestionable character of the beings that are opened up and interprets all thoughtful questioning as an attack on, an unfortunate irritation of, common sense.

However, what philosophy is according to the estimation of common sense, which is quite justified in its own domain, does not touch on the essence of philosophy, which can be determined only on the basis of relatedness to the original truth of being as such as a whole. But because the full essence of truth contains the nonessence and above all holds sway as concealing, philosophy as a questioning into this truth is intrinsically discordant. Philosophical thinking is gentle releasement that does not renounce the concealment of being as a whole. Philosophical thinking is especially the stern and resolute openness that does not disrupt the concealing but entreats its unbroken essence into the open region of understanding and thus into its own truth.

In the gentle sternness and stern gentleness with which it lets being as such be as a whole, philosophy becomes a questioning which does not cling solely to beings yet which also can allow no externally imposed decree. Kant presaged this innermost need that thinking has. For he says of philosophy:

> Here philosophy is seen in fact to be placed in a precarious position, which is supposed to be stable—although neither in heaven nor on earth is there anything on which it depends or on which it is based. It is here that it has to prove its integrity as the keeper of its laws [*Selbsthalterin ihrer Gesetze*], not as the mouthpiece of laws secretly communicated to it by some implanted sense or by who knows what tutelary nature. (*Grundlegung zur Metaphysik der Sitten. Werke*, Akademieausgabe IV, 425.)

With this essential interpretation of philosophy, Kant, whose work introduces the final turning of Western metaphysics, envisages a domain which to be sure he could understand only on the basis of his fundamental metaphysical position, founded on subjectivity, and which he had to understand as the keeping of its laws. This essential view of the determination of philosophy nevertheless goes far enough to renounce every subjugation of philosophical thinking, the most destitute kind of which lets philosophy still be of value as an "expression" of "culture" (Spengler) and as an ornament of productive mankind.

However, whether philosophy as "keeper of its laws" fulfills its primordially decisive essence, or whether it is not itself first of all kept and appointed to its task as keeper by the truth of that to which its laws pertain—this depends on the primordiality with which the original essence of truth becomes essential for thoughtful questioning.

The present undertaking takes the question of the essence of truth beyond the confines of the ordinary definition provided in the usual concept of essence and helps us to consider whether the question of the essence of truth must not be, at the same time and even first of all, the question concerning the truth of essence. But in the concept of "essence" philosophy thinks Being. In tracing the inner possibility of the correctness of statements back to the ek-sistent freedom of letting-be as its "ground," likewise in pointing to the essential commencement of this ground in concealing and in errancy, we want to show that the essence of truth is not the empty "generality" of an "abstract" universality but rather that which, self-concealing, is unique in the unremitting history of the disclosure of the "meaning" of what we call Being—what we for a long time have been accustomed to considering only as being as a whole.

## 9. Note

The question of the essence of truth arises from the question of the truth of essence. In the former question essence is understood initially in the sense of whatness (*quidditas*) or material content (*realitas*), whereas truth is understood as a characteristic of knowledge. In the question of the truth of essence, essence is understood verbally; in this word, remaining still within metaphysical presentation, Beyng is thought as the difference that holds sway between Being and beings. Truth signifies sheltering that clears [*lichtendes Bergen*] as the basic characteristic of Being. The question of the essence of truth finds its answer in the proposition *the essence of truth is the truth of essence*. After our explanation it can easily be seen that the proposition does not merely reverse the word order so as to conjure the specter of paradox. The subject of the proposition—if this unfortunate grammatical category may still be used at all—is the truth of essence. Sheltering that clears is—i.e., lets essentially unfold—accordance between knowledge and beings. The proposition is not dialectical. It is no proposition at all in the sense of a statement. The answer to the question of the essence of truth is the saying of a turning [*die Sage einer Kehre*] within the history of Being. Because sheltering that clears belongs to it, Being appears primordially in the light of concealing withdrawal. The name of this clearing [*Lichtung*] is *alētheia*.

Already in the original project, the lecture "On the Essence of Truth" was to have been completed by a second lecture, "On the Truth of Essence." The latter failed for reasons that are now indicated in the "Letter on Humanism" [Reading V].

The decisive question (in *Being and Time*, 1927) of the meaning, i.e., of the project-domain (see p. 151), i.e., of the openness, i.e., of the truth of Being and not merely of beings, remains intentionally undeveloped. Our thinking apparently remains on the path of metaphysics. Nevertheless, in its decisive steps, which lead from truth as correctness to ek-sistent freedom, and from the latter to truth as concealing and as errancy, it accomplishes a change in the questioning that belongs to the overcoming of metaphysics. The thinking attempted in the lecture comes to fulfillment in the essential experience that a nearness to the truth of Being is first prepared for historical man on the basis of the Da-sein into which man can enter. Every kind of anthropology and all subjectivity of man as subject is not merely left behind—as it was already in *Being and Time*—and the truth of Being sought as the ground of a transformed historical position; rather, the movement of the lecture is such that it sets out to think from this other ground (Dasein). The course of the questioning is intrinsically the way of a thinking which, instead of furnishing representations and concepts, experiences and tests itself as a transformation of its relatedness to Being.

# IV

# THE ORIGIN OF THE WORK OF ART

> ☙ *Only image formed keeps the vision.*
> *Yet image formed rests in the poem.*

On November 13, 1935, Heidegger delivered a public lecture in Freiburg with the title *Der Ursprung des Kunstwerkes,* "The Origin of the Work of Art." In January 1936 he repeated the lecture in Zürich, Switzerland. During the course of the year he expanded the material, and on November 17 and 24, and December 4, he presented a tripartite lecture with the same title in Frankfurt. The text of the following essay (complete in this revised edition of *Basic Writings,* yet still showing the three sections) derives from the Frankfurt lectures.

The lectures must have been difficult to listen to and understand. A reviewer for the *Frankfurter Allgemeine Zeitung* compared them to "an abandoned landscape," perhaps with some right. For Heidegger avoided the easy answer—which is really a subterfuge—that the origin of the artwork is simply the artist himself. For it is more true to say, as Merleau-Ponty does of Cézanne, that the work of art is the artist's existence and the source of his or her life. It is a series of attempts at seeing, listening, and saying, essays flowing from an inexhaustible and undiscoverable source. Whence does the artwork originate? Where does it spring from? And what springs from the work of art itself? For Heidegger these questions about the origin (*der Ursprung*) relate the matter of art to truth as *alētheia* or unconcealment. For the Being of beings, the coming to presence of things, is the original self-showing by which entities emerge from hiddenness; by the constancy of their relation to concealment beings show that they have an origin. But beings that are works of art manifest their origin in a special way. Heidegger therefore calls art the becoming of truth, the setting to work of the truth of beings.

Of course a "work" of art, whether a painting, poem, or symphony, is a "thing." Heidegger begins by trying to identify the "thingly" quality of artworks, as though "thing" were the genus to which one would add the specific difference "art" in order to make an artistic thing, i.e., a work. He examines three traditional interpretations of the "thing" stemming from ancient ontology, (1) the thing as a substance to which various accidents or properties belong, or as a subject that contains certain predicates, (2) the thing as the unity within the mind of a manifold of sense-impressions, and (3) the thing as matter invested with form. But these interpretations reflect *their* origin in a particular kind of human activity, involvement with tools or equipment (analyzed in *Being and Time,* sections 15–18); they therefore distort the character of both "thing" and "work."

Now, the inadequacy of the three traditional views of the "thing" comes to light in a curious manner: the equipmental origin of these philosophical conceptions betrays itself when Heidegger examines and describes Van Gogh's painting of some (peasant?) shoes. This work reveals things in their Being. More, it reveals the *world* of the peasant who walks in those shoes while working the *earth*. Heidegger argues that such revelation belongs to every work of art: the work erects a *world* which in turn opens a space for man and things; but this distinctive openness rests on something more stable and enduring than any world, i.e., the all-sheltering *earth*. "World" and "earth" are not to be subsumed under the categories of form and matter, nor even under the notions of unconcealment and concealment. How through the work of art we are to envisage the creative strife of world and earth is perhaps the greatest challenge in "The Origin of the Work of Art."

We find help for our effort to understand the notion of "world" when we turn to Heidegger's analysis of "worldliness" in *Being and Time,* sections 14–18. There he defines "world" as the structural whole of significant relationships that Dasein experiences—with tools, things of nature, and other human beings—as being-in-the-world. "World" is that already familiar horizon upon which everyday human existence confidently moves; it is that in which Dasein always has been and which is somehow co-disclosed in all man's projects and possibilities. "World" names the essential mystery of existence, the transcendence that makes Dasein different from all other intramundane entities, the disclosedness of beings, the openness of Being.

But if the notion of "world" is already familiar to us because of *Being and Time,* that of "earth" is strange and without precedent—except perhaps for the myth of *Cura* in section 42 of *Being and Time.* Regarding "earth" we can here provide only one hint. During the winter of 1934–35 Heidegger lectured at Freiburg University on two poems by Friedrich Hölderlin (1770–1843), "Germania" and "The Rhine." In these two poems the word "earth" appears many times, as it does throughout Hölderlin's poetry. Perhaps the *Ursprung* of Heidegger's notion of "earth" must be sought in poetry, which occupies a special place among the works of art. Indeed, "The Origin of the Work of Art" opens onto the entire question of language and poetry (see Reading X, "The Way to Language").

One of the most ancient of the Homeric Hymns, *Eis Gēn Mētera Pantōn,* "To Earth, Mother of All," stands at the source of a poetic heritage that enriched Hölderlin beyond measure. In translation Homeric Hymn number thirty reads:

*Gaia! Allmother will I sing! Revered*
*Firmgrounded nourisher of everything on earth.*
*Whatever traverses holy earth or the seas*
*Or climbs the air enjoys your dispensation.*
*From you sprout fine fruits and offspring;*
*Lady, you have power to give mortal men life*
*Or take it. But happy those you care for in*
*Your heart: all is generously present to them.*
*Fields thicken with lifegiving nourishment,*
*Herds at pasture multiply, houses fill with*
*Splendid things. Right and law rule in the city*
*Blessed by plenty and wealth among women beautiful.*
*Children beam with youth and joy, gleeful girls*
*Dance the ringdance, their hands all blossoms,*
*Leaping on flowery carpets of meadow.*
*Such pleasures your servants enjoy,*
*Goddess sublime! Generous divinity!*

In this essay, as later in "Building Dwelling Thinking" (Reading VIII), Heidegger too celebrates the protection and nourishment earth affords. In a sense all artwork and all thinking are for him participations in the creative strife of world and earth: they reveal beings and let them come to radiant appearance, but only by cultivating and safeguarding their provenance, allowing all things the darkness they require and their proper growing time. In all its work the language of art and thought houses the splendors that come to light.

# THE ORIGIN OF THE WORK OF ART

Origin here means that from which and by which something is what it is and as it is. What something is, as it is, we call its essence. The origin of something is the source of its essence. The question concerning the origin of the work of art asks about its essential source. On the usual view, the work arises out of and by means of the activity of the artist. But by what and whence is the artist what he is? By the work; for to say that the work does credit to the master means that it is the work that first lets the artist emerge as a master of his art. The artist is the origin of the work. The work is the origin of the artist. Neither is without the other. Nevertheless, neither is the sole support of the other. In themselves and in their interrelations artist and work *are* each of them by virtue of a third thing which is prior to both, namely, that which also gives artist and work of art their names—art.

As necessarily as the artist is the origin of the work in a different way than the work is the origin of the artist, so it is equally certain that, in a still different way, art is the origin of both artist and work. But can art be an origin at all? Where and how does art occur? Art—this is nothing more than a word to which nothing actual any longer corresponds. It may pass for a collective idea under which we find a place for that which alone is actual in art: works and

---

In this second edition of *Basic Writings* Heidegger's "The Origin of the Work of Art" appears *complete*, including the later "Epilogue" and the "Addendum" of 1956. The translation is by Albert Hofstadter (in *Poetry, Language, Thought*, New York: Harper & Row, 1971, pp. 17–87), with minor changes. The German text for the translation is Martin Heidegger, *Der Ursprung des Kunstwerkes*, ed. H.G. Gadamer (Stuttgart: P. Reclam, 1960). An error on p. 9 of the German text has been silently corrected.

artists. Even if the word *art* were taken to signify more than a collective notion, what is meant by the word could exist only on the basis of the actuality of works and artists. Or is the converse the case? Do works and artists exist only because art exists as their origin?

Whatever the decision may be, the question of the origin of the work of art becomes a question about the essence of art. Since the question whether and how art in general exists must still remain open, we shall attempt to discover the essence of art in the place where art undoubtedly prevails in an actual way. Art essentially unfolds in the artwork. But what and how is a work of art?

What art is should be inferable from the work. What the work of art is we can come to know only from the essence of art. Anyone can easily see that we are moving in a circle. Ordinary understanding demands that this circle be avoided because it violates logic. What art is can be gathered from a comparative examination of actual artworks. But how are we to be certain that we are indeed basing such an examination on artworks if we do not know beforehand what art is? And the essence of art can no more be arrived at by a derivation from higher concepts than by a collection of characteristics of actual artworks. For such a derivation, too, already has in view the definitions that must suffice to establish that what we in advance take to be an artwork is one in fact. But selecting characteristics from among given objects, and deriving concepts from principles, are equally impossible here, and where these procedures are practiced they are a self-deception.

Thus we are compelled to follow the circle. This is neither a makeshift nor a defect. To enter upon this path is the strength of thought, to continue on it is the feast of thought, assuming that thinking is a craft. Not only is the main step from work to art a circle like the step from art to work, but every separate step that we attempt circles in this circle.

In order to discover the essence of the art that actually prevails in the work, let us go to the actual work and ask the work what and how it is.

Works of art are familiar to everyone. Architectural and sculptural works can be seen installed in public places, in churches, and in dwellings. Artworks of the most diverse periods and peoples are housed in collections and exhibitions. If we consider the works in their untouched actuality and do not deceive ourselves, the result is that the works are as naturally present as are things. The picture hangs on the wall like a rifle or a hat. A painting, e.g., the one by Van Gogh that represents a pair of peasant shoes, travels from one exhibition to another. Works of art are shipped like coal from the Ruhr and logs from the Black Forest. During the First World War Hölderlin's hymns were packed in the soldier's knapsack together with cleaning gear. Beethoven's quartets lie in the storerooms of the publishing house like potatoes in a cellar.

All works have this thingly character. What would they be without it? But perhaps this rather crude and external view of the work is objectionable to us. Shippers or charwomen in museums may operate with such conceptions of the work of art. We, however, have to take works as they are encountered by those who experience and enjoy them. But even the much-vaunted aesthetic experience cannot get around the thingly aspect of the artwork. There is something stony in a work of architecture, wooden in a carving, colored in a painting, spoken in a linguistic work, sonorous in a musical composition. The thingly element is so irremovably present in the artwork that we are compelled rather to say conversely that the architectural work is in stone, the carving is in wood, the painting in color, the linguistic work in speech, the musical composition in sound. "Obviously," it will be replied. No doubt. But what is this self-evident thingly element in the work of art?

Presumably it becomes superfluous and confusing to inquire into this feature, since the artwork is something else over and above the thingly element. This something else in the work constitutes its artistic nature. The artwork is, to be sure, a thing that is made, but it says something other than what the mere thing itself is, *allo agoreuei*. The work makes public something other than itself; it manifests something other; it is an allegory. In the work of art something other is brought together with the thing that is made. To bring together is, in Greek, *symballein*. The work is a symbol.

Allegory and symbol provide the conceptual frame within whose channel of vision the artwork has long been characterized. But this one element in a work that manifests another, this one element that joins with another, is the thingly feature in the artwork. It seems almost as though the thingly element in the artwork is like the substructure into and upon which the other, proper element is built. And is it not this thingly feature in the work that the artist properly makes by his handicraft?

Our aim is to arrive at the immediate and full actuality of the work of art, for only in this way shall we discover actual art also within it. Hence we must first bring to view the thingly element of the work. To this end it is necessary that we should know with sufficient clarity what a thing is. Only then can we say whether the artwork is a thing, but a thing to which something else adheres; only then can we decide whether the work is at bottom something else and not a thing at all.

## Thing and Work

What in truth is the thing, so far as it is a thing? When we inquire in this way, our aim is to come to know the thing-being (thingness) of the thing. The point is to discover the thingly character of the thing. To this end we have to be acquainted with the sphere to which all those entities belong which we have long called by the name of thing.

The stone in the road is a thing, as is the clod in the field. A jug is a thing, as is the well beside the road. But what about the milk in the jug and the water in the well? These too are things if the cloud in the sky and the thistle in the field, the leaf in the autumn breeze and the hawk over the wood, are rightly called by the name of thing. All these must indeed be called things, if the name is applied even to that which does not, like those just enumerated, show itself, i.e., that which does not appear. According to Kant, the whole of the world, for example, and even God himself, is a thing of this sort, a thing that does not itself appear, namely, a "thing-in-itself." In the language of philosophy both things-in-themselves and things that appear, all beings that in any way are, are called things.

Airplanes and radio sets are nowadays among the things closest to us, but when we have in mind the last things we think of something altogether different. Death and judgment—these are the last things. On the whole the word "thing" here designates whatever is not simply nothing. In this sense the work of art is also a thing, so far as it is some sort of being. Yet this concept is of no use to us, at least immediately, in our attempt to delimit entities that have the mode of being of a thing, as against those having the mode of being of a work. And besides, we hesitate to call God a thing. In the same way we hesitate to consider the peasant in the field, the stoker at the boiler, the teacher in the school as things. A man is not a thing. It is true that we speak of a young girl who is faced with a task too difficult for her as being a young thing, still too young for it, but only because we feel that being human is in a certain way missing here and think that instead we have to do here with the factor that constitutes the thingly character of things. We hesitate even to call the deer in the forest clearing, the beetle in the grass, the blade of grass a thing. We would sooner think of a hammer as a thing, or a shoe, or an ax, or a clock. But even these are not mere things. Only a stone, a clod of earth, a piece of wood are for us such mere things. Lifeless beings of nature and objects of use. Natural things and utensils are the things commonly so called.

We thus see ourselves brought back from the widest domain, within which everything is a thing (thing = *res* = *ens* = a being), including even the highest and last things, to the narrow precinct of mere things. "Mere" here means, first, the pure thing, which is simply a thing and nothing more; but then, at the same time, it means that which is only a thing, in an almost pejorative sense. It is mere things, excluding even utensils, that count as things in the proper sense. What does the thingly character of these things, then, consist in? It is in reference to these that the thingness of things must be determinable. This determination enables us to characterize what it is that is thingly as such. Thus prepared, we are able to characterize the almost palpable actuality of works, in which something else inheres.

Now, it passes for a known fact that as far back as antiquity, no sooner was the question raised as to what beings are in general, than things in their thingness thrust themselves into prominence again and again as the standard type of beings. Consequently we are bound to meet with the definition of the thingness of things already in the traditional interpretations of beings. We thus need only to ascertain explicitly this traditional knowledge of the thing, to be relieved of the tedious labor of making our own search for the thingly character of the thing. The answers to the question "What is the thing?" are so familiar that we no longer sense anything questionable behind them.

The interpretations of the thingness of the thing which, predominant in the course of Western thought, long ago became self-evident and are now in everyday use, may be reduced to three.

This block of granite, for example, is a mere thing. It is hard, heavy, extended, bulky, shapeless, rough, colored, partly dull, partly shiny. We can take note of all these features in the stone. Thus we acknowledge its characteristics. But still, the traits signify something proper to the stone itself. They are its properties. The thing has them. The thing? What are we thinking of when we now have the thing in mind? Obviously a thing is not merely an aggregate of traits, nor an accumulation of properties by which that aggregate arises. A thing, as everyone thinks he knows, is that around which the properties have assembled. We speak in this connection of the core of things. The Greeks are supposed to have called it *to hypokeimenon*. For them, this core of the thing was something lying at the ground of the thing, something always already there. The characteristics, however, are called *ta symbebēkota*, that which has always turned up already along with the given core and occurs along with it.

These designations are no arbitrary names. Something that lies beyond the purview of this essay speaks in them, the basic Greek experience of the Being of beings in the sense of presence. It is by these determinations, however, that the interpretation of the thingness of the thing is established which henceforth becomes standard, and the Western interpretation of the Being of beings stabilized. The process begins with the appropriation of Greek words by Roman-Latin thought. *Hypokeimenon* becomes *subiectum; hypostasis* becomes *substantia; symbebēkos* becomes *accidens*. However, this translation of Greek names into Latin is in no way the innocent process it is considered to this day. Beneath the seemingly literal and thus faithful translation there is concealed, rather, a *trans*lation of Greek experience into a different way of thinking. *Roman thought takes over the Greek words without a corresponding, equally original experience of what they say, without the Greek word*. The rootlessness of Western thought begins with this translation.

According to current opinion, this definition of the thingness of the thing as the substance with its accidents seems to correspond to our natural outlook on things. No wonder that the current attitude toward things—our way of addressing ourselves to things and speaking about them—has adapted itself to this common view of the thing. A simple propositional statement consists of the subject, which is the Latin translation, hence already a reinterpretation, of *hypokeimenon*, and the predicate, in which the thing's traits are stated of it. Who would have the temerity to assail these simple fundamental relations between thing and statement, between sentence structure and thing-structure? Nevertheless, we must ask: Is the structure of a simple propositional statement (the combination of subject and predicate) the mirror image of the structure of the thing (of the union of substance with accidents)? Or could it be that even the structure of the thing as thus envisaged is a projection of the framework of the sentence?

What could be more obvious than that man transposes his propositional way of understanding things into the structure of the thing itself? Yet this view, seemingly critical yet actually rash and ill-considered, would have to explain first how such a transposition of propositional structure into the thing is supposed to be possible without the thing having already become visible. The question as to which comes first and functions as the standard, proposition-structure or thing-structure, remains to this hour undecided. It even remains doubtful whether in this form the question is at all decidable.

Actually, the sentence structure does not provide the standard for the pattern of thing-structure, nor is the latter simply mirrored in the former. Both sentence and thing-structure derive, in their typical form and their possible mutual relationship, from a common and more original source. In any case this first interpretation of the thingness of the thing, the thing as bearer of its characteristic traits, despite its currency, is not as natural as it appears to be. What seems natural to us is probably just something familiar in a long tradition that has forgotten the unfamiliar source from which it arose. And yet this unfamiliar source once struck man as strange and caused him to think and to wonder.

Our confidence in the current interpretation of the thing is only seemingly well founded. But in addition this thing-concept (the thing as bearer of its characteristics) holds not only of the mere thing in its proper sense, but also of any being whatsoever. Hence it cannot be used to set apart thingly beings from non-thingly beings. Yet even before all reflection, attentive dwelling within the sphere of things already tells us that this thing-concept does not hit upon the thingly element of the thing, its independent and self-contained character. Occasionally we still have the feeling that violence has long been done to the thingly element of things and that thought has played a part in this violence, for which reason people disavow thought instead of taking pains to make it more thoughtful. But in defining the essence of the thing, what is the use of a feeling, however certain, if thought alone has the right to speak here? Perhaps, however, what we call feeling or mood, here and in similar instances, is more reasonable—that is, more intelligently perceptive—because more open to Being than all that reason which, having meanwhile become *ratio*, was misinterpreted as being rational. The hankering after the irrational, as abortive offspring of the unthought rational, therewith performed a curious service. To be sure, the current thing-concept always fits each thing. Nevertheless, it does not lay hold of the thing as it is in its own being, but makes an assault upon it.

Can such an assault perhaps be avoided—and how? Only, certainly, by granting the thing, as it were, a free field to display its thingly character directly. Everything that might interpose itself between the thing and us in apprehending and talking about it must first be set aside. Only then do we yield ourselves to the undistorted presencing of the thing. But we do not need first to call or arrange for this situation in which we let things encounter us without mediation. The situation always prevails. In what the senses of sight, hearing, and touch convey, in the sensations of color, sound, roughness, hardness, things move us bodily, in the literal meaning of the word. The thing is the *aisthēton*, that which is perceptible by sensations in the senses belonging to sensibility. Hence the concept later becomes a commonplace according to which a thing is nothing but the unity of a manifold of what is given in the senses. Whether this unity is conceived as sum or as totality or as *Gestalt* alters nothing in the standard character of this thing-concept.

Now this interpretation of the thingness of the thing is as correct and demonstrable in every case as the previous one. This already suffices to cast doubt on its truth. If we consider moreover what we are searching for, the thingly character of the thing, then this thing-concept again leaves us at a loss. We never really first perceive a throng of sensations, e.g., tones and noises, in the appearance of things—as this thing-concept alleges; rather we hear the storm whistling in the chimney, we hear the three-motored plane, we hear the Mercedes in immediate distinction from the Volkswagen. Much closer to us than all sensations are the things themselves. We hear the door shut in the house and never hear acoustical sensations or even mere sounds. In order to hear a bare sound we have to listen away from things, divert our ear from them, i.e., listen abstractly.

In the thing-concept just mentioned there is not so much an assault upon the thing as rather an inordinate attempt to bring it into the greatest possible proximity to us. But a thing never reaches that position as long as we assign as its thingly feature what is perceived by the senses. Whereas the first interpretation keeps the thing at arm's length from us, as it were, and sets it too far off, the second makes it press too physically upon us. In both interpretations the thing vanishes. It is therefore necessary to avoid the exaggerations of both. The thing itself must be allowed to remain in its self-containment. It must be accepted in its own steadfastness. This the third interpretation seems to do, which is just as old as the first two.

That which gives things their constancy and pith but is also at the same time the source of their particular mode of sensuous pressure—colored, resonant, hard, massive—is the matter in things. In this analysis of the thing as matter (*hyle*), form (*morphē*) is already coposited. What is constant in a thing, its consistency, lies in the fact that matter stands together with a form. The thing is formed matter. This interpretation appeals to the immediate view with which the thing solicits us by its outward appearance (*eidos*). In this synthesis of matter and form a thing-concept has finally been found which applies equally to things of nature and to utensils.

This concept puts us in a position to answer the question concerning the thingly element in the work of art. The thingly element is manifestly the matter of which it consists. Matter is the substrate and field for the artist's formative action. But we could have advanced this obvious and well-known definition of the thingly element at the very outset. Why do we make a detour through other applicable thing-concepts? Because we also mistrust this concept of the thing, which represents it as formed matter.

But is not precisely this pair of concepts, matter-form, usually employed in the domain in which we are supposed to be moving? To be sure. The distinction of matter and form is *the conceptual schema which is used, in the greatest variety of ways, quite generally for all art theory and aesthetics*. This incontestable fact, however, proves neither that the distinction of matter and form is adequately founded, nor that it belongs originally to the domain of art and the artwork. Moreover, the range of application of this pair of concepts has long extended far beyond the field of aesthetics. Form and content are the most hackneyed concepts under which anything and everything may be subsumed. And if form is correlated with the rational and matter with the irrational; if the rational is taken to be the logical and the irrational the alogical; if in addition the subject-object relation is coupled with the conceptual pair form-matter; then representation has at its command a conceptual machinery that nothing is capable of withstanding.

If, however, it is thus with the distinction between matter and form, how then shall we make use of it to lay hold of the particular domain of mere things by contrast with all other entities? But perhaps this characterization in terms of matter and form would recover its defining power if only we reversed the process of expanding and emptying these concepts. Certainly, but this presupposes that we know in what sphere of beings they realize their genuine defining power. That this is the domain of mere things is so far only an assumption. Reference to the copious use made of this conceptual framework in aesthetics might sooner lead to the idea that matter and form are specifications stemming from the essence of the artwork and were in the first place transferred from it back to the thing. Where does the matter-form structure have its origin—in the thingly character of the thing or in the workly character of the artwork?

The self-contained block of granite is something material in a definite if unshapely form. Form means here the distribution and arrangement of the material parts in spatial locations, resulting in a particular shape, namely, that of a block. But a jug, an ax, a shoe are also matter occurring in a form. Form as shape is not the consequence here of a prior distribution of the matter. The form, on the contrary, determines the arrangement of the matter. Even more, it prescribes in each case the kind and selection of the matter—impermeable for a jug, sufficiently hard for an ax, firm yet flexible for shoes. The interfusion of form and matter prevailing here is, moreover, controlled beforehand by the purposes served by jug, ax, shoes. Such usefulness is never assigned or added on afterward to a being of the type of a jug, ax, or pair of shoes. But neither is it something that floats somewhere above it as an end.

Usefulness is the basic feature from which this being regards us, that is, flashes at us and thereby is present and thus is this being. Both the formative act and the choice of material—a choice given with the act—and therewith the dominance of the conjunction of matter and form, are all grounded in such usefulness. A being that falls under usefulness is always the product of a process of making. It is made as a piece of equipment for something. As determinations of beings, accordingly, matter and form have their proper place in the essential nature of equipment. This name designates what is produced expressly for employment and use. Matter and form are in no case original determinations of the thingness of the mere thing.

A piece of equipment, a pair of shoes for instance, when finished, is also self-contained like the mere thing, but it does not have the character of having taken shape by itself like the granite boulder. On the other hand, equipment displays an affinity with the artwork insofar as it is something produced by the human hand. However, by its self-sufficient presencing the work of art is similar rather to the mere thing which has taken shape by itself and is self-contained. Nevertheless we do not count such works among mere things. As a rule it is the use-objects around us that are the nearest and the proper things. Thus the piece of equipment is half thing, because characterized by thingliness, and yet it is something more; at the same time it is half artwork and yet something less, because lacking the self-sufficiency of the artwork. Equipment has a peculiar position intermediate between thing and work, assuming that such a calculated ordering of them is permissible.

The matter-form structure, however, by which the Being of a piece of equipment is first determined, readily presents itself as the immediately intelligible constitution of every being, because here man himself as maker participates in the way in which the piece of equipment comes into being. Because equipment takes an intermediate place between mere thing and work, the suggestion is that nonequipmental beings—things and works and ultimately all beings—are to be comprehended with the help of the Being of equipment (the matter-form structure).

The inclination to treat the matter-form structure as *the* constitution of every being receives an additional impulse from the fact that on the basis of a religious faith, namely, the biblical faith, the totality of all beings is represented in advance as something created, which here means made. The philosophy of this faith can of course assure us that all of God's creative work is to be thought of as different from the action of a craftsman. Nevertheless, if at the same time or even beforehand, in accordance with a presumed predetermination of Thomistic philosophy for interpreting the Bible, the *ens creatum* is conceived as a unity of *materia* and *forma*, then faith is expounded by way of a philosophy whose truth lies in an unconcealedness of beings which differs in kind from the world believed in by faith.

The idea of creation, grounded in faith, can lose its guiding power for knowledge of beings as a whole. But the theological interpretation of all beings, the view of the world in terms of matter and form borrowed from an alien philosophy, having once been instituted, can still remain a force. This happens in the transition from the

Middle Ages to modern times. The metaphysics of the modern period rests on the form-matter structure devised in the medieval period, which itself merely recalls in its words the buried natures of *eidos* and *hyle*. Thus the interpretation of "thing" by means of matter and form, whether it remains medieval or becomes Kantian-transcendental, has become current and self-evident. But for that reason, no less than the other interpretations mentioned of the thingness of the thing, it is an encroachment upon the thing-being of the thing.

The situation stands revealed as soon as we speak of things in the proper sense as mere things. The "mere," after all, means the removal of the character of usefulness and of being made. The mere thing is a sort of equipment, albeit equipment denuded of its equipmental being. Thing-being consists in what is then left over. But this remnant is not actually defined in its ontological character. It remains doubtful whether the thingly character comes to view at all in the process of stripping off everything equipmental. Thus the third mode of interpretation of the thing, that which follows the lead of the matter-form structure, also turns out to be an assault upon the thing.

These three modes of defining thingness conceive of the thing as a bearer of traits, as the unity of a manifold of sensations, as formed matter. In the course of the history of truth about beings, the interpretations mentioned have also entered into combinations, a matter we may now pass over. In such combinations, they have further strengthened their innate tendency to expand so as to apply in the same way to thing, to equipment, and to work. Thus they give rise to a mode of thought by which we think not only about thing, equipment, and work but about all beings in general. This long-familiar mode of thought preconceives all immediate experience of beings. The preconception shackles reflection on the Being of any given being. Thus it comes about that prevailing thing-concepts obstruct the way toward the thingly character of the thing as well as toward the equipmental character of equipment, and all the more toward the workly character of the work.

That is why it is necessary to know about these thing-concepts, in order thereby to take heed of their provenance and their boundless presumption, but also of their semblance of self-evidence. This knowledge becomes all the more necessary when we risk the attempt to bring to view and express in words the thingly character of the thing, the equipmental character of equipment, and the workly character of the work. To this end, however, only one element is needful: to keep at a distance all the preconceptions and assaults of the above modes of thought, to leave the thing to rest in its own self, for instance, in its thing-being. What seems easier than to let a thing be just the being that it is? Or does this turn out to be the most difficult of tasks, particularly if such an intention—to let a being be as it is—represents the opposite of the indifference that simply turns its back upon the being itself in favor of an unexamined concept of Being? We ought to turn toward the being, think about it in regard to its Being, but by means of this thinking at the same time let it rest upon itself in its very own essence.

This exertion of thought seems to meet with its greatest resistance in defining the thingness of the thing; for where else could the cause lie of the failure of the efforts mentioned? The unpretentious thing evades thought most stubbornly. Or can it be that this self-refusal of the mere thing, this self-contained, irreducible spontaneity, belongs precisely to the essence of the thing? Must not this strange and uncommunicative feature of the essence of the thing become intimately familiar to thought that tries to think the thing? If so, then we should not force our way to its thingly character.

That the thingness of the thing is particularly difficult to express and only seldom expressible is infallibly documented by the history of its interpretation indicated above. This history coincides with the destiny in accordance with which Western thought has hitherto thought the Being of beings. However, not only do we now establish this point; at the same time we discover a clue in this history. Is it an accident that in the interpretation of the thing the view that takes matter and form as guide attains special dominance? This definition of the thing derives from an interpretation of the equipmental being of equipment. And equipment, having come into being through human making, is a being particularly familiar to human thinking. At the same time, this being that is so familiar in its Being has a peculiar intermediate position between thing and work. We shall follow this clue and search first for the equipmental character of equipment. Perhaps this will suggest something to us about the thingly character of the thing and the workly character of the work. We must only avoid making thing and work prematurely into subspecies of equipment. We are disregarding the possibility, however, that differences relating to the essential history of Being may yet also be present in the way equipment *is*.

But what path leads to the equipmental quality of equipment? How shall we discover what a piece of equipment truly is? The procedure necessary at present must plainly avoid any attempts that again immediately entail the encroachments of the usual interpretations. We are most easily insured against this if we simply describe some equipment without any philosophical theory.

We choose as example a common sort of equipment—a pair of peasant shoes. We do not even need to exhibit actual pieces of this sort of useful article in order to describe them. Everyone is acquainted with them. But since it is a matter here of direct description, it may be well to facilitate the visual realization of them. For this purpose a pictorial representation suffices. We shall choose a well-known painting by Van Gogh, who painted such shoes several times. But what is there to see here? Everyone knows what shoes consist of. If they are not wooden or bast shoes, there will be leather soles and uppers, joined together by thread and nails. Such gear serves to clothe the feet. Depending on the use to which the shoes are to be put, whether for work in the field or for dancing, matter and form will differ.

Such statements, no doubt correct, only explicate what we already know. The equipmental quality of equipment consists in its usefulness. But what about this usefulness itself? In conceiving it, do we already conceive along with it the equipmental character of equipment? In order to succeed in doing this, must we not look out for useful equipment in its use? The peasant woman wears her shoes in the field. Only here are they what they are. They are all the more genuinely so, the less the peasant woman thinks about the shoes while she is at work, or looks at them at all, or is even aware of them. She stands and walks in them. That is how shoes actually serve. It is in this process of the use of equipment that we must actually encounter the character of equipment.

As long as we only imagine a pair of shoes in general, or simply look at the empty, unused shoes as they merely stand there in the picture, we shall never discover what the equipmental being of the equipment in truth is. From Van Gogh's painting we cannot even tell where these shoes stand. There is nothing surrounding this pair of peasant shoes in or to which they might belong—only an undefined space. There are not even clods of soil from the field or the field-path sticking to them, which would at least hint at their use. A pair of peasant shoes and nothing more. And yet.

From the dark opening of the worn insides of the shoes the toilsome tread of the worker stares forth. In the stiffly rugged heaviness of the shoes there is the accumulated tenacity of her slow trudge through the far-spreading and ever-uniform furrows of the field swept by a raw wind. On the leather lie the dampness and richness of the soil. Under the soles stretches the loneliness of the field-path as evening falls. In the shoes vibrates the silent call of the earth, its quiet gift of the ripening grain and its unexplained self-refusal in the fallow desolation of the wintry field. This equipment is pervaded by uncomplaining worry as to the certainty of bread, the wordless joy of having once more withstood want, the trembling before the impending childbed and shivering at the surrounding menace of death. This equipment belongs to the *earth*, and it is protected in the *world* of the peasant woman. From out of this protected belonging the equipment itself rises to its resting-within-itself.

But perhaps it is only in the picture that we notice all this about the shoes. The peasant woman, on the other hand, simply wears them. If only this simple wearing were so simple. When she takes off her shoes late in the evening, in deep but healthy fatigue, and reaches out for them again in the still dim dawn, or passes them by on the day of rest, she knows all this without noticing or reflecting. The equipmental being of the equipment consists indeed in its usefulness. But this usefulness itself rests in the abundance of an essential Being of the equipment. We call it reliability. By virtue of this reliability the peasant woman is made privy to the silent call of the earth; by virtue of the reliability of the equipment she is sure of her world. World and earth exist for her, and for those who are with her in her mode of being, only thus—in the equipment. We say "only" and therewith fall into error; for the reliability of the equipment first gives to the simple world its security and assures to the earth the freedom of its steady thrust.

The equipmental being of equipment, reliability, keeps gathered within itself all things according to their manner and extent. The usefulness of equipment is nevertheless only the essential consequence of reliability. The former vibrates in the latter and would be nothing without it. A single piece of equipment is worn out and used up; but at the same time the use itself also falls into disuse, wears away, and becomes usual. Thus equipmentality wastes away, sinks into mere stuff. In such wasting, reliability vanishes. This dwindling, however, to which use-things owe their boringly obtrusive usualness, is only one more testimony to the original essence of equipmental being. The worn-out usualness of the equipment then obtrudes itself as the sole mode of being, apparently peculiar to it exclusively. Only blank usefulness now remains visible. It awakens the impression that the origin of equipment lies in a mere fabricating that impresses a form upon some matter. Nevertheless, in its genuinely equipmental being, equipment stems from a more distant source. Matter and form and their distinction have a deeper origin.

The repose of equipment resting within itself consists in its reliability. Only in this reliability do we discern what equipment in truth is. But we still know nothing of what we first sought: the thing's thingly character. And we know nothing at all of what we really and solely seek: the workly character of the work in the sense of the work of art.

Or have we already learned something unwittingly—in passing, so to speak—about the work-being of the work?

The equipmental quality of equipment was discovered. But how? Not by a description and explanation of a pair of shoes actually present; not by a report about the process of making shoes; and also not by the observation of the actual use of shoes occurring here and there; but only by bringing ourselves before Van Gogh's painting. This painting spoke. In the nearness of the work we were suddenly somewhere else than we usually tend to be.

The artwork lets us know what shoes are in truth. It would be the worst self-deception to think that our description, as a subjective action, had first depicted everything thus and then projected it into the painting. If anything is questionable here, it is rather that we experienced too little in the nearness of the work and that we expressed the experience too crudely and too literally. But above all, the work did not, as it might seem at first, serve merely for a better visualizing of what a piece of equipment is. Rather, the equipmentality of equipment first expressly comes to the fore through the work and only in the work.

What happens here? What is at work in the work? Van Gogh's painting is the disclosure of what the equipment, the pair of peasant shoes, *is* in truth. This being emerges into the unconcealment of its Being. The Greeks called the unconcealment of beings *alētheia*. We say "truth" and think little enough in using this word. If there occurs in the work a disclosure of a particular being, disclosing what and how it is, then there is here an occurring, a happening of truth at work.

In the work of art the truth of beings has set itself to work. "To set" means here "to bring to stand." Some particular being, a pair of peasant shoes, comes in the work to stand in the light of its Being. The Being of beings comes into the steadiness of its shining.

The essence of art would then be this: the truth of beings setting itself to work. But until now art presumably has had to do with the beautiful and beauty, and not with truth. The arts that produce such works are called the fine arts, in contrast with the applied or industrial arts that manufacture equipment. In fine art the art itself is not beautiful, but is called so because it produces the beautiful. Truth, in contrast, belongs to logic. Beauty, however, is reserved for aesthetics.

But perhaps the proposition that art is truth setting itself to work intends to revive the fortunately obsolete view that art is an imitation and depiction of something actual? The reproduction of something at hand requires, to be sure, agreement with the actual being, adaptation to it; the Middle Ages called it *adaequatio*; Aristotle already spoke of *homoiōsis*. Agreement with what *is* has long been taken to be the essence of truth. But then, is it our opinion that this painting by Van Gogh depicts a pair of peasant shoes somewhere at hand, and is a work of art because it does so successfully? Is it our opinion that the painting draws a likeness from something actual and transposes it into a product of artistic—production? By no means.

The work, therefore, is not the reproduction of some particular entity that happens to be at hand at any given time; it is, on the contrary, the reproduction of things' general essence. But then where and how is this general essence, so that artworks are able to agree with it? With what essence of what thing should a Greek temple agree? Who could maintain the impossible view that the Idea of Temple is represented in the building? And yet, truth is set to work in such a work, if it is a work. Or let us think of Hölderlin's hymn, "The Rhine." What is pregiven to the poet, and how is it given, so that it can then be regiven in the poem? And if in the case of this hymn and similar poems the idea of a copy-relation between something already actual and the artwork clearly fails, the view that the work is a copy is confirmed in the best possible way by a work of the kind presented in C. F. Meyer's poem "Roman Fountain."

*Roman Fountain*

The jet ascends and falling fills
The marble basin circling round;
This, veiling itself over, spills
Into a second basin's ground.
The second in such plenty lives,
Its bubbling flood a third invests,
And each at once receives and gives
And streams and rests.

This is neither a poetic painting of a fountain actually present nor a reproduction of the general essence of a Roman fountain. Yet truth is set into the work. What truth is happening in the work? Can truth happen [*geschehen*] at all and thus be historical [*geschichtlich*]? Yet truth, people say, is something timeless and supertemporal.

We seek the actuality of the artwork in order actually to find there the art prevailing within it. The thingly substructure is what proved to be the most immediate actuality in the work. But to comprehend this thingly feature the traditional thing-concepts are not adequate; for they themselves fail to grasp the essence of the thing. The currently predominant thing-concept, thing as formed matter, is not even derived from the essence of the thing but from the essence of equipment. It also turned out that equipmental being generally has long since occupied a peculiar preeminence in the interpretation of beings. This preeminence of equipmentality, which, however, has never been expressly thought, suggested that we pose the question of equipment anew while avoiding the current interpretations.

We allowed a work to tell us what equipment is. By this means, almost clandestinely, it came to light what is at work in the work: the disclosure of the particular being in its Being, the happening of truth. If, however, the actuality of the work can be defined solely by means of what is at work in the work, then what about our intention to seek out the actual artwork in its actuality? As long as we supposed that the actuality of the work lay primarily in its thingly substructure we were going astray. We are now confronted by a remarkable result of our considerations—if it still deserves to be called a result at all. Two points become clear:

First, the dominant thing-concepts are inadequate as means of grasping the thingly aspect of the work.

Second, what we tried to treat as the most immediate actuality of the work, its thingly substructure, does not belong to the work in that way at all.

As soon as we look for such a thingly substructure in the work, we have unwittingly taken the work as equipment, to which we then also ascribe a superstructure supposed to contain its artistic quality. But the work is not a piece of equipment that is fitted out in addition with an aesthetic value that adheres to it. The work is no more anything of the kind than the bare thing is a piece of equipment that merely lacks the specific equipmental characteristics of usefulness and being made.

Our formulation of the question of the work has been shaken because we asked, not about the work, but half about a thing and half about equipment. Still, this formulation of the question was not first developed by us. It is the formulation native to aesthetics. The way in which aesthetics views the artwork from the outset is dominated by the traditional interpretation of all beings. Yet the shaking of this accustomed formulation is not the essential point. What matters is a first opening of our vision to the fact that what is workly in the work, equipmental in equipment, and thingly in the thing comes closer to us only when we think the Being of beings. To this end it is necessary beforehand that the barriers of our preconceptions fall away and that the current pseudo-concepts be set aside. That is why we had to take this detour. But it brings us directly to a road that may lead to a determination of the thingly feature in the work. The thingly feature in the work should not be denied; but if it belongs admittedly to the work-being of the work, it must be conceived by way of the work's workly nature. If this is so, then the road toward the determination of the thingly reality of the work leads not from thing to work but from work to thing.

The artwork opens up in its own way the Being of beings. This opening up, i.e., this revealing, i.e., the truth of beings, happens in the work. In the artwork, the truth of beings has set itself to work. Art is truth setting itself to work. What is truth itself, that it sometimes propriates as art? What is this setting-itself-to-work?

## The Work and Truth

The origin of the artwork is art. But what is art? Art is actual in the artwork. Hence we first seek the actuality of the work. In what does it consist? Artworks universally display a thingly character, albeit in a wholly distinct way. The attempt to interpret this thing-character of the work with the aid of the usual thing-concepts failed—not only because these concepts do not lay hold of the thingly feature, but because, in raising the question of its thingly substructure, we force the work into a preconceived framework by which we obstruct our own access to the work-being of the work. Nothing can be discovered about the thingly aspect of the work so long as the pure self-subsistence of the work has not distinctly displayed itself.

Yet is the work ever in itself accessible? To gain access to the work, it would be necessary to remove it from all relations to something other than itself, in order to let it stand on its own for itself alone. But the artist's most peculiar intention already aims in this direction. The work is to be released by the artist to its pure self-subsistence. It is precisely in great art—and only such art is under consideration here—that the artist remains inconsequential as compared with the work, almost like a passageway that destroys itself in the creative process for the work to emerge.

Well, then, the works themselves stand and hang in collections and exhibitions. Yet are they here in themselves as the works they themselves are, or are they not rather here as objects of the art industry? Works are made available for public and private art appreciation. Official agencies assume the care and maintenance of works. Connoisseurs and critics busy themselves with them. Art dealers supply the market. Art-historical study makes the works the objects of a science. Yet in all this busy activity do we encounter the work itself?

The Aegina sculptures in the Munich collection, Sophocles' *Antigone* in the best critical edition, are, as the works they are, torn out of their own native sphere. However high their quality and power of impression, however good their state of preservation, however certain their interpretation, placing them in a collection has withdrawn them from their own world. But even when we make an effort to cancel or avoid such displacement of works—when, for instance, we visit the temple in Paestum at its own site or the Bamberg cathedral on its own square—the world of the work that stands there has perished.

World-withdrawal and world-decay can never be undone. The works are no longer the works they were. It is they themselves, to be sure, that we encounter there, but they themselves are gone by. As bygone works they stand over against us in the realm of tradition and conservation. Henceforth they remain merely such objects. Their standing before us is still indeed a consequence of, but no longer the same as, their former self-subsistence. This self-subsistence has fled from them. The whole art industry, even if carried to the extreme and exercised in every way for the sake of works themselves, extends only to the object-being of the works. But this does not constitute their work-being.

However, does the work still remain a work if it stands outside all relations? Is it not essential for the work to stand in relations? Yes, of course—except that it remains to ask in what relations it stands.

Where does a work belong? The work belongs, as work, uniquely within the realm that is opened up by itself. For the work-being of the work occurs essentially and only in such opening up. We said that in the work there was a happening of truth at work. The reference to Van Gogh's picture tried to point to this happening. With regard to it there arose the question as to what truth is and how truth can happen.

We now ask the question of truth with a view to the work. But in order to become more familiar with what the question involves, it is necessary to make visible once more the happening of truth in the work. For this attempt let us deliberately select a work that cannot be ranked as representational art.

A building, a Greek temple, portrays nothing. It simply stands there in the middle of the rock-cleft valley. The building encloses the figure of the god, and in this concealment lets it stand out into the holy precinct through the open portico. By means of the temple, the god is present in the temple. This presence of the god is in itself the extension and delimitation of the precinct as a holy precinct. The temple and its precinct, however, do not fade away into the indefinite. It is the temple-work that first fits together and at the same time gathers around itself the unity of those paths and relations in which birth and death, disaster and blessing, victory and disgrace, endurance and decline acquire the shape of destiny for human being. The all-governing expanse of this open relational context is the world of this historical people. Only from and in this expanse does the nation first return to itself for the fulfillment of its vocation.

Standing there, the building rests on the rocky ground. This resting of the work draws up out of the rock the obscurity of that rock's bulky yet spontaneous support. Standing there, the building holds its ground against the storm raging above it and so first makes the storm itself manifest in its violence. The luster and gleam of the stone, though itself apparently glowing only by the grace of the sun, first brings to radiance the light of the day, the breadth of the sky, the darkness of the night. The temple's firm towering makes visible the invisible space of air. The steadfastness of the work contrasts with the surge of the surf, and its own repose brings out the raging of the sea. Tree and grass, eagle and bull, snake and cricket first enter into their distinctive shapes and thus come to appear as what they are. The Greeks early called this emerging and rising in itself and in all things *physis*. It illuminates also that on which and in which man bases his dwelling. We call this ground the *earth*. What this word says is not to be associated with the idea of a mass of matter deposited somewhere, or with the merely astronomical idea of a planet. Earth is that whence the arising brings back and shelters everything that arises as such. In the things that arise, earth occurs essentially as the sheltering agent.

The temple-work, standing there, opens up a world and at the same time sets this world back again on earth, which itself only thus emerges as native ground. But men and animals, plants and things, are never present and familiar as unchangeable objects, only to represent incidentally also a fitting environment for the temple, which one fine day is added to what is already there. We shall get closer to what *is*, rather, if we think of all this in reverse order, assuming of course that we have, to begin with, an eye for how differently everything then faces us. Mere reversing, done for its own sake, reveals nothing.

The temple, in its standing there, first gives to things their look and to men their outlook on themselves. This view remains open as long as the work is a work, as long as the god has not fled from it. It is the same with the sculpture of the god, a votive offering of the victor in the athletic games. It is not a portrait whose purpose is to make it easier to realize how the god looks; rather, it is a work that lets the god himself be present and thus *is* the god himself. The same holds for the linguistic work. In the tragedy nothing is staged or displayed theatrically, but the battle of the new gods against the old is being fought. The linguistic work, originating in the speech of the people, does not refer to this battle; it transforms the people's saying so that now every living word fights the battle and puts up for decision what is holy and what unholy, what great and what small, what brave and what cowardly, what lofty and what flighty, what master and what slave (see Heraclitus, Fragment 53).

In what, then, does the work-being of the work consist? Keeping steadily in view the points just crudely enough indicated, two essential features of the work may for the moment be brought out more distinctly. We set out here, from the long familiar foreground of the work's being, the thingly character which gives support to our customary attitude toward the work.

When a work is brought into a collection or placed in an exhibition we say also that it is "set up." But this setting up differs essentially from setting up in the sense of erecting a building, raising a statue, presenting a tragedy at a holy festival. The latter setting up is erecting in the sense of dedication and praise. Here "setting up" no longer means a bare placing. To dedicate means to consecrate, in the sense that in setting up the work the holy is opened up as holy and the god is invoked into the openness of his presence. Praise belongs to dedication as doing honor to the dignity and splendor of the god. Dignity and splendor are not properties beside and behind which the god, too, stands as something distinct, but it is rather in the dignity, in the splendor that the god comes to presence. In the reflected glory of this splendor there glows, i.e., there clarifies, what we called the world. To e-rect means: to open the right in the sense of a guiding measure, a form in which what is essential gives guidance. But why is the setting up of a work an erecting that consecrates and praises? Because the work, in its work-being, demands it. How is it that the work comes to demand such a setting up? Because it itself, in its own work-being, is something that sets up. What does the work, as work, set up? Towering up within itself, the work opens up a *world* and keeps it abidingly in force.

To be a work means to set up a world. But what is it to be a world? The answer was hinted at when we referred to the temple. On the path we must follow here, the essence of world can only be indicated. What is more, this indication limits itself to warding off anything that might at first distort our view of the essential.

The world is not the mere collection of the countable or uncountable, familiar and unfamiliar things that are at hand. But neither is it a merely imagined framework added by our representation to the sum of such given things. The *world worlds*, and is more fully in being than the tangible and perceptible realm in which we believe ourselves to be at home. World is never an object that stands before us and can be seen. World is the ever-nonobjective to which we are subject as long as the paths of birth and death, blessing and curse keep us transported into Being. Wherever those utterly essential decisions of our history are made, are taken up and abandoned by us, go unrecognized and are rediscovered by new inquiry, there the world worlds. A stone is worldless. Plant and animal likewise have no world; but they belong to the covert throng of a surrounding into which they are linked. The peasant woman, on the other hand, has a world because she dwells in the overtness of beings. Her equipment, in its reliability, gives to this world a necessity and nearness of its own. By the opening up of a world, all things gain their lingering and hastening, their remoteness and nearness, their scope and limits. In a world's worlding is gathered that spaciousness out of which the protective grace of the gods is granted or withheld. Even this doom, of the god remaining absent, is a way in which world worlds.

A work, by being a work, makes space for that spaciousness. "To make space for" means here especially to liberate the free space of the open region and to establish it in its structure. This installing occurs through the erecting mentioned earlier. The work as work sets up a world. The work holds open the open region of the world. But the setting up of a world is only the first essential feature in the work-being of a work to be referred to here. Starting again from the foreground of the work, we shall attempt to make clear in the same way the second essential feature that belongs with the first.

When a work is created, brought forth out of this or that work material—stone, wood, metal, color, language, tone—we say also that it is made, set forth out of it. But just as the work requires a setting up in the sense of a consecrating-praising erection, because the work's work-being consists in the setting up of a world, so a setting forth is needed because the work-being of the work itself has the character of setting forth. The work as work, in its essence, is a setting forth. But what does the work set forth? We come to know about this only when we explore what comes to the fore and is customarily spoken of as the production [*Herstellung*, literally, "setting forth"] of works.

To work-being there belongs the setting up of a world. Thinking of it within this perspective, what is the essence of that in the work which is usually called the work material? Because it is determined by usefulness and serviceability, equipment takes into its service that of which it consists: the matter. In fabricating equipment—e.g., an ax—stone is used, and used up. It disappears into usefulness. The material is all the better and more suitable the less it resists vanishing in the equipmental being of the equipment. By contrast the temple-work, in setting up a world, does not cause the material to disappear, but rather causes it to come forth for the very first time and to come into the open region of the work's world. The rock comes to bear and rest and so first becomes rock; metals come to glitter and shimmer, colors to glow, tones to sing, the word to say. All this comes forth as the work sets itself back into the massiveness and heaviness of stone, into the firmness and pliancy of wood, into the hardness and luster of metal, into the brightening and darkening of color, into the clang of tone, and into the naming power of the word.

That into which the work sets itself back and which it causes to come forth in this setting back of itself we called the earth. Earth is that which comes forth and shelters. Earth, irreducibly spontaneous, is effortless and untiring. Upon the earth and in it, historical man grounds his dwelling in the world. In setting up a world, the work sets forth the earth. This setting forth must be thought here in the strict sense of the word. The work moves the earth itself into the open region of a world and keeps it there. *The work lets the earth be an earth.*

But why must this setting forth of the earth happen in such a way that the work sets itself back into it? What is the earth that it attains to the unconcealed in just such a manner? A stone presses downward and manifests its heaviness. But while this heaviness exerts an opposing pressure upon us it denies us any penetration into it. If we attempt such a penetration by breaking open the rock, it still does not display in its fragments anything inward that has been opened up. The stone has instantly withdrawn again into the same dull pressure and bulk of its fragments. If we try to lay hold of the stone's heaviness in another way, by placing the stone on a balance, we merely bring the heaviness into the form of a calculated weight. This perhaps very precise determination of the stone remains a number, but the weight's burden has escaped us. Color shines and wants only to shine. When we analyze it in rational terms by measuring its wavelengths, it is gone. It shows itself only when it remains undisclosed and unexplained. Earth thus shatters every attempt to penetrate it. It causes every merely calculating importunity upon it to turn into a destruction. This destruction may herald itself under the appearance of mastery and of progress in the form of the technical-scientific objectivation of nature, but this mastery nevertheless remains an impotence of will. The earth appears openly cleared as itself only when it is perceived and preserved as that which is essentially undisclosable, that which shrinks from every disclosure and constantly keeps itself closed up. All things of earth, and the earth itself as a whole, flow together into a reciprocal accord. But this confluence is not a blurring of their outlines. Here there flows the bordering stream, restful within itself, which delimits everything present in its presencing. Thus in each of the self-secluding things there is the same not-knowing-of-one-another. The earth is essentially self-secluding. To set forth the earth means to bring it into the open region as the self-secluding.

This setting forth of the earth is achieved by the work as it sets itself back into the earth. The self-seclusion of earth, however, is not a uniform, inflexible staying under cover, but unfolds itself in an inexhaustible variety of simple modes and shapes. To be sure, the sculptor uses stone just as the mason uses it, in his own way. But he does not use it up. That happens in a certain way only where the work miscarries. To be sure, the painter also uses pigment, but in such a way that color is not used up but rather only now comes to shine forth. To be sure, the poet also uses the word—not, however, like ordinary speakers and writers who have to use them up, but rather in such a way that the word only now becomes and remains truly a word.

Nowhere in the work is there any trace of a work material. It even remains doubtful whether, in the essential definition of equipment, what the equipment consists of is properly described in its equipmental essence as matter.

The setting up of a world and the setting forth of earth are two essential features in the work-being of the work. They belong together, however, in the unity of work-being. This is the unity we seek when we ponder the self-subsistence of the work and try to tell of this closed, unitary repose of self-support.

However, in the essential features just mentioned, if our account has any validity at all, we have indicated in the work rather a happening and in no sense a repose. For what is rest if not the opposite of motion? It is at any rate not an opposite that excludes motion from itself, but rather includes it. Only what is in motion can rest. The mode of rest varies with the kind of motion. In motion as the mere displacement of a physical body, rest is, to be sure, only the limiting case of motion. Where rest includes motion, there can exist a repose which is an inner concentration of motion, hence supreme agitation, assuming that the mode of motion requires such a rest.

Now, the repose of the work that rests in itself is of this sort. We shall therefore come nearer to this repose if we can succeed in grasping the state of movement of the happening in work-being in its unity. We ask. What relation do the setting up of a world and the setting forth of the earth exhibit in the work itself?

The world is the self-opening openness of the broad paths of the simple and essential decisions in the destiny of a historical people. The earth is the spontaneous forthcoming of that which is continually self-secluding and to that extent sheltering and concealing. World and earth are essentially different from one another and yet are never separated. The world grounds itself on the earth, and earth juts through world. Yet the relation between world and earth does not wither away into the empty unity of opposites unconcerned with one another. The world, in resting upon the earth, strives to surmount it. As self-opening it cannot endure anything closed. The earth, however, as sheltering and concealing, tends always to draw the world into itself and keep it there.

The opposition of world and earth is strife. But we would surely all too easily falsify its essence if we were to confound strife with discord and dispute, and thus see it only as disorder and destruction. In essential strife, rather, the opponents raise each other into the self-assertion of their essential natures. Self-assertion of essence, however, is never a rigid insistence upon some contingent state, but surrender to the concealed originality of the provenance of one's own Being. In strife, each opponent carries the other beyond itself. Thus the strife becomes ever more intense as striving, and more properly what it is. The more strife, for its part, outdoes itself, the more inflexibly do the opponents let themselves go into the intimacy of simple belonging to one another. The earth cannot dispense with the open region of the world if it itself is to appear as earth in the liberated surge of its self-seclusion. The world in turn cannot soar out of the earth's sight if, as the governing breadth and path of all essential destiny, it is to ground itself on something decisive.

In setting up a world and setting forth the earth, the work is an instigating of this strife. This does not happen so that the work should at the same time settle and put an end to strife by an insipid agreement, but so that the strife may remain a strife. Setting up a world and setting forth the earth, the work accomplishes this strife. The work-being of the work consists in the instigation of strife between world and earth. It is because the strife arrives at its high point in the simplicity of intimacy that the unity of the work comes about in the instigation of strife. The latter is the continually self-overreaching gathering of the work's agitation. The repose of the work that rests in itself thus has its essence in the intimacy of strife.

From this repose of the work we can now first see what is at work in the work. Until now it was a merely provisional assertion that in an artwork truth is set to work. In what way does truth happen in the work-being of the work, which now means to say, how does truth happen in the instigation of strife between world and earth? What is truth?

How slight and stunted our knowledge of the essence of truth is, is shown by the laxity we permit ourselves in using this basic word. By truth is usually meant this or that particular truth. That means: something true. A cognition articulated in a proposition can be of this sort. However, we call not only a proposition true, but also a thing, true gold in contrast with sham gold. True here means genuine, actual gold. What does the expression "actual" mean here? To us it is what is in truth. The true is what corresponds to the actual, and the actual is what is in truth. The circle has closed again.

What does "in truth" mean? Truth is the essence of the true. What do we have in mind when speaking of essence? Usually it is thought to be those features held in common by everything that is true. The essence is discovered in the generic and universal concept, which represents the one feature that holds indifferently for many things. This indifferent essence (essentiality in the sense of *essentia*) is, however, only the unessential essence. What does the essential essence of something consist in? Presumably it lies in what the entity *is* in truth. The true essence of a thing is determined by way of its true Being, by way of the truth of the given being. But we are now seeking not the truth of essence but the essence of truth. There thus appears a curious tangle. Is it only a curiosity or even merely the empty sophistry of a conceptual game, or is it—an abyss?

Truth means the essence of the true. We think this essence in recollecting the Greek word *alētheia*, the unconcealment of beings. Yet is this enough to define the essence of truth? Are we not passing off a mere change of word usage—unconcealment instead of truth—as a characterization of the matter at issue? Certainly we do not get beyond an interchange of names as long as we do not come to know what must have happened in order to be compelled to say the *essence* of truth in the word "unconcealment."

Does this require a revival of Greek philosophy? Not at all. A revival, even if such an impossibility were possible, would be of no help to us; for the hidden history of Greek philosophy consists from its beginning in this, that it does not remain in conformity with the essence of truth that flashes out in the word *alētheia*, and has to misdirect its knowing and its speaking about the essence of truth more and more into the discussion of a derivative essence of truth. The essence of truth as *alētheia* was not thought out in the thinking of the Greeks, and certainly not in the philosophy that followed after. Unconcealment is, for thought, the most concealed thing in Greek existence, although from early times it determines the presencing of everything present.

Yet why should we not be satisfied with the essence of truth that has by now been familiar to us for centuries? Truth means today and has long meant the conformity of knowledge with the matter. However, the matter must show itself to be such if knowledge and the proposition that forms and expresses knowledge are to be able to conform to it; otherwise the matter cannot become binding on the proposition. How can the matter show itself if it cannot itself stand forth out of concealment, if it does not itself stand in the unconcealed? A proposition is true by conforming to the unconcealed, to what is true. Propositional truth is always, and always exclusively, this correctness. The critical concepts of truth which, since Descartes, start out from truth as certainty, are merely variations of the definition of truth as correctness. The essence of truth which is familiar to us—correctness in representation—stands and falls with truth as unconcealment of beings.

If here and elsewhere we conceive of truth as unconcealment, we are not merely taking refuge in a more literal translation of a Greek word. We are reminding ourselves of what, unexperienced and unthought, underlies our familiar and therefore outworn essence of truth in the sense of correctness. We do, of course, occasionally take the trouble to concede that naturally, in order to understand and verify the correctness (truth) of a proposition, one really should go back to something that is already evident, and that this presupposition is indeed unavoidable. As long as we talk and believe in this way, we always understand truth merely as correctness, which of course still requires a further presupposition, that we ourselves just happen to make, heaven knows how or why.

But it is not we who presuppose the unconcealment of beings; rather, the unconcealment of beings (Being) puts us into such a condition of being that in our representation we always remain installed within and in attendance upon unconcealment. Not only must that in *conformity* with which a cognition orders itself be already in some way unconcealed. The entire *realm* in which this "conforming to something" goes on must already occur as a whole in the unconcealed; and this holds equally of that *for* which the conformity of a proposition to a matter becomes manifest. With all our correct representations we would get nowhere, we could not even presuppose that there already is manifest something to which we can conform ourselves, unless the unconcealment of beings had already exposed us to, placed us in that cleared realm in which every being stands for us and from which it withdraws.

But how does this take place? How does truth happen as this unconcealment? First, however, we must say more clearly what this unconcealment itself is.

Things are, and human beings, gifts, and sacrifices are, animals and plants are, equipment and works are. The particular being stands in Being. Through Being there passes a veiled fatality that is ordained between the godly and the countergodly. There is much in being that man cannot master. There is but little that comes to be known. What is known remains inexact, what is mastered insecure. Beings are never of our making, or even merely our representations, as it might all too easily seem. When we contemplate this whole as one, then we apprehend, so it appears, all that is—though we grasp it crudely enough.

And yet—beyond beings, not away from them but before them, there is still something else that happens. In the midst of beings as a whole an open place occurs. There is a clearing. Thought of in reference to beings, this clearing is more in being than are beings. This open center is therefore not surrounded by beings; rather, the clearing center itself encircles all that is, as does the nothing, which we scarcely know.

Beings can be as beings only if they stand within and stand out within what is cleared in this clearing. Only this clearing grants and guarantees to us humans a passage to those beings that we ourselves are not, and access to the being that we ourselves are. Thanks to this clearing, beings are unconcealed in certain changing degrees. And yet a being can be *concealed*, as well, only within the sphere of what is cleared. Each being we encounter and which encounters us keeps to this curious opposition of presencing, in that it always withholds itself at the same time in a concealment. The clearing in which beings stand is in itself at the same time concealment. Concealment, however, prevails in the midst of beings in a twofold way.

Beings refuse themselves to us down to that one and seemingly least feature which we touch upon most readily when we can say no more of beings than that they are. Concealment as refusal is not simply and only the limit of knowledge in any given circumstance, but the beginning of the clearing of what is cleared. But concealment, though of another sort, to be sure, at the same time also occurs within what is cleared. One being places itself in front of another being, the one helps to hide the other, the former obscures the latter, a few obstruct many, one denies all. Here concealment is not simple refusal. Rather, a being appears, but presents itself as other than it is.

This concealment is dissembling. If one being did not simulate another, we could not make mistakes or act mistakenly in regard to beings; we could not go astray and transgress, and especially could never overreach ourselves. That a being should be able to deceive as semblance is the condition for our being able to be deceived, not conversely.

Concealment can be a refusal or merely a dissembling. We are never fully certain whether it is the one or the other. Concealment conceals and dissembles itself. This means that the open place in the midst of beings, the clearing, is never a rigid stage with a permanently raised curtain on which the play of beings runs its course. Rather, the clearing happens only as this double concealment. The unconcealment of beings—this is never a merely existent state, but a happening. Unconcealment (truth) is neither an attribute of matters in the sense of beings, nor one of propositions.

We believe we are at home in the immediate circle of beings. Beings are familiar, reliable, ordinary. Nevertheless, the clearing is pervaded by a constant concealment in the double form of refusal and dissembling. At bottom, the ordinary is not ordinary; it is extraordinary. The essence of truth, that is, of unconcealment, is dominated throughout by a denial. Yet this denial is not a defect or a fault, as though truth were an unalloyed unconcealment that has rid itself of everything concealed. If truth could accomplish this, it would no longer be itself. *This denial, in the form of a double concealment, belongs to the essence of truth as unconcealment*. Truth, in its essence, is un-truth. We put the matter this way in order to serve notice, with a possibly surprising trenchancy, that denial in the manner of concealment belongs to unconcealment as clearing. The proposition "the essence of truth is un-truth" is not, however, intended to state that truth is at bottom falsehood. Nor does it mean that truth is never itself but, viewed dialectically, is also its opposite.

Truth occurs precisely as itself in that the concealing denial, as refusal, provides the steady provenance of all clearing, and yet, as dissembling, metes out to all clearing the indefeasible severity of error. Concealing denial is intended to denote that opposition in the essence of truth which subsists between clearing and concealing. It is the opposition of the original strife. The essence of truth is, in itself, the primal strife in which that open center is won within which beings stand and from which they set themselves back into themselves.

This open region happens in the midst of beings. It exhibits an essential feature that we have already mentioned. To the open region there belongs a world and the earth. But the world is not simply the open region that corresponds to clearing, and the earth is not simply the closed region that corresponds to concealment. Rather, the world is the clearing of the paths of the essential guiding directions with which all decision complies. Every decision, however, bases itself on something not mastered, something concealed, confusing; else it would never be a decision. The earth is not simply the closed region but rather that which rises up as self-closing. World and earth are always intrinsically and essentially in conflict, belligerent by nature. Only as such do they enter into the strife of clearing and concealing.

Earth juts through the world and world grounds itself on the earth only so far as truth happens as the primal strife between clearing and concealing. But how does this happen? We answer: it happens in a few essential ways. One of these ways in which truth happens is the work-being of the work. Setting up a world and setting forth the earth, the work is the instigation of the strife in which the unconcealment of beings as a whole, or truth, is won.

Truth happens in the temple's standing where it is. This does not mean that something is correctly represented and rendered here, but that beings as a whole are brought into unconcealment and held therein. To hold [*halten*] originally means to take into protective heed [*hüten*]. Truth happens in Van Gogh's painting. This does not mean that something at hand is correctly portrayed, but rather that in the revelation of the equipmental being of the shoes beings as a whole—world and earth in their counterplay—attain to unconcealment.

Thus in the work *it* is truth, not merely something true, that is at work. The picture that shows the peasant shoes, the poem that says the Roman fountain, do not simply make manifest what these isolated beings as such are—if indeed they manifest anything at all; rather, they make unconcealment as such happen in regard to beings as a whole. The more simply and essentially the shoes are engrossed in their essence, the more directly and engagingly do all beings attain a greater degree of being along with them. That is how self-concealing Being is cleared. Light of this kind joins its shining to and into the work. This shining, joined in the work, is the beautiful. *Beauty is one way in which truth essentially occurs as unconcealment.*

We now, indeed, grasp the essence of truth more clearly in certain respects. What is at work in the work may accordingly have become clearer. But the work's now visible work-being still does not tell us anything about the work's closest and most obtrusive reality, about the thingly aspect of the work. Indeed it almost seems as though, in pursuing the exclusive aim of grasping the work's independence as purely as possible, we had completely overlooked the one thing, that a work is always a work, which means that it is something effected. If there is anything that distinguishes the work as work, it is that the work has been created. Since the work is created, and creation requires a medium out of which and in which it creates, the thingly element, too, enters into the work. This is incontestable. Still the question remains: How does being created belong to the work? This can be elucidated only if two points are cleared up:

1. What do being created and creation mean here in distinction from making and being made?
2. What is the inmost essence of the work itself, from which alone can be gauged how far createdness belongs to the work and how far it determines the work-being of the work?

Creation is here always thought in reference to the work. To the essence of the work there belongs the happening of truth. From the outset we define the essence of creating by its relation to the essence of truth as the unconcealment of beings. The pertinence of createdness to the work can be elucidated only by way of a more fundamental clarification of the essence of truth. The question of truth and its essence returns again.

We must raise that question once more, if the proposition that truth is at work in the work is not to remain a mere assertion.

We must now first ask in a more essential way: To what extent does the impulse toward such a thing as a work lie in the essence of truth? Of what essence is truth, that it can be set into work, or even under certain conditions must be set into work, in order to be *as* truth? But we defined the setting-into-a-work of truth as the essence of art. Hence our last question becomes:

What is truth, that it can happen as, or even must happen as, art? How is it that *there is* art at all?

## Truth and Art

Art is the origin of the artwork and of the artist. Origin is the provenance of the essence in which the Being of a being essentially unfolds. What is art? We seek its essence in the actual work. The actuality of the work has been defined by that which is at work in the work, by the happening of truth. This happening we think of as the instauration of strife between world and earth. Repose occurs in the concentrated agitation of this striving. The self-composure of the work is grounded here.

In the work, the happening of truth is at work. But what is thus at work is so *in* the work. This means that the actual work is here already presupposed as the bearer of this happening. At once the problem of the thingly feature of the work at hand confronts us again. One thing thus finally becomes clear: however zealously we inquire into the work's self-sufficiency, we shall still fail to find its actuality as long as we do not also agree to take the work as something worked, effected. To take it thus lies closest to us, for in the word "work" we hear what is worked. The workly character of the work consists in its having been created by the artist. It may seem curious that this most obvious and all-clarifying definition of the work is mentioned only now.

The work's createdness, however, can obviously be grasped only in terms of the process of creation. Thus, constrained by the matter at issue, we must consent after all to go into the activity of the artist in order to arrive at the origin of the work of art. The attempt to define the work-being of the work purely in terms of the work itself proves to be unfeasible.

In turning away now from the work to examine the essence of the creative process, we should like nevertheless to keep in mind what was said first of the picture of the peasant shoes and later of the Greek temple.

We think of creation as a bringing forth. But the making of equipment, too, is a bringing forth. Handicraft—a remarkable play of language—does not, to be sure, create works, not even when we contrast, as we must, the handmade with the factory product. But what is it that distinguishes bringing forth as creation from bringing forth in the mode of making? It is as difficult to track down the essential features of the creation of works and the making of equipment as it is easy to distinguish verbally between the two modes of bringing forth. Going along with first appearances, we find the same procedure in the activity of potter and sculptor, of joiner and painter. The creation of a work requires craftsmanship. Great artists prize craftsmanship most highly. They are the first to call for its painstaking cultivation, based on complete mastery. They above all others constantly take pains to educate themselves ever anew in thorough craftsmanship. It has often enough been pointed out that the Greeks, who knew a few things about works of art, use the same word, *technē*, for craft and art and call the craftsman and the artist by the same name: *technitēs*.

It thus seems advisable to define the essence of creative work in terms of its craft aspect. But reference to the linguistic usage of the Greeks, with their experience of the matter, must give us pause. However usual and convincing the reference may be to the Greek practice of naming craft and art by the same name, *technē*, it nevertheless remains oblique and superficial; for *technē* signifies neither craft nor art, and not at all the technical in our present-day sense; it never means a kind of practical performance.

The word *technē* denotes rather a mode of knowing. To know means to have seen, in the widest sense of seeing, which means to apprehend what is present, as such. For Greek thought the essence of knowing consists in *alētheia*, that is, in the revealing of beings. It supports and guides all comportment toward beings. *Technē*, as knowledge experienced in the Greek manner, is a bringing forth of beings in that it *brings forth* what is present as such *out of* concealment and specifically *into* the unconcealment of its appearance; *technē* never signifies the action of making.

The artist is a *technitēs* not because he is also a craftsman, but because both the setting forth of works and the setting forth of equipment occur in a bringing forth that causes beings in the first place to come forward and be present in assuming an outward aspect. Yet all this happens in the midst of the being that surges upward, growing of its own accord, *physis*. Calling art *technē* does not at all imply that the artist's action is seen in the light of craft. What looks like craft in the creation of a work is of a different sort.

Such doing is determined and pervaded by the essence of creation, and indeed remains contained within that creating.

What then, if not craft, is to guide our thinking about the essence of creation? What else than a view of what is to be created—the work? Although it becomes actual only as the creative act is performed, and thus depends for its actuality upon this act, the essence of creation is determined by the essence of the work. Even though the work's createdness has a relation to creation, nevertheless both createdness and creation must be defined in terms of the work-being of the work. By now it can no longer seem strange that we first and at length dealt with the work alone, to bring its createdness into view only at the end. If createdness belongs to the work as essentially as the word "work" makes it sound, then we must try to understand even more essentially what so far could be defined as the work-being of the work.

In the light of the essential definition of the work we have reached at this point, according to which the happening of truth is at work in the work, we are able to characterize creation as follows: to create is to let something emerge as a thing that has been brought forth. The work's becoming a work is a way in which truth becomes and happens. It all rests in the essence of truth. But what is truth, that it has to happen in such a thing as something created? How does truth have an impulse toward a work grounded in its very essence? Is this intelligible in terms of the essence of truth as thus far elucidated?

Truth is un-truth, insofar as there belongs to it the reservoir of the not-yet-revealed, the un-uncovered, in the sense of concealment. In unconcealment, as truth, there occurs also the other "un-" of a double restraint or refusal. Truth essentially occurs as such in the opposition of clearing and double concealing. Truth is the primal strife in which, always in some particular way, the open region is won within which everything stands and from which everything withholds itself that shows itself and withdraws itself as a being. Whenever and however this strife breaks out and happens, the opponents, clearing and concealing, move apart because of it. Thus the open region of the place of strife is won. The openness of this open region, that is, truth, can be what it is, namely, *this* openness, only if and as long as it establishes itself within its open region. Hence there must always be some being in this open region in which the openness takes its stand and attains its constancy. In thus taking possession of the open region, openness holds it open and sustains it. Setting and taking possession are here everywhere drawn from the Greek sense of *thesis*, which means a setting up in the unconcealed.

In referring to this self-establishing of openness in the open region, thinking touches on a sphere that cannot yet be explicated here. Only this much should be noted, that if the essence of the unconcealment of beings belongs in any way to Being itself (see *Being and Time*, section 44), then Being, by way of its own essence, lets the free space of openness (the clearing of the There) happen, and introduces it as a place *of the sort* in which each being emerges in its own way.

Truth happens only by establishing itself in the strife and the free space opened up by truth itself. Because truth is the opposition of clearing and concealing, there belongs to it what is here to be called *establishing*. But truth does not exist in itself beforehand, somewhere among the stars, only subsequently to descend elsewhere among beings. This is impossible for the reason alone that it is after all only the openness of beings that first affords the possibility of a somewhere and of sites filled by present beings. Clearing of openness and establishment in the open region belong together. They are the same single essence of the happening of truth. This happening is historical in multiple ways.

One essential way in which truth establishes itself in the beings it has opened up is truth setting itself into work. Another way in which truth occurs is the act that founds a political state. Still another way in which truth comes to shine forth is the nearness of that which is not simply a being, but the being that is most in being. Still another way in which truth grounds itself is the essential sacrifice. Still another way in which truth becomes is the thinker's questioning, which, as the thinking of Being, names Being in its question-worthiness. By contrast, science is not an original happening of truth, but always the cultivation of a domain of truth already opened, specifically by apprehending and confirming that which shows itself to be possibly and necessarily correct within that field. When and insofar as a science passes beyond correctness and goes on to a truth, which means that it arrives at the essential disclosure of beings as such, it is philosophy.

Because it is in the essence of truth to establish itself within beings, in order thus first to become truth, the *impulse toward the work* lies in the essence of truth as one of truth's distinctive possibilities, by which it can itself occur as being in the midst of beings.

The establishing of truth in the work is the bringing forth of a being such as never was before and will never come to be again. The bringing forth places this being in the open region in such a way that what is to be brought forth first clears the openness of the open region into which it comes forth. Where this bringing forth expressly brings the openness of beings, or truth, that which is brought forth is a work. Creation is such a bringing forth. As such a bringing, it is rather a receiving and removing within the relation to unconcealment. What, accordingly, does the createdness consist in? It may be elucidated by two essential determinations.

Truth establishes itself in the work. Truth essentially occurs only as the strife between clearing and concealing in the opposition of world and earth. Truth wills to be established in the work as this strife of world and earth. The strife is not to be resolved in a being brought forth for that purpose, nor is it to be merely housed there; the strife, on the contrary, is started by it. This being must therefore contain within itself the essential traits of the strife. In the strife the unity of world and earth is won. As a world opens itself, it submits to the decision of a historical humanity the question of victory and defeat, blessing and curse, mastery and slavery. The dawning world brings out what is as yet undecided and measureless, and thus discloses the hidden necessity of measure and decisiveness.

Yet as a world opens itself the earth comes to tower. It stands forth as that which bears all, as that which is sheltered in its own law and always wrapped in itself. World demands its decisiveness and its measure and lets beings attain to the open region of their paths. Earth, bearing and jutting, endeavors to keep itself closed and to entrust everything to its law. Strife is not a rift [*Riss*], as a mere cleft is ripped open; rather, it is the intimacy with which opponents belong to each other. This rift carries the opponents into the provenance of their unity by virtue of their common ground. It is a basic design, an outline sketch, that draws the basic features of the upsurgence of the clearing of beings. This rift does not let the opponents break apart; it brings what opposes measure and boundary into its common outline.*

Truth establishes itself as strife within a being that is to be brought forth only in such a way that the strife opens up in this being; that is, this being is itself brought into the rift. The rift is the drawing together, into a unity, of sketch and basic design, breach and outline. Truth establishes itself in a being in such a way, indeed, that this being itself takes possession of the open region of truth. This occupying, however, can happen only if what is to be brought forth, the rift, entrusts itself to the self-secluding element that juts into the open region. The rift must set itself back into the gravity of stone, the mute hardness of wood, the dark glow of colors. As the earth takes the rift back into itself, the rift is first set

*In German *der Riss* is a crack, tear, laceration, cleft, or rift; but it is also a plan or design in drawing. The verb *reissen* from which it derives is cognate with the English word *writing*. *Der Riss* is incised or inscribed as a rune or letter. Heidegger here employs a series of words (*Abriss*, *Aufriss*, *Umriss*, and especially *Grundriss*) to suggest that the rift of world and earth releases a sketch, outline, profile, blueprint, or ground plan. The rift is writ.—Ed.

forth into the open region and thus placed, that is, set, within that which towers into the open region as self-secluding and sheltering.

The strife that is brought into the rift and thus set back into the earth and thus fixed in place is the *figure* [*Gestalt*]. Createdness of the work means truth's being fixed in place in the figure. Figure is the structure in whose shape the rift composes itself. This composed rift is the fugue of truth's shining. What is here called figure [*Gestalt*] is always to be thought in terms of the particular placing [*Stellen*] and enframing [*Ge-stell*] as which the *work* occurs when it sets itself up and sets itself forth.

In the creation of a work, the strife, as rift, must be set back into the earth, and the earth itself must be set forth and put to use as self-secluding. Such use, however, does not use up or misuse the earth as matter, but rather sets it free to be nothing but itself. This use of the earth is a working with it that, to be sure, looks like the employment of matter in handicraft. Hence the illusion that artistic creation is also an activity of handicraft. It never is. But it is at all times a use of the earth in the fixing in place of truth in the figure. In contrast, the making of equipment never directly effects the happening of truth. The production of equipment is finished when a material has been so formed as to be ready for use. For equipment to be ready means that it is released beyond itself, to be used up in usefulness.

Not so when a work is created. This becomes clear in the light of the second characteristic, which may be introduced here.

The readiness of equipment and the createdness of the work agree in this, that in each case something is produced. Yet in contrast to all other modes of production, the work is distinguished by being created so that its createdness is part of the created work. But does not this hold true for everything brought forth, indeed for anything that has in any way come to be? Everything brought forth surely has this endowment of having been brought forth, if it has any endowment at all. Certainly. However, in the work, createdness is expressly created into the created being, so that it stands out from it, from the being thus brought forth, in an expressly particular way. If this is how matters stand, then we must also be able to discover and experience the createdness explicitly in the work.

The emergence of createdness from the work does not mean that the work is to give the impression of having been made by a great artist. The point is not that the created being be certified as the performance of a capable person, so that the producer is thereby brought to public notice. It is not the *N. N. fecit* that is to be made known. Rather, the simple *factum est* is to be held forth into the open region by the work: namely this, that unconcealment of a being has happened here, and that as this happening it happens here for the first time; or, that such a work *is* at all rather than is not. The thrust that the work, as this work is, and the uninterruptedness of this plain thrust, constitute the steadfastness of the work's self-subsistence. Precisely where the artist and the process and the circumstances of the genesis of the work remain unknown, this thrust, this "*that* it is" of createdness, emerges into view most purely from the work.

To be sure, "that" it is made is a property also of all equipment that is available and in use. But this "that" does not become prominent in the equipment; it disappears in usefulness. The more handy a piece of equipment is, the more inconspicuous it remains that, for example, this particular hammer is, and the more exclusively does the equipment keep itself in its equipmentality. In general, of everything present to us, we can note that it *is*; but this also, if it is noted at all, is noted only soon to fall into oblivion, as is the wont of everything commonplace. And what is more commonplace than this, that a being is? In a work, by contrast, this fact, that it *is* as a work, is just what is unusual. The event of its being created does not simply reverberate through the work; rather, the work casts before itself the eventful fact that the work is as this work, and it has constantly this fact about itself. The more essentially the work opens itself, the more luminous becomes the uniqueness of the fact that it is rather than is not. The more essentially this thrust comes into the open region, the more strange and solitary the work becomes. In the bringing forth of the work there lies this offering "that it be."

The question of the work's createdness ought to have brought us nearer to its workly character and therewith to its actuality. Createdness revealed itself as strife being fixed in place in the figure by means of the rift. Createdness here is itself expressly created into the work and stands as the silent thrust into the open region of the "that." But the work's actuality does not exhaust itself even in the createdness. On the contrary, this view of the essence of the work's createdness now enables us to take the step toward which everything thus far said tends.

The more solitarily the work, fixed in the figure, stands on its own and the more cleanly it seems to cut all ties to human beings, the more simply does the thrust come into the open that such a work *is*, and the more essentially is the extraordinary thrust to the surface and what is long-familiar thrust down. But this multiple thrusting is nothing violent, for the more purely the work is itself transported into the openness of beings—an openness opened by itself—the more simply does it transport us into this openness and thus at the same time transport us out of the realm of the ordinary. To submit to this displacement means to transform our accustomed ties to world and earth and henceforth to restrain all usual doing and prizing, knowing and looking, in order to stay within the truth that is happening in the work. Only the restraint of this staying lets what is created be the work that it is. This letting the work be a work we call preserving the work. It is only for such preserving that the work yields itself in its createdness as actual, which now means, present in the manner of a work.

Just as a work cannot be without being created, but is essentially in need of creators, so what is created cannot itself come into being without those who preserve it.

However, if a work does not find preservers, does not immediately find them capable of responding to the truth happening in the work, this does not at all mean that the work may also be a work without preservers. Being a work, it always remains tied to preservers, even and particularly when it is still only waiting for them, only pleading and persevering for them to enter into its truth. Even the oblivion into which the work can sink is not nothing; it is still a preservation. It feeds on the work. Preserving the work means standing within the openness of beings that happens in the work. This "standing-within" of preservation, however, is a knowing. Yet knowing does not consist in mere information and notions about something. He who truly knows beings knows what he wills to do in the midst of them.

The willing here referred to, which neither merely applies knowledge nor decides beforehand, is thought in terms of the basic experience of thinking in *Being and Time*. Knowing that remains a willing, and willing that remains a knowing, is the existing human being's ecstatic entry into the unconcealment of Being. The resoluteness intended in *Being and Time* is not the deliberate action of a subject but the opening up of human being, out of its captivity in beings, to the openness of Being.* However, in existence, man does not proceed from some inside to some outside; rather, the essence of *Existenz* is out-standing standing-within the essential sunderance of the clearing of beings. Neither in the creation mentioned before nor in the willing mentioned now do we think of the performance or act of a subject striving toward himself as his self-posited goal.

Willing is the sober unclosedness of that existential self-transcendence which exposes itself to the openness of beings as it is set into the work. In this way, standing-within is brought under law. Preserving the work, as knowing, is a sober standing-within the awesomeness of the truth that is happening in the work.

This knowledge, which as a willing makes its home in the work's truth, and only thus remains a knowing, does not deprive the work

*The word for resoluteness, *Entschlossenheit*, if taken literally, would mean "unclosedness."—Tr.

of its independence, does not drag it into the sphere of mere lived experience, and does not degrade it to the role of a stimulator of such experience.* Preserving the work does not reduce people to their private experiences, but brings them into affiliation with the truth happening in the work. Thus it grounds being for and with one another as the historical standing-out of human existence in relation to unconcealment. Most of all, knowledge in the manner of preserving is far removed from that merely aestheticizing connoisseurship of the work's formal aspects, its qualities and charms. Knowing as having seen is a being resolved; it is standing within the strife that the work has fitted into the rift.

The proper way to preserve the work is co-created and prescribed only and exclusively by the work. Preserving occurs at different levels of knowledge, with always differing degrees of scope, constancy, and lucidity. When works are offered for sheer artistic enjoyment, this does not yet prove that they stand in preservation as works.

As soon as the thrust into the awesome is parried and captured by the sphere of familiarity and connoisseurship, the art business has begun. Even a painstaking transmission of works to posterity, all scientific efforts to regain them, no longer reach the work's own being, but only a remembrance of it. But even this remembrance may still offer to the work a place from which it joins in shaping history. The work's own peculiar actuality, on the other hand, is brought to bear only where the work is preserved in the truth that happens through the work itself.

The work's actuality is determined in its basic features by the essence of the work's being. We can now return to our opening question: How do matters stand with the work's thingly feature that is to guarantee its immediate actuality? They stand so that now we no longer raise this question about the work's thingly element; for

*This is precisely the complaint that Heidegger levels against Nietzsche's notion of will to power as art—not in the 1936–37 lecture course on Nietzsche but in notes sketched during the year 1939. To some extent the whole of the present essay may be viewed as a response to the Nietzschean *Wille zur Macht als Kunst*.—Ed.

as long as we ask it, we take the work directly and as a foregone conclusion, as an object that is simply at hand. In that way we never question in terms of the work, but in our own terms. In our terms—we, who then do not let the work be a work but view it as an object that is supposed to produce this or that state of mind in us.

But what looks like the thingly element, in the sense of our usual thing-concepts, in the work taken as object, is, seen from the perspective of the work, its earthy character. The earth juts up within the work because the work essentially unfolds as something in which truth is at work and because truth essentially unfolds only by installing itself within a particular being. In the earth, however, as essentially self-secluding, the openness of the open region finds that which most intensely resists it; it thereby finds the site of its constant stand, the site in which the figure must be fixed in place.

Was it then superfluous, after all, to enter into the question of the thingly character of the thing? By no means. To be sure, the work's work-character cannot be defined in terms of its thingly character, but as against that the question about the thing's thingly character can be brought onto the right course by way of a knowledge of the work's work-character. This is no small matter, if we recollect that those ancient, traditional modes of thought attack the thing's thingly character and make it subject to an interpretation of beings as a whole, an interpretation that remains unfit to apprehend the essence of equipment and of the work, and which makes us equally blind to the original essence of truth.

To determine the thing's thingness, neither consideration of the bearer of properties, nor that of the manifold of sense data in their unity, and least of all that of the matter-form structure regarded by itself, which is derived from equipment, is adequate. Anticipating a meaningful and weighty interpretation of the thingly character of things, we must aim at the thing's belonging to the earth. The essence of the earth, in its free and unhurried bearing and self-seclusion, reveals itself, however, only in the earth's jutting into a world, only in the opposition of the two. This strife is fixed in place in the figure of the work and becomes manifest by it. What holds true of equipment—namely, that we come to know its equipmental character specifically only through the work itself—also holds of the thingly character of the thing. The fact that we never know thingness directly, and if we know it at all, then only vaguely, and thus require the work—this fact proves indirectly that in the work's work-being the happening of truth, the opening up of beings, is at work.

But, we might finally object, if the work is indeed to bring thingness cogently into the open region, must it not then itself—and indeed before its own creation and for the sake of its creation—have been brought into a relation with the things of earth, with nature? Someone who must have known all about this, Albrecht Dürer, did after all make the well-known remark: "For in truth, art lies hidden within nature; he who can wrest it from her, has it." "Wrest" here means to draw out the rift and to draw the design with the drawing-pen on the drawing-board.* But we at once raise the counterquestion: How can the rift be drawn out if it is not brought into the open region by the creative projection as a rift, which is to say, brought out beforehand as strife of measure and unmeasure? True, there lies hidden in nature a rift-design, a measure and a boundary and, tied to it, a capacity for bringing forth—that is, art. But it is equally certain that this art hidden in nature becomes manifest only through the work, because it lies originally in the work.

The trouble we are taking over the actuality of the work is intended as spadework for discovering art and the essence of art in the actual work. The question concerning the essence of art, the way toward knowledge of it, is first to be placed on a firm ground again. The answer to the question, like every genuine answer, is only the final result of the last step in a long series of questions. Each answer remains in force as an answer only as long as it is rooted in questioning.

*"*Reissen heisst hier Herausholen des Risses und den Riss reissen mit der Reissfeder auf dem Reissbrett.*"

The actuality of the work has become not only clearer for us in the light of its work-being, but also essentially richer. The preservers of a work belong to its createdness with an essentiality equal to that of the creators. But it is the work that makes the creators possible in their essence, the work that by its own essence is in need of preservers. If art is the origin of the work, this means that art lets those who essentially belong together at work, the creator and the preserver, originate, each in his own essence. What, however, is art itself that we call it rightly an origin?

In the work, the happening of truth is at work and, indeed, at work according to the manner of a work. Accordingly the essence of art was defined to begin with as the setting-into-work of truth. Yet this definition is intentionally ambiguous. It says on the one hand: art is the fixing in place of self-establishing truth in the figure. This happens in creation as the bringing forth of the unconcealment of beings. Setting-into-work, however, also means the bringing of work-being into movement and happening. This happens as preservation. Thus art is the creative preserving of truth in the work. *Art then is a becoming and happening of truth*. Does truth, then, arise out of nothing? It does indeed if by *nothing* we mean the sheer "not" of beings, and if we here think of the being as something at hand in the ordinary way, which thereafter comes to light and is challenged by the existence of the work as only presumptively a true being. Truth is never gathered from things at hand, never from the ordinary. Rather, the opening up of the open region, and the clearing of beings, happens only when the openness that makes its advent in thrownness is projected.*

*Thrownness, *Geworfenheit*, is understood in *Being and Time* as an existential characteristic of Dasein, human being, its thatness, its "that it is," and it refers to the facticity of human being's being handed over to itself, its being on its own responsibility; as long as human being is what it is, it is thrown, cast, *im Wurf*. Projection, *Entwurf*, on the other hand, is a second existential character of human being, referring to its driving forward toward its own possibility of being. It takes the form of understanding, which the author speaks of as the mode of being of human being in which human being *is* in its possibilities *as* possibilities. It is not the mere having of a

Truth, as the clearing and concealing of beings, happens in being composed. *All art*, as the letting happen of the advent of the truth of beings, is as such, *in essence, poetry*. The essence of art, on which both the artwork and the artist depend, is the setting-itself-into-work of truth. It is due to art's poetic essence that, in the midst of beings, art breaks open an open place, in whose openness everything is other than usual. By virtue of the projection set into the work of the unconcealment of beings, which casts itself toward us, everything ordinary and hitherto existing becomes an unbeing. This unbeing has lost the capacity to give and keep Being as measure. The curious fact here is that the work in no way affects hitherto existing beings by causal connections. The working of the work does not consist in the taking effect of a cause. It lies in a change, happening from out of the work, of the unconcealment of beings, and this means, of Being.

Poetry, however, is not an aimless imagining of whimsicalities and not a flight of mere notions and fancies into the realm of the unreal. What poetry, as clearing projection, unfolds of unconcealment and projects ahead into the rift-design of the figure, is the open region which poetry lets happen, and indeed in such a way that only now, in the midst of beings, the open region brings beings to shine and ring out. If we fix our vision on the essence of the work and its connection with the happening of the truth of beings, it becomes questionable whether the essence of poetry, and this means at the same time the essence of projection, can be adequately thought of in terms of the power of imagination.

The essence of poetry, which has now been ascertained very broadly—but not on that account vaguely—may here be kept firmly in mind as something worthy of questioning, something that still has to be thought through.

---

preconceived plan, but is the projecting of possibility in human being that occurs antecedently to all plans and makes planning possible. Human being is both thrown and projected; it is a "thrown project," a factical directedness toward possibilities of being.—Tr.

If all art is in essence poetry, then the arts of architecture, painting, sculpture, and music must be traced back to poesy. That is pure arbitrariness. It certainly is, as long as we mean that those arts are varieties of the art of language, if it is permissible to characterize poesy by that easily misinterpretable title. But poesy is only one mode of the clearing projection of truth, i.e., of poetic composition in this wider sense. Nevertheless, the linguistic work, poetry in the narrower sense, has a privileged position in the domain of the arts.

To see this, only the right concept of language is needed. In the current view, language is held to be a kind of communication. It serves for verbal exchange and agreement, and in general for communicating. But language is not only and not primarily an audible and written expression of what is to be communicated. It not only puts forth in words and statements what is overtly or covertly intended to be communicated; language alone brings beings as beings into the open for the first time. Where there is no language, as in the Being of stone, plant, and animal, there is also no openness of beings, and consequently no openness of nonbeing and of the empty.

Language, by naming beings for the first time, first brings beings to word and to appearance. Only this naming nominates beings *to* their Being *from out of* their Being. Such saying is a projecting of clearing, in which announcement is made of what it is that beings come into the open *as*. Projecting is the release of a throw by which unconcealment infuses itself into beings as such. This projective announcement forthwith becomes a renunciation of all the dim confusion in which a being veils and withdraws itself.

Projective saying is poetry: the saying of world and earth, the saying of the arena of their strife and thus of the place of all nearness and remoteness of the gods. Poetry is the saying of the unconcealment of beings. Actual language at any given moment is the happening of this saying, in which a people's world historically arises for it and the earth is preserved as that which remains closed.

Projective saying is saying which, in preparing the sayable, simultaneously brings the unsayable as such into a world. In such saying, the concepts of a historical people's essence, i.e., of its belonging to world history, are preformed for that people.

Poetry is thought of here in so broad a sense and at the same time in such intimate essential unity with language and word, that we must leave open whether art in all its modes, from architecture to poesy, exhausts the essence of poetry.

Language itself is poetry in the essential sense. But since language is the happening in which beings first disclose themselves to man each time as beings, poesy—or poetry in the narrower sense—is the most original form of poetry in the essential sense. Language is not poetry because it is the primal poesy; rather, poesy propriates in language because language preserves the original essence of poetry. Building and plastic creation, on the other hand, always happen already, and happen only, in the open region of saying and naming. It is the open region that pervades and guides them. But for this very reason they remain their own ways and modes in which truth directs itself into work. They are an ever special poetizing within the clearing of beings which has already happened unnoticed in language.

Art, as the setting-into-work of truth, is poetry. Not only the creation of the work is poetic, but equally poetic, though in its own way, is the preserving of the work; for a work is in actual effect as a work only when we remove ourselves from our commonplace routine and move into what is disclosed by the work, so as to bring our own essence itself to take a stand in the truth of beings.

The essence of art is poetry. The essence of poetry, in turn, is the founding of truth. We understand founding here in a triple sense: founding as bestowing, founding as grounding, and founding as beginning. Founding, however, is actual only in preserving. Thus to each mode of founding there corresponds a mode of preserving. We can do no more now than to present this structure of the essence of art in a few strokes, and even this only to the extent that the earlier characterization of the essence of the work offers an initial hint.

The setting-into-work of truth thrusts up the awesome and at the same time thrusts down the ordinary and what we believe to be such. The truth that discloses itself in the work can never be proved or derived from what went before. What went before is refuted in its exclusive actuality by the work. What art founds can therefore never be compensated and made up for by what is already at hand and available. Founding is an overflow, a bestowal.

The poetic projection of truth that sets itself into work as figure is also never carried out in the direction of an indeterminate void. Rather, in the work, truth is thrown toward the coming preservers, that is, toward a historical group of human beings. What is thus cast forth is, however, never an arbitrary demand. Truly poetic projection is the opening up of that into which human being as historical is already cast. This is the earth and, for a historical people, its earth, the self-secluding ground on which it rests together with everything that it already is, though still hidden from itself. But this is also its world, which prevails in virtue of the relation of human being to the unconcealment of Being. For this reason, everything with which man is endowed must, in the projection, be drawn up from the closed ground and expressly set upon this ground. In this way the ground is first grounded as the bearing ground.

All creation, because it is such a drawing-up, is a drawing, as of water from a spring. Modern subjectivism, to be sure, immediately misinterprets creation, taking it as the sovereign subject's performance of genius. The founding of truth is a founding not only in the sense of free bestowal, but at the same time foundation in the sense of this ground-laying grounding. Poetic projection comes from nothing in this respect, that it never takes its gift from the ordinary and traditional. Yet it never comes from nothing in that what is projected by it is only the withheld determination of historical Dasein itself.

Bestowing and grounding have in themselves the unmediated character of what we call a beginning. Yet this unmediated character of a beginning, the peculiarity of a leap out of the unmediable, does not exclude but rather includes the fact that the beginning prepares itself for the longest time and wholly inconspicuously. A genuine beginning, as a leap, is always a head start, in which everything to come is already leaped over, even if as something still veiled. The beginning already contains the end latent within itself. A genuine beginning, however, has nothing of the neophyte character of the primitive. The primitive, because it lacks the bestowing, grounding leap and head start, is always futureless. It is not capable of releasing anything more from itself because it contains nothing more than that in which it is caught.

A beginning, on the contrary, always contains the undisclosed abundance of the awesome, which means that it also contains strife with the familiar and ordinary. Art as poetry is founding, in the third sense of instigation of the strife of truth: founding as beginning. Always when beings as a whole, as beings themselves, demand a grounding in openness, art attains to its historical essence as foundation. This foundation happened in the West for the first time in Greece. What was in the future to be called Being was set into work, setting the standard. The realm of beings thus opened up was then transformed into a being in the sense of God's creation. This happened in the Middle Ages. This kind of being was again transformed at the beginning and during the course of the modern age. Beings became objects that could be controlled and penetrated by calculation. At each time a new and essential world irrupted. At each time the openness of beings had to be established in beings themselves by the fixing in place of truth in figure. At each time there happened unconcealment of beings. Unconcealment sets itself into work, a setting which is accomplished by art.

Whenever art happens—that is, whenever there is a beginning—a thrust enters history; history either begins or starts over again. History here means not a sequence in time of events, of whatever sort, however important. History is the transporting of a people into its appointed task as entry into that people's endowment.

Art is the setting-into-work of truth. In this proposition an essential ambiguity lies hidden, in which truth is at once the subject and the object of the setting. But subject and object are unsuitable names here. They keep us from thinking precisely this ambiguous essence, a task that no longer belongs to the present consideration. Art is historical, and as historical it is the creative preserving of truth in the work. Art happens as poetry. Poetry is founding in the triple sense of bestowing, grounding, and beginning. Art, as founding, is essentially historical. This means not simply that art has a history in the extrinsic sense that in the course of time it, too, appears along with many other things, and in the process changes and passes away and offers changing aspects for historiology. Art is history in the essential sense that it grounds history.

Art lets truth originate. Art, founding preserving, is the spring that leaps to the truth of beings in the work. To originate something by a leap, to bring something into being from out of its essential source in a founding leap—this is what the word "origin" [*Ursprung*, literally, primal leap] means.

The origin of the work of art—that is, the origin of both the creators and the preservers, which is to say of a people's historical existence—is art. This is so because art is in its essence an origin: a distinctive way in which truth comes into being, that is, becomes historical.

We inquire into the essence of art. Why do we inquire in this way? We inquire in this way in order to be able to ask more properly whether art is or is not an origin in our historical existence, whether and under what conditions it can and must be an origin.

Such reflection cannot force art and its coming-to-be. But this reflective knowledge is the preliminary and therefore indispensable preparation for the becoming of art. Only such knowledge prepares its space for art, their way for the creators, their location for the preservers.

In such knowledge, which can only grow slowly, the question is decided whether art can be an origin and then must be a forward spring, or whether it is to remain a mere appendix and then can only be carried along as a routine cultural phenomenon.

Are we in our existence historically at the origin? Do we know, which means do we give heed to, the essence of the origin? Or, in our relation to art, do we still merely make appeal to a cultivated acquaintance with the past?

For this either-or and its decision there is an infallible sign. Hölderlin, the poet—whose work still confronts the Germans as a test to be stood—named it in saying:

*Schwer verlässt*
*was nahe dem Ursprung wohnet, den Ort.*

Reluctantly
that which dwells near its origin abandons the site.

—"The Journey," verses 18–19

# EPILOGUE

The foregoing reflections are concerned with the riddle of art, the riddle that art itself is. They are far from claiming to solve the riddle. The task is to see the riddle.

Almost from the time when specialized thinking about art and the artist began, this thought was called aesthetic. Aesthetics takes the work of art as an object, the object of *aisthēsis*, of sensuous apprehension in the wide sense. Today we call this apprehension lived experience. The way in which man experiences art is supposed to give information about its essence. Lived experience is the source that is standard not only for art appreciation and enjoyment but also for artistic creation. Everything is an experience. Yet perhaps lived experience is the element in which art dies. The dying occurs so slowly that it takes a few centuries.

To be sure, people speak of immortal works of art and of art as an eternal value. Speaking this way means using that language which does not trouble with precision in all essential matters, for fear that in the end to be precise would call for—thinking. And is there any greater anxiety today than that in the face of thinking? Does this talk about immortal works and the eternal value of art have any content or substance? Or are these merely the half-baked clichés of an age when great art, together with its essence, has departed from among human beings?

In the most comprehensive reflection on the essence of art that the West possesses—comprehensive because it stems from metaphysics—namely, Hegel's *Lectures on Aesthetics*, the following propositions occur:

> Art no longer counts for us as the highest manner in which truth obtains existence for itself.

> One may well hope that art will continue to advance and perfect itself, but its form has ceased to be the highest need of spirit.

> In all these relationships art is and remains for us, on the side of its highest vocation, something past.*

The judgment that Hegel passes in these statements cannot be evaded by pointing out that since Hegel's lectures on aesthetics were given for the last time during the winter of 1828–29 at the University of Berlin we have seen the rise of many new artworks and new art movements. Hegel never meant to deny this possibility. But the question remains: Is art still an essential and necessary way in which that truth happens which is decisive for our historical existence, or is art no longer of this character? If, however, it is such no longer, then there remains the question as to why this is so. The truth of Hegel's judgment has not yet been decided; for behind this verdict there stands Western thought since the Greeks. Such thought corresponds to a truth of beings that has already happened. Decision upon the judgment will be made, when it is made, from and about this truth of beings. Until then the judgment remains in force. But for that very reason the question is necessary as to whether the truth that the judgment declares is final and conclusive, and what follows if it is.

Such questions, which solicit us more or less definitely, can be asked only after we have first taken into consideration the essence of art. We attempt to take a few steps by posing the question of the origin of the artwork. The problem is to bring to view the work-character of the work. What the word "origin" here means is thought by way of the essence of truth.

The truth of which we have spoken does not coincide with that which is generally recognized under the name and assigned to cognition and science as a quality, in order to distinguish from it the

*In the original pagination of the *Vorlesungen*, which is repeated in the Jubiläum edition edited by H. Glockner (Stuttgart: Frommanns Verlag, 1953), these passages occur at X, 1, 134; 135; 16. In the "Theorie Werkausgabe" (Frankfurt am Main: Suhrkamp, 1970) they are to be found in vol. 13, 141, 142, and 25.—Ed.

beautiful and the good, which function as names for the values of nontheoretical activities.

Truth is the unconcealment of beings as beings. Truth is the truth of Being. Beauty does not occur apart from this truth. When truth sets itself into the work, it appears. Appearance—as this being of truth in the work and as work—is beauty. Thus the beautiful belongs to truth's propriative event. It does not exist merely relative to pleasure and purely as its object. The beautiful does lie in form, but only because the *forma* once took its light from Being as the beingness of beings. Being at that time was appropriated as *eidos*. The *idea* fits itself into the *morphē*. The *synolon*, the unitary whole of *morphē* and *hylē*, namely the *ergon*, *is* in the manner of *energeia*. This mode of presence becomes the *actualitas* of the *ens actu*. The *actualitas* becomes reality. Actuality becomes objectivity. Objectivity becomes lived experience. In the way in which, for the world determined by the West, beings are as the actual, there is concealed a peculiar confluence of beauty with truth. The history of the essence of Western art corresponds to the change in the essence of truth. Western art is no more intelligible in terms of beauty taken for itself than it is in terms of lived experience, supposing that the metaphysical concept of art reaches into the essence of art.

# ADDENDUM

On pages 189 and 197 an essential difficulty will force itself on the attentive reader: it looks as if the remarks about the "fixing in place of truth" and the "letting happen of the advent of truth" could never be brought into accord. For "fixing in place" implies a willing that blocks and thus prevents the advent of truth. In *letting*-happen on the other hand, there is manifested a compliance and thus, as it were, a nonwilling, that clears the way for the advent of truth.

The difficulty is resolved if we understand fixing in place in the sense intended throughout the entire text of the essay, above all in the key specification "*setting*-into-work." Also correlated with "to place" and "to set" is "to lay"; all three meanings are still intended jointly by the Latin *ponere*.

We must think "to place" in the sense of *thesis*. Thus on page 186 the statement is made, "Setting and taking possession are here everywhere (!) thought on the basis of the Greek sense of *thesis*, which means a setting up in the unconcealed." The Greek "setting" means placing, as for instance, letting a statue be set up. It means laying, laying down an oblation. Placing and laying have the sense of bringing *here* into the unconcealed, bringing *forth* into what is present, that is, letting lie forth. Setting and placing here never mean the modern concept of commandeering things to be placed over against the self (the ego-subject). The standing of the statue (i.e., the presencing of the radiance facing us) is different from the standing of what stands over against us in the sense of the object. "Standing" (See p. 162) is constancy of shining. By contrast, thesis, anti-thesis, and synthesis in the dialectic of Kant and German Idealism mean a placing or putting within the sphere of subjectivity of consciousness. Accordingly, Hegel—correctly in terms of his position—interpreted the Greek *thesis* in the sense of the immediate positing of the object. Setting in this sense, therefore, is for him still untrue, because it is not yet mediated by antithesis and synthesis. (See "Hegel und die Griechen" in the *Festschrift* for H. G. Gadamer, 1960.)*

But if, in the context of our essay on the work of art, we keep in mind the Greek sense of *thesis*—to let lie forth in its shining and presencing—then the "fix" in "fix in place" can never have the sense of rigid, motionless, and secure.

"Fixed" means outlined, admitted into the boundary (*peras*), brought into the outline (See p. 188). The boundary in the Greek sense does not block off; rather, being itself brought forth, it first brings to its radiance what is present. Boundary sets free into the unconcealed; by its contour in the Greek light the mountain stands in its towering and repose. The boundary that fixes and consolidates is in this repose—repose in the fullness of motion. All this holds of the work in the Greek sense of *ergon*; this work's "Being" is *energeia*, which gathers infinitely more movement within itself than do the modern "energies."

Thus the "fixing in place" of truth, rightly understood, can never run counter to the "letting happen." For one thing, this "letting" is nothing passive but a doing in the highest degree (see "Wissenschaft und Besinnung" in *Vorträge und Aufsätze*, p. 49†) in the sense of *thesis*, a "working" and "willing" that in the present essay (p. 192) is characterized as the "existing human being's ecstatic entry into the unconcealment of Being." For another thing, the "happen" in the

*The reference was added to the Reclam edition in 1960. The essay appears also in Martin Heidegger, *Wegmarken* (Frankfurt-am-Main: V. Klostermann, 1967), pp. 255–72.—Ed.

†The reference is to a discussion of the German *Tun*, doing, which points to the core of its meaning as a laying forth, placing here, bringing here, and bringing forth—"working," in the sense either of something bringing itself forth out of itself into presence or of man performing the bringing here and bringing forth of something. Both are ways in which something that is present presences.—Tr.

letting happen of truth is the movement that prevails in the clearing *and* concealing, or more precisely in their union, that is to say, the movement of the clearing of self-concealment as such, from which in turn all self-clearing stems. What is more, this "movement" even requires a fixing in place in the sense of a bringing forth, where the bringing is to be understood in the sense given it on page 187, in that the creative bringing forth "is rather a receiving and an incorporating of a relation to unconcealment."

In accordance with what has so far been explained, the meaning of the noun *Ge-Stell* [enframing] used on page 189, is thus defined: the gathering of the bringing-forth, of the letting-come-forth-here into the rift-design as bounding outline (*peras*). The Greek sense of *morphē* as *Gestalt* is made clear by *Ge-Stell* so understood. Now, the word *Ge-Stell*, which we used in later writings as the explicit key expression for the essence of modern technology, was indeed conceived in reference to that broader sense of *Ge-Stell* (*not* in reference to such other senses as bookshelf or montage). That context is essential, because related to the destiny of Being. Enframing, as the essence of modern technology, derives from the Greek way of experiencing letting-lie-forth, *logos*, from the Greek *poiēsis* and *thesis*. In setting up the frame—which now means in commandeering [*Herausfordern*] everything into assured availability—there sounds the claim of the *ratio reddenda*, i.e., of the *logon didonai* [the reasons, grounds, or accounts to be rendered], but in such a way that today this claim that is made in enframing takes control of the absolute, and the process of representation [*Vor-stellen*, literally, putting forth], on the basis of the Greek sense of apprehending, devotes itself to securing and fixing in place.

When we hear the words "fix in place" and "enframing" in "The Origin of the Work of Art," we must, on the one hand, put out of mind the modern meaning of placing or enframing, and yet at the same time we must not fail to note that, and in what way, the Being that defines the modern period—Being as enframing—stems from the Western destiny of Being and has not been thought *up* by philosophers but rather thought *to* thinkers (see *Vorträge und Aufsätze*, pp. 28 and 49*).

It is still our burden to discuss the specifications given briefly on pages 185–87 about the "establishing" and "self-establishing of truth in beings." Here again we must avoid understanding "establish" [*einrichten*] in the modern sense and in the manner of the lecture on technology as "organize" and "finish or complete." Rather, "establishing" recalls the "impulse of truth toward the work," mentioned on page 187 the impulse that, in the midst of beings, truth should itself come to be in the manner of work, should itself come to be as being.

If we recollect how truth as unconcealment of beings means nothing but the presencing of beings as such, that is, *Being*—see page 197—then talk about the self-establishing of truth, that is, of Being, in beings, touches on the problem of the ontological difference (See *Identitität und Differenz* [1957], pp. 37ff).† For this reason there is the note of caution on page 186 of "The Origin of the Work of Art": "In referring to this self-establishing of openness in the open region, thinking touches on a sphere that cannot yet be explicated here." The whole essay, "The Origin of the Work of Art," deliberately yet tacitly moves on the path of the question of the essence of Being. Reflection on what *art* may be is completely and decidedly determined only in regard to the question of *Being*. Art is considered neither an area of cultural achievement nor an appearance of spirit; it belongs to the *propriative event* [Ereignis] by way of which the "meaning of Being" (See *Being and Time*) can alone be defined. What art may be is one of the questions to which

---

*The reference to p. 49 is to the conception of doing, as given in the previous note. The passage on p. 28 of *Vorträge und Aufsätze* appears in these *Basic Writings*, pp. 325–26.—Tr.

†See Martin Heidegger, *Identity and Difference*, trans. Joan Stambaugh (New York: Harper & Row, 1969), pp. 50ff., 116ff.—Tr.

no answers are given in the essay. Whatever gives the impression of such an answer are directives for questioning. (See the first sentences of the Epilogue.)

Among these directives there are two *important hints,* on pages 196 and 202. In both places mention is made of an "ambiguity." On page 202 an "essential ambiguity" is noted in regard to the definition of art as "the setting-into-work of truth." In this ambiguity, truth is "subject" on the one hand and "object" on the other. *Both* descriptions remain "unsuitable." If truth is the "subject," then the definition "the setting-into-work of truth" means: "truth's setting *itself* into work" (compare pages 196 and 162). Art is then conceived in terms of the propriative event. Being, however, is a call to man and is not without man. Accordingly, art is at the same time defined as the setting-into-work of truth, where truth *now* is "object" and art is human creating and preserving.

Within the *human* relation to art there results the second ambiguity of the setting-into-work of truth, which on page 191 was called creation and preservation. According to pages 196 and 182 the art*work* and the art*ist* rest "simultaneously" in what goes on in art. In the rubric "the setting-into-work of truth," in which it remains undecided but decid*able* who does the setting or in what way it occurs, there is concealed *the relation of Being and human being,* a relation that is unsuitably conceived even in this version—a distressing difficulty, which has been clear to me since *Being and Time* and has since been expressed in a variety of versions (See, most recently, "Zur Seinsfrage" and the present essay, p. 186, "Only this much should be noted, that. . . .").

The problematic context that prevails here then comes together at the proper place in the discussion, where the essence of language and of poetry is touched on, all this again only in regard to the belonging together of Being and saying.

There remains the unavoidable quandary that the reader, who naturally comes to the essay from without, will refrain at first and for a long time from perceiving and interpreting the matters at issue here in terms of the reticent domain that is the source of what has to be thought. For the author himself, however, there remains the quandary of always having to speak in the language most opportune for each of the various stations on his way.

# V

# LETTER ON HUMANISM

*To think is to confine yourself to a single thought that one day stands still like a star in the world's sky.*

In Brussels during the spring of 1845, not long after his expulsion from Paris, Karl Marx jotted down several notes on the German philosopher Ludwig Feuerbach. The second of these reads: "The question whether human thought achieves objective truth is not a question of theory but a *practical* question. . . . Dispute over the actuality or non-actuality of any thinking that isolates itself from *praxis* is a purely *scholastic* question." Ever since that time—especially in France, which Marx exalted as the heart of the Revolution—the relation of philosophy to political practice has been a burning issue. It is not surprising that the impulse for Heidegger's reflections on action, Marxism, existentialism, and humanism in the "Letter on Humanism" came from a French colleague.

On November 10, 1946, a century after Marx sketched his theses on Feuerbach, Jean Beaufret addressed a number of questions to Heidegger, who responded to Beaufret's letter in December with the following piece. (Actually Heidegger reworked and expanded the letter for publication in 1947.) Both Beaufret's inquiry and Heidegger's response refer to a brief essay by Jean-Paul Sartre, originally a public address, with the title *Existentialism Is a Humanism* (Paris: Nagel, 1946). There Sartre defined existentialism as the conviction "that existence precedes essence, or . . . that one must take subjectivity as one's point of departure" (p. 17). In Sartre's view no objectively definable "human nature" underlies man conceived as *existence:* man is nothing more than how he *acts,* what he *does*. This because he has lost all otherworldly underpinnings, has been abandoned to a realm where there are only human beings who have no choice but to make choices. For Sartre man is in the predicament of having to choose and to act without appeal to any concept of human nature that would guarantee the rightness of his choice and the efficacy of his action. "There is reality only in action," Sartre insists (p. 55), and existentialism "defines man by action" (p. 62), which is to say, "in connection with an *engagement*" (p. 78). Nevertheless, Sartre reaffirms (pp. 64ff.)

that man's freedom to act is rooted in subjectivity, which alone grants man his dignity, so that the Cartesian *cogito* becomes the only possible *point de départ* for existentialism and the only possible basis for a humanism (p. 93).

Heidegger responds by keeping open the question of action but strongly criticizing the tradition of subjectivity, which celebrates the "I think" as the font of liberty. Much of the "Letter" is taken up with renewed insistence that Dasein or existence is and remains beyond the pale of Cartesian subjectivism. Again Heidegger writes *Existenz* as *Ek-sistenz,* in order to stress man's "standing out" into the "truth of Being." Humanism underestimates man's unique position in the clearing of Being (*Lichtung des Seins*), Heidegger argues, conceding that to this extent he rejects the humanistic tradition. For it remains stamped in the mold of metaphysics, engrossed in beings, oblivious to Being.

But any opposition to humanism sounds like a rejection of humanity and of humane values. Heidegger therefore discusses the meaning of "values" and of the "nihilism" that ostensibly results when such things are put in question. He finds—as Nietzsche did—that not the denial of such values but their installation in the first place is the source of nihilism. For establishment of values anticipates eventual disestablishment, both actions amounting to a willful self-congratulation of the representing subject.

As Sartre tries to clear a path between the leading competitive "humanisms," those of Christianity and Communism, Heidegger attempts to distinguish his understanding of ek-sistence from man as *imago dei* or *homo faber.* He tries to prevent the question of the clearing of Being from collapsing into the available answers of divine or human light. In so doing he comments on basic questions of religion and ethics. He rejects Sartre's "over-hasty" identification with atheism, not in order to embrace theism but to reflect freely on the nature of the holy and the hale, as of malignancy and the rage of evil. His reflections remain highly relevant at a time when discourses on ethics abound—whether avowedly "metaphysical" or professedly "nonmetaphysical," whether as "practical reason" or "applied ethics."

Returning at the end to the question of action, Heidegger claims that thought of Being occurs prior to the distinction between theory and practice or contemplation and deed. Such thinking seems of the highest importance to Heidegger—yet he warns us not to overestimate it in terms of practical consequences.

Hannah Arendt was fond of calling the "Letter" Heidegger's *Prachtstück,* his most splendid effort. Yet a number of questions might continue to plague us. Is Heidegger's self-interpretation, his account of the "turning," adequate here, even when we note that it is part of an ongoing "immanent critique" (see Reading XI) of *Being and Time?* More important, are the motivations of Heidegger's critique of humanism and of the *animal rationale* altogether clear? Why, for instance, insist that there be an "abyss of essence" separating humanity from animality? Perhaps most disturbing, can Heidegger invoke "malignancy" and "the rage of evil" without breaking his silence and offering some kind of reflection on the Extermination? And how can Heidegger's thought help us to think about those evils that continue to be so very much at home in *our* world? However splendid the "Letter on Humanism," it should only serve to call *us* to *thinking.*

# LETTER ON HUMANISM

We are still far from pondering the essence of action decisively enough. We view action only as causing an effect. The actuality of the effect is valued according to its utility. But the essence of action is accomplishment. To accomplish means to unfold something into the fullness of its essence, to lead it forth into this fullness—*producere*. Therefore only what already is can really be accomplished. But what "is" above all is Being. Thinking accomplishes the relation of Being to the essence of man. It does not make or cause the relation. Thinking brings this relation to Being solely as something handed over to it from Being. Such offering consists in the fact that in thinking Being comes to language. Language is the house of Being. In its home man dwells. Those who think and those who create with words are the guardians of this home. Their guardianship accomplishes the manifestation of Being insofar as they bring the manifestation to language and maintain it in language through their speech. Thinking does not become action only because some effect issues from it or because it is applied. Thinking acts insofar as it thinks. Such action is presumably the simplest and at the same time the highest, because it concerns the relation of Being to man. But all working or effecting lies in Being and is di-

---

This new translation of *Brief über den Humanismus* by Frank A. Capuzzi in collaboration with J. Glenn Gray appears here in its entirety. I have edited it with reference to the helpful French bilingual edition, Martin Heidegger, *Lettre sur l'humanisme*, translated by Roger Munier, revised edition (Paris: Aubier Montaigne, 1964). A previous English translation by Edgar Lohner is included in *Philosophy in the Twentieth Century*, edited by William Barrett and Henry D. Aiken (New York: Random House, 1962), III, 271–302. The German text was first published in 1947 by A. Francke Verlag, Bern; the present translation is based on the text in Martin Heidegger, *Wegmarken* (Frankfurt am Main: Vittorio Klostermann Verlag, 1967), pp. 145–194.

rected toward beings. Thinking, in contrast, lets itself be claimed by Being so that it can say the truth of Being. Thinking accomplishes this letting. Thinking is *l'engagement par l'Être pour l'Être* [engagement by Being for Being]. I do not know whether it is linguistically possible to say both of these ("*par*" and "*pour*") at once, in this way: *penser, c'est l'engagement de l'Être* [thinking is the engagement of Being]. Here the possessive form "*de l'* . . ." is supposed to express both subjective and objective genitives. In this regard "subject" and "object" are inappropriate terms of metaphysics, which very early on in the form of Occidental "logic" and "grammar" seized control of the interpretation of language. We today can only begin to descry what is concealed in that occurrence. The liberation of language from grammar into a more original essential framework is reserved for thought and poetic creation. Thinking is not merely *l'engagement dans l'action* for and by beings, in the sense of the actuality of the present situation. Thinking is *l'engagement* by and for the truth of Being. The history of Being is never past but stands ever before; it sustains and defines every *condition et situation humaine*. In order to learn how to experience the aforementioned essence of thinking purely, and that means at the same time to carry it through, we must free ourselves from the technical interpretation of thinking. The beginnings of that interpretation reach back to Plato and Aristotle. They take thinking itself to be a *technē*, a process of reflection in service to doing and making. But here reflection is already seen from the perspective of *praxis* and *poiēsis*. For this reason thinking, when taken for itself, is not "practical." The characterization of thinking as *theōria* and the determination of knowing as "theoretical" behavior occur already within the "technical" interpretation of thinking. Such characterization is a reactive attempt to rescue thinking and preserve its autonomy over against acting and doing. Since then "philosophy" has been in the constant predicament of having to justify its existence before the "sciences." It believes it can do that most effectively by elevating itself to the rank of a science. But such an effort is the abandonment of the essence of thinking. Philosophy is hounded by the fear that it loses prestige and validity if it is not a science. Not to be a science is taken as a failing that is equivalent to being unscientific. Being, as the element of thinking, is abandoned by the technical interpretation of thinking. "Logic," beginning with the Sophists and Plato, sanctions this explanation. Thinking is judged by a standard that does not measure up to it. Such judgment may be compared to the procedure of trying to evaluate the essence and powers of a fish by seeing how long it can live on dry land. For a long time now, all too long, thinking has been stranded on dry land. Can then the effort to return thinking to its element be called "irrationalism"?

Surely the questions raised in your letter would have been better answered in direct conversation. In written form thinking easily loses its flexibility. But in writing it is difficult above all to retain the multidimensionality of the realm peculiar to thinking. The rigor of thinking, in contrast to that of the sciences, does not consist merely in an artificial, that is, technical-theoretical exactness of concepts. It lies in the fact that speaking remains purely in the element of Being and lets the simplicity of its manifold dimensions rule. On the other hand, written composition exerts a wholesome pressure toward deliberate linguistic formulation. Today I would like to grapple with only one of your questions. Perhaps its discussion will also shed some light on the others.

You ask: *Comment redonner un sens au mot 'Humanisme'?* [How can we restore meaning to the word "humanism"?] This question proceeds from your intention to retain the word "humanism." I wonder whether that is necessary. Or is the damage caused by all such terms still not sufficiently obvious? True, "-isms" have for a long time now been suspect. But the market of public opinion continually demands new ones. We are always prepared to supply the demand. Even such names as "logic," "ethics," and "physics" begin to flourish only when original thinking comes to an end. During the time of their greatness the Greeks thought without such headings. They did not even call thinking "philosophy." Thinking comes to an end when it slips out of its element. The element is what enables thinking to be a thinking. The element is what properly enables: it is the enabling [*das Vermögen*]. It embraces thinking and so brings it into its essence. Said plainly, thinking is the thinking of Being. The genitive says something twofold. Thinking is of Being inasmuch as thinking, propriated by Being, belongs to Being. At the same time thinking is of Being insofar as thinking, belonging to Being, listens to Being. As the belonging to Being that listens, thinking is what it is according to its essential origin. Thinking *is*—this says: Being has fatefully embraced its essence. To embrace a "thing" or a "person" in its essence means to love it, to favor it. Thought in a more original way such favoring [*Mögen*] means to bestow essence as a gift. Such favoring is the proper essence of enabling, which not only can achieve this or that but also can let something essentially unfold in its provenance, that is, let it be. It is on the "strength" of such enabling by favoring that something is properly able to be. This enabling is what is properly "possible" [*das "Mögliche"*], whose essence resides in favoring. From this favoring Being enables thinking. The former makes the latter possible. Being is the enabling-favoring, the "may be" [*das "Mög-liche"*]. As the element, Being is the "quiet power" of the favoring-enabling, that is, of the possible. Of course, our words *möglich* [possible] and *Möglichkeit* [possibility], under the dominance of "logic" and "metaphysics," are thought solely in contrast to "actuality"; that is, they are thought on the basis of a definite—the metaphysical—interpretation of Being as *actus* and *potentia*, a distinction identified with the one between *existentia* and *essentia*. When I speak of the "quiet power of the possible" I do not mean the *possibile* of a merely represented *possibilitas*, nor *potentia* as the *essentia* of an *actus* of *existentia*; rather, I mean Being itself, which in its favoring presides over thinking and hence over the essence of humanity, and that means over its relation to Being. To enable something here means to preserve it in its essence, to maintain it in its element.

When thinking comes to an end by slipping out of its element it replaces this loss by procuring a validity for itself as *technē*, as an instrument of education and therefore as a classroom matter and later a cultural concern. By and by philosophy becomes a technique for explaining from highest causes. One no longer thinks; one occupies oneself with "philosophy." In competition with one another, such occupations publicly offer themselves as "-isms" and try to offer more than the others. The dominance of such terms is not accidental. It rests above all in the modern age upon the peculiar dictatorship of the public realm. However, so-called "private existence" is not really essential, that is to say free, human being. It simply insists on negating the public realm. It remains an offshoot that depends upon the public and nourishes itself by a mere withdrawal from it. Hence it testifies, against its own will, to its subservience to the public realm. But because it stems from the dominance of subjectivity the public realm itself is the metaphysically conditioned establishment and authorization of the openness of individual beings in their unconditional objectification. Language thereby falls into the service of expediting communication along routes where objectification—the uniform accessibility of everything to everyone—branches out and disregards all limits. In this way language comes under the dictatorship of the public realm, which decides in advance what is intelligible and what must be rejected as unintelligible. What is said in *Being and Time* (1927), sections 27 and 35, about the "they" in no way means to furnish an incidental contribution to sociology.* Just as little does the "they"

*The preparatory fundamental analysis of Dasein tries to define concrete structures of human being in its predominant state, "average everydayness." For the most part Dasein is absorbed in the public realm (*die Öffentlichkeit*), which dictates the range of possibilities that shall obtain for it in all dimensions of its life: "We enjoy ourselves and take our pleasures as *they* do; we read, see, and judge works of literature and art as *they* do; but we also shrink back in revulsion from the 'masses' of men just as *they* do; and are '*scandalized*' by what *they* find shocking" (*Sein und Zeit*, pp. 126–27). Heidegger argues that the public realm—the neutral, impersonal "they"—tends to level off genuine possibilities and force individuals to keep their distance from one another and from themselves. It holds Dasein in subservience and hinders knowledge mean merely the opposite, understood in an ethical-existentiell way, of the selfhood of persons. Rather, what is said there contains a reference, thought in terms of the question of the truth of Being, to the word's primordial belongingness to Being. This relation remains concealed beneath the dominance of subjectivity that presents itself as the public realm. But if the truth of Being has become thought-provoking for thinking, then reflection on the essence of language must also attain a different rank. It can no longer be a mere philosophy of language. That is the only reason *Being and Time* (section 34) contains a reference to the essential dimension of language and touches upon the simple question as to what mode of Being language as language in any given case has.* The widely and rapidly spreading devastation of language not only undermines aesthetic and moral responsibility in every use of language; it arises from a threat to the essence of humanity. A merely cultivated use of language is still no proof that we have as yet escaped the danger to our essence. These days, in fact, such usage might sooner testify that we have not yet seen and cannot see the danger because we have never yet placed ourselves in view of it. Much bemoaned of late, and much too lately, the downfall of language is, however, not the grounds for, but already a consequence of, the state of affairs in which language under the dominance of the modern metaphysics

---

of the self and the world. It allows the life-and-death issues of existence proper to dissolve in "chatter," which is "the possibility of understanding everything without prior dedication to, and appropriation of, the matter at stake" (*Sein und Zeit*, p. 169). (All references to *Being and Time* in this essay and throughout the book cite the pagination of the German edition.)—ED.

*In section 34 of *Being and Time* Heidegger defines the existential-ontological foundation of language as speech or talk (*die Rede*). It is as original a structure of being-in-the-world as mood or understanding, of which it is the meaningful articulation. To it belong not only speaking out and asserting but also hearing and listening, heeding and being silent and attentive. As the Greeks experienced it, Dasein is living being that speaks, not so much in producing vocal sounds as in discovering the world, and this by letting beings come to appear as they are. Cf. the analysis of *logos* in section 7 B of Reading I, above; on the crucial question of the "mode of Being" of language, see Reading X, "The Way to Language."—ED.

of subjectivity almost irremediably falls out of its element. Language still denies us its essence: that it is the house of the truth of Being. Instead, language surrenders itself to our mere willing and trafficking as an instrument of domination over beings. Beings themselves appear as actualities in the interaction of cause and effect. We encounter beings as actualities in a calculative businesslike way, but also scientifically and by way of philosophy, with explanations and proofs. Even the assurance that something is inexplicable belongs to these explanations and proofs. With such statements we believe that we confront the mystery. As if it were already decided that the truth of Being lets itself at all be established in causes and explanatory grounds or, what comes to the same, in their incomprehensibility.

But if man is to find his way once again into the nearness of Being he must first learn to exist in the nameless. In the same way he must recognize the seductions of the public realm as well as the impotence of the private. Before he speaks man must first let himself be claimed again by Being, taking the risk that under this claim he will seldom have much to say. Only thus will the pricelessness of its essence be once more bestowed upon the word, and upon man a home for dwelling in the truth of Being.

But in the claim upon man, in the attempt to make man ready for this claim, is there not implied a concern about man? Where else does "care" tend but in the direction of bringing man back to his essence?* What else does that in turn betoken but that man

*In the final chapter of division one of *Being and Time* Heidegger defines "care" as the Being of Dasein. It is a name for the structural whole of existence in all its modes and for the broadest and most basic possibilities of discovery and disclosure of self and world. Most poignantly experienced in the phenomenon of anxiety—which is not fear of anything at hand but awareness of my being-in-the-world as such—"care" describes the sundry ways I get involved in the issue of my birth, life, and death, whether by my projects, inclinations, insights, or illusions. "Care" is the all-inclusive name for my concern for other people, preoccupations with things, and awareness of my proper Being. It expresses the movement of my life out of a past, into a future, through the present. In section 65 the ontological meaning of the Being of care proves to be *temporality*.—ED.

(*homo*) become human (*humanus*)? Thus *humanitas* really does remain the concern of such thinking. For this is humanism: meditating and caring, that man be human and not inhumane, "inhuman," that is, outside his essence. But in what does the humanity of man consist? It lies in his essence.

But whence and how is the essence of man determined? Marx demands that "man's humanity" be recognized and acknowledged.* He finds it in "society." "Social" man is for him "natural" man. In "society" the "nature" of man, that is, the totality of "natural needs" (food, clothing, reproduction, economic sufficiency) is equably secured. The Christian sees the humanity of man, the *humanitas* of *homo*, in contradistinction to *Deitas*. He is the man of the history of redemption who as a "child of God" hears and accepts the call of the Father in Christ. Man is not of this world, since the "world," thought in terms of Platonic theory, is only a temporary passage to the beyond.

*Humanitas*, explicitly so called, was first considered and striven for in the age of the Roman Republic. *Homo humanus* was opposed to *homo barbarus*. *Homo humanus* here means the Romans, who exalted and honored Roman *virtus* through the "embodiment" of the *paideia* [education] taken over from the Greeks. These were the Greeks of the Hellenistic age, whose culture was acquired in the schools of philosophy. It was concerned with *eruditio et institutio in bonas artes* [scholarship and training in good conduct]. *Paideia* thus understood was translated as *humanitas*. The genuine *romanitas* of *homo romanus* consisted in such *humanitas*. We encounter the first humanism in Rome: it therefore remains in essence a specifically Roman phenomenon, which emerges from the encounter of Roman civilization with the culture of late Greek civilization.

*The phrase *der menschliche Mensch* appears in Karl Marx, *Economic-philosophic Manuscripts of 1844*, the so-called "Paris Manuscripts," third MS, p. IV. Cf. *Marx-Engels-Werke* (Berlin, 1973), Ergänzungsband I, 536. This third manuscript is perhaps the best source for Marx's syncretic "humanism," based on man's natural, social, practical, and conscious species-existence.—Ed.

The so-called Renaissance of the fourteenth and fifteenth centuries in Italy is a *renascentia romanitatis*. Because *romanitas* is what matters, it is concerned with *humanitas* and therefore with Greek *paideia*. But Greek civilization is always seen in its later form and this itself is seen from a Roman point of view. The *homo romanus* of the Renaissance also stands in opposition to *homo barbarus*. But now the in-humane is the supposed barbarism of gothic Scholasticism in the Middle Ages. Therefore a *studium humanitatis*, which in a certain way reaches back to the ancients and thus also becomes a revival of Greek civilization, always adheres to historically understood humanism. For Germans this is apparent in the humanism of the eighteenth century supported by Winckelmann, Goethe, and Schiller. On the other hand, Hölderlin does not belong to "humanism," precisely because he thought the destiny of man's essence in a more original way than "humanism" could.

But if one understands humanism in general as a concern that man become free for his humanity and find his worth in it, then humanism differs according to one's conception of the "freedom" and "nature" of man. So too are there various paths toward the realization of such conceptions. The humanism of Marx does not need to return to antiquity any more than the humanism which Sartre conceives existentialism to be. In this broad sense Christianity too is a humanism, in that according to its teaching everything depends on man's salvation (*salus aeterna*); the history of man appears in the context of the history of redemption. However different these forms of humanism may be in purpose and in principle, in the mode and means of their respective realizations, and in the form of their teaching, they nonetheless all agree in this, that the *humanitas* of *homo humanus* is determined with regard to an already established interpretation of nature, history, world, and the ground of the world, that is, of beings as a whole.

Every humanism is either grounded in a metaphysics or is itself made to be the ground of one. Every determination of the essence of man that already presupposes an interpretation of beings without asking about the truth of Being, whether knowingly or not, is metaphysical. The result is that what is peculiar to all metaphysics, specifically with respect to the way the essence of man is determined, is that it is "humanistic." Accordingly, every humanism remains metaphysical. In defining the humanity of man humanism not only does not ask about the relation of Being to the essence of man; because of its metaphysical origin humanism even impedes the question by neither recognizing nor understanding it. On the contrary, the necessity and proper form of the question concerning the truth of Being, forgotten in and through metaphysics, can come to light only if the question "What is metaphysics?" is posed in the midst of metaphysics' domination. Indeed every inquiry into Being, even the one into the truth of Being, must at first introduce its inquiry as a "metaphysical" one.

The first humanism, Roman humanism, and every kind that has emerged from that time to the present, has presupposed the most universal "essence" of man to be obvious. Man is considered to be an *animal rationale*. This definition is not simply the Latin translation of the Greek *zōon logon echon* but rather a metaphysical interpretation of it. This essential definition of man is not false. But it is conditioned by metaphysics. The essential provenance of metaphysics, and not just its limits, became questionable in *Being and Time*. What is questionable is above all commended to thinking as what is to be thought, but not at all left to the gnawing doubts of an empty skepticism.

Metaphysics does indeed represent beings in their Being, and so it thinks the Being of beings. But it does not think the difference of both.[1] Metaphysics does not ask about the truth of Being itself. Nor does it therefore ask in what way the essence of man belongs to the truth of Being. Metaphysics has not only failed up to now to ask

1. Cf. Martin Heidegger, *Vom Wesen des Grundes* (1929), p. 8; *Kant and the Problem of Metaphysics*, trans. Richard Taft (Bloomington: Indiana University Press, 1990), section 43; and *Being and Time*, section 44, p. 230.

this question, the question is inaccessible to metaphysics as such. Being is still waiting for the time when it will become thought-provoking to man. With regard to the definition of man's essence, however one may determine the *ratio* of the *animal* and the reason of the living being, whether as a "faculty of principles" or a "faculty of categories" or in some other way, the essence of reason is always and in each case grounded in this: for every apprehending of beings in their Being, Being itself is already illumined and propriated in its truth. So too with *animal*, *zōon*, an interpretation of "life" is already posited that necessarily lies in an interpretation of beings as *zōē* and *physis*, within which what is living appears. Above and beyond everything else, however, it finally remains to ask whether the essence of man primordially and most decisively lies in the dimension of *animalitas* at all. Are we really on the right track toward the essence of man as long as we set him off as one living creature among others in contrast to plants, beasts, and God? We can proceed in that way; we can in such fashion locate man within being as one being among others. We will thereby always be able to state something correct about man. But we must be clear on this point, that when we do this we abandon man to the essential realm of *animalitas* even if we do not equate him with beasts but attribute a specific difference to him. In principle we are still thinking of *homo animalis*—even when *anima* [soul] is posited as *animus sive mens* [spirit or mind], and this in turn is later posited as subject, person, or spirit [*Geist*]. Such positing is the manner of metaphysics. But then the essence of man is too little heeded and not thought in its origin, the essential provenance that is always the essential future for historical mankind. Metaphysics thinks of man on the basis of *animalitas* and does not think in the direction of his *humanitas*.

Metaphysics closes itself to the simple essential fact that man essentially occurs only in his essence, where he is claimed by Being. Only from that claim "has" he found that wherein his essence dwells. Only from this dwelling "has" he "language" as the home that preserves the ecstatic for his essence.* Such standing in the clearing of Being I call the ek-sistence of man. This way of Being is proper only to man. Ek-sistence so understood is not only the ground of the possibility of reason, *ratio*, but is also that in which the essence of man preserves the source that determines him.

Ek-sistence can be said only of the essence of man, that is, only of the human way "to be." For as far as our experience shows, only man is admitted to the destiny of ek-sistence. Therefore ek-sistence can also never be thought of as a specific kind of living creature among others—granted that man is destined to think the essence of his Being and not merely to give accounts of the nature and history of his constitution and activities. Thus even what we attribute to man as *animalitas* on the basis of the comparison with "beasts" is itself grounded in the essence of ek-sistence. The human body is something essentially other than an animal organism. Nor is the error of biologism overcome by adjoining a soul to the human body, a mind to the soul, and the existentiell to the mind, and then louder than before singing the praises of the mind—only to let everything relapse into "life-experience," with a warning that thinking by its inflexible concepts disrupts the flow of life and that thought of Being distorts existence. The fact that physiology and physiological chemistry can scientifically investigate man as an organism is no proof that in this "organic" thing, that is, in the body scientifically explained, the essence of man consists. That has as little validity as the notion that the essence of nature has been discovered in atomic energy. It could even be that nature, in the face it turns toward man's technical mastery, is simply concealing its

*In *Being and Time* "ecstatic" (from the Greek *ekstasis*) means the way Dasein "stands out" in the various moments of the temporality of care, being "thrown" out of a past and "projecting" itself toward a future by way of the present. The word is closely related to another Heidegger introduces now to capture the unique sense of man's Being—*ek-sistence*. This too means the way man "stands out" into the truth of Being and so is exceptional among beings that are at hand only as things of nature or human production. Cf. Heidegger's definition of "existence" in Reading I, section 4, above, and his use of ek-sistence in Reading III.—ED.

essence. Just as little as the essence of man consists in being an animal organism can this insufficient definition of man's essence be overcome or offset by outfitting man with an immortal soul, the power of reason, or the character of a person. In each instance essence is passed over, and passed over on the basis of the same metaphysical projection.

What man is—or, as it is called in the traditional language of metaphysics, the "essence" of man—lies in his ek-sistence. But ek-sistence thought in this way is not identical with the traditional concept of *existentia*, which means actuality in contrast to the meaning of *essentia* as possibility. In *Being and Time* (p. 42) this sentence is italicized: "The 'essence' of Dasein lies in its existence." However, here the opposition between *existentia* and *essentia* is not under consideration, because neither of these metaphysical determinations of Being, let alone their relationship, is yet in question. Still less does the sentence contain a universal statement about *Dasein*, since the word came into fashion in the eighteenth century as a name for "object," intending to express the metaphysical concept of the actuality of the actual. On the contrary, the sentence says: man occurs essentially in such a way that he is the "there" [*das "Da"*], that is, the clearing of Being. The "Being" of the *Da*, and only it, has the fundamental character of ek-sistence, that is, of an ecstatic inherence in the truth of Being. The ecstatic essence of man consists in ek-sistence, which is different from the metaphysically conceived *existentia*. Medieval philosophy conceives the latter as *actualitas*. Kant represents *existentia* as actuality in the sense of the objectivity of experience. Hegel defines *existentia* as the self-knowing Idea of absolute subjectivity. Nietzsche grasps *existentia* as the eternal recurrence of the same. Here it remains an open question whether through *existentia*—in these explanations of it as actuality, which at first seem quite different—the Being of a stone or even life as the Being of plants and animals is adequately thought. In any case living creatures are as they are without standing outside their Being as such and within the truth of Being, preserving in such standing the essential nature of their Being. Of all the beings that are, presumably the most difficult to think about are living creatures, because on the one hand they are in a certain way most closely akin to us, and on the other are at the same time separated from our ek-sistent essence by an abyss. However, it might also seem as though the essence of divinity is closer to us than what is so alien in other living creatures, closer, namely, in an essential distance which, however distant, is nonetheless more familiar to our ek-sistent essence than is our scarcely conceivable, abysmal bodily kinship with the beast. Such reflections cast a strange light upon the current and therefore always still premature designation of man as *animal rationale*. Because plants and animals are lodged in their respective environments but are never placed freely in the clearing of Being which alone is "world," they lack language. But in being denied language they are not thereby suspended worldlessly in their environment. Still, in this word "environment" converges all that is puzzling about living creatures. In its essence, language is not the utterance of an organism; nor is it the expression of a living thing. Nor can it ever be thought in an essentially correct way in terms of its symbolic character, perhaps not even in terms of the character of signification. Language is the clearing-concealing advent of Being itself.

Ek-sistence, thought in terms of *ecstasis*, does not coincide with *existentia* in either form or content. In terms of content ek-sistence means standing out into the truth of Being. *Existentia* (*existence*) means in contrast *actualitas*, actuality as opposed to mere possibility as Idea. Ek-sistence identifies the determination of what man is in the destiny of truth. *Existentia* is the name for the realization of something that is as it appears in its Idea. The sentence "Man ek-sists" is not an answer to the question of whether man actually is or not; rather, it responds to the question concerning man's "essence." We are accustomed to posing this question with equal impropriety whether we ask what man is or who he is. For in the *Who?* or the *What?* we are already on the lookout for something like a person or an object. But the personal no less than the objective misses and misconstrues the essential unfolding of ek-sistence in the history of Being. That is why the sentence cited from *Being and Time* (p. 42) is careful to enclose the word "essence" in quotation marks. This indicates that "essence" is now being defined from neither *esse essentiae* nor *esse existentiae* but rather from the ek-static character of Dasein. As ek-sisting, man sustains Da-sein in that he takes the *Da*, the clearing of Being, into "care." But Da-sein itself occurs essentially as "thrown." It unfolds essentially in the throw of Being as the fateful sending.

But it would be the ultimate error if one wished to explain the sentence about man's ek-sistent essence as if it were the secularized transference to human beings of a thought that Christian theology expresses about God (*Deus est suum esse* [God is His Being]); for ek-sistence is not the realization of an essence, nor does ek-sistence itself even effect and posit what is essential. If we understand what *Being and Time* calls "projection" as a representational positing, we take it to be an achievement of subjectivity and do not think it in the only way the "understanding of Being" in the context of the "existential analysis" of "being-in-the-world" can be thought—namely, as the ecstatic relation to the clearing of Being. The adequate execution and completion of this other thinking that abandons subjectivity is surely made more difficult by the fact that in the publication of *Being and Time* the third division of the first part, "Time and Being," was held back (cf. *Being and Time*, p. 87, above). Here everything is reversed. The division in question was held back because thinking failed in the adequate saying of this turning [*Kehre*] and did not succeed with the help of the language of metaphysics. The lecture "On the Essence of Truth," thought out and delivered in 1930 but not printed until 1943, provides a certain insight into the thinking of the turning from "Being and Time" to "Time and Being." This turning is not a change of standpoint from *Being and Time*, but in it the thinking that was sought first arrives at the location of that dimension out of which *Being and Time* is experienced, that is to say, experienced from the fundamental experience of the oblivion of Being.

By way of contrast, Sartre expresses the basic tenet of existentialism in this way: Existence precedes essence.* In this statement he is taking *existentia* and *essentia* according to their metaphysical meaning, which from Plato's time on has said that *essentia* precedes *existentia*. Sartre reverses this statement. But the reversal of a metaphysical statement remains a metaphysical statement. With it he stays with metaphysics in oblivion of the truth of Being. For even if philosophy wishes to determine the relation of *essentia* and *existentia* in the sense it had in medieval controversies, in Leibniz's sense, or in some other way, it still remains to ask first of all from what destiny of Being this differentiation in Being as *esse essentiae* and *esse existentiae* comes to appear to thinking. We have yet to consider why the question about the destiny of Being was never asked and why it could never be thought. Or is the fact that this is how it is with the differentiation of *essentia* and *existentia* not at all a sign of forgetfulness of Being? We must presume that this destiny does not rest upon a mere failure of human thinking, let alone upon a lesser capacity of early Western thinking. Concealed in its essential provenance, the differentiation of *essentia* (essentiality) and *existentia* (actuality) completely dominates the destiny of Western history and of all history determined by Europe.

Sartre's key proposition about the priority of *existentia* over *essentia* does, however, justify using the name "existentialism" as an appropriate title for a philosophy of this sort. But the basic tenet of "existentialism" has nothing at all in common with the statement from *Being and Time*—apart from the fact that in *Being and Time* no statement about the relation of *essentia* and *existentia* can yet be expressed, since there it is still a question of preparing something precursory. As is obvious from what we have just said, that happens

*See Jean-Paul Sartre, *L'Existentialisme est un humanisme* (Paris: Nagel, 1946), pp. 17, 21, and elsewhere.—Ed.

clumsily enough. What still today remains to be said could perhaps become an impetus for guiding the essence of man to the point where it thoughtfully attends to that dimension of the truth of Being which thoroughly governs it. But even this could take place only to the honor of Being and for the benefit of Da-sein, which man ek-sistingly sustains; not, however, for the sake of man, so that civilization and culture through man's doings might be vindicated.

But in order that we today may attain to the dimension of the truth of Being in order to ponder it, we should first of all make clear how Being concerns man and how it claims him. Such an essential experience happens to us when it dawns on us that man is in that he ek-sists. Were we now to say this in the language of the tradition, it would run: the ek-sistence of man is his substance. That is why in *Being and Time* the sentence often recurs, "The 'substance' of man is existence" (pp. 117, 212, 314). But "substance," thought in terms of the history of Being, is already a blanket translation of *ousia*, a word that designates the presence of what is present and at the same time, with puzzling ambiguity, usually means what is present itself. If we think the metaphysical term "substance" in the sense already suggested in accordance with the "phenomenological destructuring" carried out in *Being and Time* (cf. p. 63, above), then the statement "The 'substance' of man is ek-sistence" says nothing else but that the way that man in his proper essence becomes present to Being is ecstatic inherence in the truth of Being. Through this determination of the essence of man the humanistic interpretations of man as *animal rationale*, as "person," as spiritual-ensouled-bodily being, are not declared false and thrust aside. Rather, the sole implication is that the highest determinations of the essence of man in humanism still do not realize the proper dignity of man. To that extent the thinking in *Being and Time* is against humanism. But this opposition does not mean that such thinking aligns itself against the humane and advocates the inhuman, that it promotes the inhumane and deprecates the dignity of man. Humanism is opposed because it does not set the *human-*

*itas* of man high enough. Of course the essential worth of man does not consist in his being the substance of beings, as the "Subject" among them, so that as the tyrant of Being he may deign to release the beingness of beings into an all too loudly bruited "objectivity."

Man is rather "thrown" from Being itself into the truth of Being, so that ek-sisting in this fashion he might guard the truth of Being, in order that beings might appear in the light of Being as the beings they are. Man does not decide whether and how beings appear, whether and how God and the gods or history and nature come forward into the clearing of Being, come to presence and depart. The advent of beings lies in the destiny of Being. But for man it is ever a question of finding what is fitting in his essence that corresponds to such destiny; for in accord with this destiny man as ek-sisting has to guard the truth of Being. Man is the shepherd of Being. It is in this direction alone that *Being and Time* is thinking when ecstatic existence is experienced as "care" (cf. section 44 C, pp. 226ff.).

Yet Being—what is Being? It is It itself. The thinking that is to come must learn to experience that and to say it. "Being"—that is not God and not a cosmic ground. Being is farther than all beings and is yet nearer to man than every being, be it a rock, a beast, a work of art, a machine, be it an angel or God. Being is the nearest. Yet the near remains farthest from man. Man at first clings always and only to beings. But when thinking represents beings as beings it no doubt relates itself to Being. In truth, however, it always thinks only of beings as such; precisely not, and never, Being as such. The "question of Being" always remains a question about beings. It is still not at all what its elusive name indicates: the question in the direction of Being. Philosophy, even when it becomes "critical" through Descartes and Kant, always follows the course of metaphysical representation. It thinks from beings back to beings with a glance in passing toward Being. For every departure from beings and every return to them stands already in the light of Being.

But metaphysics recognizes the clearing of Being either solely as the view of what is present in "outward appearance" (*idea*) or critically as what is seen as a result of categorial representation on the part of subjectivity. This means that the truth of Being as the clearing itself remains concealed for metaphysics. However, this concealment is not a defect of metaphysics but a treasure withheld from it yet held before it, the treasure of its own proper wealth. But the clearing itself is Being. Within the destiny of Being in metaphysics the clearing first affords a view by which what is present comes into touch with man, who is present to it, so that man himself can in apprehending (*noein*) first touch upon Being (*thigein*, Aristotle, *Met.* IX, 10). This view first gathers the aspect to itself. It yields to such aspects when apprehending has become a setting-forth-before-itself in the *perceptio* of the *res cogitans* taken as the *subiectum* of *certitudo*.

But how—provided we really ought to ask such a question at all—how does Being relate to ek-sistence? Being itself is the relation to the extent that It, as the location of the truth of Being amid beings, gathers to itself and embraces ek-sistence in its existential, that is, ecstatic, essence. Because man as the one who ek-sists comes to stand in this relation that Being destines for itself, in that he ecstatically sustains it, that is, in care takes it upon himself, he at first fails to recognize the nearest and attaches himself to the next nearest. He even thinks that this is the nearest. But nearer than the nearest and at the same time for ordinary thinking farther than the farthest is nearness itself: the truth of Being.

Forgetting the truth of Being in favor of the pressing throng of beings unthought in their essence is what ensnarement [*Verfallen*] means in *Being and Time*.* This word does not signify the Fall of

*In *Being and Time* (see esp. sections 25–27, 38, and 68 C) *Verfallen*, literally a "falling" or "lapsing," serves as a third constitutive moment of being-in-the-world. Dasein is potentiality for Being, directed toward a future in which it can realize its possibilities: this is its "existentiality." But existence is always "thrown" out of a past that determines its trajectory: this is its "facticity." Meanwhile, Dasein usually busies

Man understood in a "moral-philosophical" and at the same time secularized way; rather, it designates an essential relationship of man to Being within Being's relation to the essence of man. Accordingly, the terms "authenticity" and "inauthenticity," which are used in a provisional fashion, do not imply a moral-existentiell or an "anthropological" distinction but rather a relation which, because it has been hitherto concealed from philosophy, has yet to be thought for the first time, an "ecstatic" relation of the essence of man to the truth of Being. But this relation is as it is not by reason of ek-sistence; on the contrary, the essence of ek-sistence derives existentially-ecstatically from the essence of the truth of Being.

The one thing thinking would like to attain and for the first time tries to articulate in *Being and Time* is something simple. As such, Being remains mysterious, the simple nearness of an unobtrusive governance. The nearness occurs essentially as language itself. But language is not mere speech, insofar as we represent the latter at best as the unity of phoneme (or written character), melody, rhythm, and meaning (or sense). We think of the phoneme and written character as a verbal body for language, of melody and rhythm as its soul, and whatever has to do with meaning as its mind. We usually think of language as corresponding to the essence of man represented as *animal rationale*, that is, as the unity of body-soul-mind. But just as ek-sistence—and through it the relation of the truth of Being to man—remains veiled in the *humanitas* of *homo animalis*, so does the metaphysical-animal explanation of language cover up the essence of language in the history of Being. According to *this* essence, language is the house of Being, which is

---

itself in quotidian affairs, losing itself in the present, forgetting what is most its own: this is its *Verfallensein*. (The last-named is not simply a matter of "everyday" dealings, however, since the tendency to let theoretical problems slip into the ready-made solutions of a tradition affects interpretation itself.) To forget what is most its own is what Heidegger means by *Uneigentlichkeit*, usually rendered as "inauthenticity" but perhaps better understood as "inappropriateness."—ED.

propriated by Being and pervaded by Being. And so it is proper to think the essence of language from its correspondence to Being and indeed as this correspondence, that is, as the home of man's essence.

But man is not only a living creature who possesses language along with other capacities. Rather, language is the house of Being in which man ek-sists by dwelling, in that he belongs to the truth of Being, guarding it.

So the point is that in the determination of the humanity of man as ek-sistence what is essential is not man but Being—as the dimension of the *ecstasis* of ek-sistence. However, the dimension is not something spatial in the familiar sense. Rather, everything spatial and all space-time occur essentially in the dimensionality that Being itself is.

Thinking attends to these simple relationships. It tries to find the right word for them within the long-traditional language and grammar of metaphysics. But does such thinking—granted that there is something in a name—still allow itself to be described as humanism? Certainly not so far as humanism thinks metaphysically. Certainly not if humanism is existentialism and is represented by what Sartre expresses: *précisément nous sommes sur un plan où il y a seulement des hommes* [We are precisely in a situation where there are only human beings].* Thought from *Being and Time,* this should say instead: *précisément nous sommes sur un plan où il y a principalement l'Être* [We are precisely in a situation where principally there is Being]. But where does *le plan* come from and what

*Heidegger cites Sartre's *L'Existentialisme est un humanisme*, p. 36. The context of Sartre's remark is as follows. He is arguing (pp. 33ff.) "that God does not exist, and that it is necessary to draw the consequences to the end." To those who assert that the death of God leaves traditional values and norms untouched—and humanism is one such value—Sartre rejoins "that it is very distressing that God does not exist, because with him vanishes every possibility of finding values in some intelligible heaven; we can no longer locate an *a priori* Good since there is no infinite and perfect consciousness to think it; it is nowhere written that the Good exists, that we must be honest, that we mustn't lie, precisely because we are in a situation where there are only human beings."—Ed.

is it? *L'Être et le plan* are the same. In *Being and Time* (p. 212) we purposely and cautiously say, *il y a l'Être*: "there is / it gives" ["*es gibt*"] Being. *Il y a* translates "it gives" imprecisely. For the "it" that here "gives" is Being itself. The "gives" names the essence of Being that is giving, granting its truth. The self-giving into the open, along with the open region itself, is Being itself.

At the same time "it gives" is used preliminarily to avoid the locution "Being is"; for "is" is commonly said of some thing that is. We call such a thing a being. But Being "is" precisely not "a being." If "is" is spoken without a closer interpretation of Being, then Being is all too easily represented as a "being" after the fashion of the familiar sorts of beings that act as causes and are actualized as effects. And yet Parmenides, in the early age of thinking, says, *esti gar einai*, "for there is Being." The primal mystery for all thinking is concealed in this phrase. Perhaps "is" can be said only of Being in an appropriate way, so that no individual being ever properly "is." But because thinking should be directed only toward saying Being in its truth, instead of explaining it as a particular being in terms of beings, whether and how Being is must remain an open question for the careful attention of thinking.

The *esti gar einai* of Parmenides is still unthought today. That allows us to gauge how things stand with the progress of philosophy. When philosophy attends to its essence it does not make forward strides at all. It remains where it is in order constantly to think the Same. Progression, that is, progression forward from this place, is a mistake that follows thinking as the shadow that thinking itself casts. Because Being is still unthought, *Being and Time* too says of it, "there is / it gives." Yet one cannot speculate about this *il y a* precipitately and without a foothold. This "there is / it gives" rules as the destiny of Being. Its history comes to language in the words of essential thinkers. Therefore the thinking that thinks into the truth of Being is, as thinking, historical. There is not a "systematic" thinking and next to it an illustrative history of past opinions. Nor is there, as Hegel thought, only a systematics that can fashion the law of its thinking into the law of history and simultaneously subsume history into the system. Thought in a more primordial way, there is the history of Being to which thinking belongs as recollection of this history, propriated by it. Such recollective thought differs essentially from the subsequent presentation of history in the sense of an evanescent past. History does not take place primarily as a happening. And its happening is not evanescence. The happening of history occurs essentially as the destiny of the truth of Being and from it.[2] Being comes to destiny in that It, Being, gives itself. But thought in terms of such destiny this says: it gives itself and refuses itself simultaneously. Nonetheless, Hegel's definition of history as the development of "Spirit" is not untrue. Neither is it partly correct and partly false. It is as true as metaphysics, which through Hegel first brings to language its essence—thought in terms of the absolute—in the system. Absolute metaphysics, with its Marxian and Nietzschean inversions, belongs to the history of the truth of Being. Whatever stems from it cannot be countered or even cast aside by refutations. It can only be taken up in such a way that its truth is more primordially sheltered in Being itself and removed from the domain of mere human opinion. All refutation in the field of essential thinking is foolish. Strife among thinkers is the "lovers' quarrel" concerning the matter itself. It assists them mutually toward a simple belonging to the Same, from which they find what is fitting for them in the destiny of Being.

Assuming that in the future man will be able to think the truth of Being, he will think from ek-sistence. Man stands ek-sistingly in the destiny of Being. The ek-sistence of man is historical as such, but not only or primarily because so much happens to man and to things human in the course of time. Because it must think the ek-sistence of Da-sein, the thinking of *Being and Time* is essentially concerned that the historicity of Dasein be experienced.

2. See the lecture on Hölderlin's hymn, "Wie wenn am Feiertage . . ." in Martin Heidegger, *Erläuterungen zu Hölderlins Dichtung*, fourth, expanded ed. (Frankfurt am Main: V. Klostermann, 1971), p. 76.

But does not *Being and Time* say on p. 212, where the "there is / it gives" comes to language, "Only so long as Dasein is, is there [*gibt es*] Being"? To be sure. It means that only so long as the clearing of Being propriates does Being convey itself to man. But the fact that the *Da*, the clearing as the truth of Being itself, propriates is the dispensation of Being itself. This is the destiny of the clearing. But the sentence does not mean that the Dasein of man in the traditional sense of *existentia*, and thought in modern philosophy as the actuality of the *ego cogito*, is that being through which Being is first fashioned. The sentence does not say that Being is the product of man. The "Introduction" to *Being and Time* (p. 85, above) says simply and clearly, even in italics, "Being is the *transcendens* pure and simple." Just as the openness of spatial nearness seen from the perspective of a particular thing exceeds all things near and far, so is Being essentially broader than all beings, because it is the clearing itself. For all that, Being is thought on the basis of beings, a consequence of the approach—at first unavoidable—within a metaphysics that is still dominant. Only from such a perspective does Being show itself in and as a transcending.

The introductory definition, "Being is the *transcendens* pure and simple," articulates in one simple sentence the way the essence of Being hitherto has illumined man. This retrospective definition of the essence of Being from the clearing of beings as such remains indispensable for the prospective approach of thinking toward the question concerning the truth of Being. In this way thinking attests to its essential unfolding as destiny. It is far from the arrogant presumption that wishes to begin anew and declares all past philosophy false. But whether the definition of Being as the *transcendens* pure and simple really does express the simple essence of the truth of Being—this and this alone is the primary question for a thinking that attempts to think the truth of Being. That is why we also say (p. 230) that how Being *is* is to be understood chiefly from its "meaning" ["*Sinn*"], that is, from the truth of Being. Being is illumined for man in the ecstatic projection [*Entwurf*]. But this projection does not create Being.

Moreover, the projection is essentially a thrown projection. What throws in projection is not man but Being itself, which sends man into the ek-sistence of Da-sein that is his essence. This destiny propriates as the clearing of Being—which it is. The clearing grants nearness to Being. In this nearness, in the clearing of the *Da*, man dwells as the ek-sisting one without yet being able properly to experience and take over this dwelling. In the lecture on Hölderlin's elegy "Homecoming" (1943) this nearness "of" Being, which the *Da* of Dasein is, is thought on the basis of *Being and Time*; it is perceived as spoken from the minstrel's poem; from the experience of the oblivion of Being it is called the "homeland." The word is thought here in an essential sense, not patriotically or nationalistically, but in terms of the history of Being. The essence of the homeland, however, is also mentioned with the intention of thinking the homelessness of contemporary man from the essence of Being's history. Nietzsche was the last to experience this homelessness. From within metaphysics he was unable to find any other way out than a reversal of metaphysics. But that is the height of futility. On the other hand, when Hölderlin composes "Homecoming" he is concerned that his "countrymen" find their essence. He does not at all seek that essence in an egoism of his nation. He sees it rather in the context of a belongingness to the destiny of the West. But even the West is not thought regionally as the Occident in contrast to the Orient, nor merely as Europe, but rather world-historically out of nearness to the source. We have still scarcely begun to think of the mysterious relations to the East that found expression in Hölderlin's poetry.[3] "German" is not spoken to the world so that the

3. Cf. "The Ister" and "The Journey" [*Die Wanderung*], third stanza and ff. [In the translations by Michael Hamburger (Ann Arbor: University of Michigan Press, 1966), pp. 492ff. and 392ff.]

world might be reformed through the German essence; rather, it is spoken to the Germans so that from a fateful belongingness to the nations they might become world-historical along with them.[4] The homeland of this historical dwelling is nearness to Being.

In such nearness, if at all, a decision may be made as to whether and how God and the gods withhold their presence and the night remains, whether and how the day of the holy dawns, whether and how in the upsurgence of the holy an epiphany of God and the gods can begin anew. But the holy, which alone is the essential sphere of divinity, which in turn alone affords a dimension for the gods and for God, comes to radiate only when Being itself beforehand and after extensive preparation has been illuminated and is experienced in its truth. Only thus does the overcoming of homelessness begin from Being, a homelessness in which not only man but the essence of man stumbles aimlessly about.

Homelessness so understood consists in the abandonment of Being by beings. Homelessness is the symptom of oblivion of Being. Because of it the truth of Being remains unthought. The oblivion of Being makes itself known indirectly through the fact that man always observes and handles only beings. Even so, because man cannot avoid having some notion of Being, it is explained merely as what is "most general" and therefore as something that encompasses beings, or as a creation of the infinite being, or as the product of a finite subject. At the same time "Being" has long stood for "beings" and, inversely, the latter for the former, the two of them caught in a curious and still unraveled confusion.

As the destiny that sends truth, Being remains concealed. But the world's destiny is heralded in poetry, without yet becoming manifest as the history of Being. The world-historical thinking of Hölderlin that speaks out in the poem "Remembrance" is therefore essentially more primordial and thus more significant for the future

4. Cf. Hölderlin's poem "Remembrance" [*Andenken*] in the *Tübingen Memorial* (1943), p. 322. [Hamburger, pp. 488ff.]

than the mere cosmopolitanism of Goethe. For the same reason Hölderlin's relation to Greek civilization is something essentially other than humanism. When confronted with death, therefore, those young Germans who knew about Hölderlin lived and thought something other than what the public held to be the typical German attitude.

Homelessness is coming to be the destiny of the world. Hence it is necessary to think that destiny in terms of the history of Being. What Marx recognized in an essential and significant sense, though derived from Hegel, as the estrangement of man has its roots in the homelessness of modern man.* This homelessness is specifically evoked from the destiny of Being in the form of metaphysics, and through metaphysics is simultaneously entrenched and covered up as such. Because Marx by experiencing estrangement attains an essential dimension of history, the Marxist view of history is superior to that of other historical accounts. But since neither Husserl nor—so far as I have seen till now—Sartre recognizes the essential importance of the historical in Being, neither phenomenology nor existentialism enters that dimension within which a productive dialogue with Marxism first becomes possible.

For such dialogue it is certainly also necessary to free oneself from naive notions about materialism, as well as from the cheap refutations that are supposed to counter it. The essence of materialism does not consist in the assertion that everything is simply matter but rather in a metaphysical determination according to which every being appears as the material of labor. The modern metaphysical essence of labor is anticipated in Hegel's *Phenomenology of Spirit* as the self-establishing process of unconditioned production, which is the objectification of the actual through man experienced as subjectivity. The essence of materialism is concealed

*On the notion of *Entfremdung*, estrangement or alienation, see Marx's *first* Paris MS, pp. XXIIff., *Werke*, Ergänzungsband I, 510–22. The relation of estrangement to the "world-historical" developments that Heidegger here stresses is perhaps more clearly stated in Marx-Engels, *The German Ideology*, *Werke*, III, 34–36.—Ed.

in the essence of technology, about which much has been written but little has been thought. Technology is in its essence a destiny within the history of Being and of the truth of Being, a truth that lies in oblivion. For technology does not go back to the *technē* of the Greeks in name only but derives historically and essentially from *technē* as a mode of *alētheuein*, a mode, that is, of rendering beings manifest [*Offenbarmachen*]. As a form of truth technology is grounded in the history of metaphysics, which is itself a distinctive and up to now the only perceptible phase of the history of Being. No matter which of the various positions one chooses to adopt toward the doctrines of communism and to their foundation, from the point of view of the history of Being it is certain that an elemental experience of what is world-historical speaks out in it. Whoever takes "communism" only as a "party" or a "Weltanschauung" is thinking too shallowly, just as those who by the term "Americanism" mean, and mean derogatorily, nothing more than a particular life-style. The danger into which Europe as it has hitherto existed is ever more clearly forced consists presumably in the fact above all that its thinking—once its glory—is falling behind in the essential course of a dawning world destiny which nevertheless in the basic traits of its essential provenance remains European by definition. No metaphysics, whether idealistic, materialistic, or Christian, can in accord with its essence, and surely not in its own attempts to explicate itself, "get a hold on" this destiny yet, and that means thoughtfully to reach and gather together what in the fullest sense of Being now is.

In the face of the essential homelessness of man, man's approaching destiny reveals itself to thought on the history of Being in this, that man find his way into the truth of Being and set out on this find. Every nationalism is metaphysically an anthropologism, and as such subjectivism. Nationalism is not overcome through mere internationalism; it is rather expanded and elevated thereby into a system. Nationalism is as little brought and raised to *humanitas* by internationalism as individualism is by an ahistorical collectivism. The latter is the subjectivity of man in totality. It completes subjectivity's unconditioned self-assertion, which refuses to yield. Nor can it be even adequately experienced by a thinking that mediates in a one-sided fashion. Expelled from the truth of Being, man everywhere circles round himself as the *animal rationale*.

But the essence of man consists in his being more than merely human, if this is represented as "being a rational creature." "More" must not be understood here additively, as if the traditional definition of man were indeed to remain basic, only elaborated by means of an existentiell postscript. The "more" means: more originally and therefore more essentially in terms of his essence. But here something enigmatic manifests itself: man is in thrownness. This means that man, as the ek-sisting counter-throw [*Gegenwurf*] of Being, is more than *animal rationale* precisely to the extent that he is less bound up with man conceived from subjectivity. Man is not the lord of beings. Man is the shepherd of Being. Man loses nothing in this "less"; rather, he gains in that he attains the truth of Being. He gains the essential poverty of the shepherd, whose dignity consists in being called by Being itself into the preservation of Being's truth. The call comes as the throw from which the thrownness of Da-sein derives. In his essential unfolding within the history of Being, man is the being whose Being as ek-sistence consists in his dwelling in the nearness of Being. Man is the neighbor of Being.

But—as you no doubt have been wanting to rejoin for quite a while now—does not such thinking think precisely the *humanitas* of *homo humanus?* Does it not think *humanitas* in a decisive sense, as no metaphysics has thought it or can think it? Is this not "humanism" in the extreme sense? Certainly. It is a humanism that thinks the humanity of man from nearness to Being. But at the same time it is a humanism in which not man but man's historical essence is at stake in its provenance from the truth of Being. But then does not the ek-sistence of man also stand or fall in this game of stakes? Indeed it does.

In *Being and Time* (p. 85, above) it is said that every question of philosophy "recoils upon existence." But existence here is not the actuality of the *ego cogito*. Neither is it the actuality of subjects who act with and for each other and so become who they are. "Ek-sistence," in fundamental contrast to every *existentia* and "*existence*," is ecstatic dwelling in the nearness of Being. It is the guardianship, that is, the care for Being. Because there is something simple to be thought in this thinking it seems quite difficult to the representational thought that has been transmitted as philosophy. But the difficulty is not a matter of indulging in a special sort of profundity and of building complicated concepts; rather, it is concealed in the step back that lets thinking enter into a questioning that experiences—and lets the habitual opining of philosophy fall away.

It is everywhere supposed that the attempt in *Being and Time* ended in a blind alley. Let us not comment any further upon that opinion. The thinking that hazards a few steps in *Being and Time* has even today not advanced beyond that publication. But perhaps in the meantime it has in one respect come farther into its own matter. However, as long as philosophy merely busies itself with continually obstructing the possibility of admittance into the matter for thinking, i.e., into the truth of Being, it stands safely beyond any danger of shattering against the hardness of that matter. Thus to "philosophize" about being shattered is separated by a chasm from a thinking that is shattered. If such thinking were to go fortunately for a man, no misfortune would befall him. He would receive the only gift that can come to thinking from Being.

But it is also the case that the matter of thinking is not achieved in the fact that talk about the "truth of Being" and the "history of Being" is set in motion. Everything depends upon this alone, that the truth of Being come to language and that thinking attain to this language. Perhaps, then, language requires much less precipitate expression than proper silence. But who of us today would want to imagine that his attempts to think are at home on the path of silence? At best, thinking could perhaps point toward the truth of Being, and indeed toward it as what is to be thought. It would thus be more easily weaned from mere supposing and opining and directed to the now rare handicraft of writing. Things that really matter, although they are not defined for all eternity, even when they come very late still come at the right time.

Whether the realm of the truth of Being is a blind alley or whether it is the free space in which freedom conserves its essence is something each one may judge after he himself has tried to go the designated way, or even better, after he has gone a better way, that is, a way befitting the question. On the penultimate page of *Being and Time* (p. 437) stand the sentences: "The *conflict* with respect to the interpretation of Being (that is, therefore, not the interpretation of beings or of the Being of man) cannot be settled, *because it has not yet been kindled*. And in the end it is not a question of 'picking a quarrel,' since the kindling of the conflict does demand some preparation. To this end alone the foregoing investigation is under way." Today after two decades these sentences still hold. Let us also in the days ahead remain as wanderers on the way into the neighborhood of Being. The question you pose helps to clarify the way.

You ask, *Comment redonner un sens au mot 'Humanisme'*? "How can some sense be restored to the word 'humanism'?" Your question not only presupposes a desire to retain the word "humanism" but also contains an admission that this word has lost its meaning.

It has lost it through the insight that the essence of humanism is metaphysical, which now means that metaphysics not only does not pose the question concerning the truth of Being but also obstructs the question, insofar as metaphysics persists in the oblivion of Being. But the same thinking that has led us to this insight into the questionable essence of humanism has likewise compelled us to think the essence of man more primordially. With regard to this more essential *humanitas* of *homo humanus* there arises the possibility of restoring to the word "humanism" a historical sense that is older than its oldest meaning chronologically reckoned. The restoration is not to be understood as though the word "humanism" were wholly without meaning and a mere *flatus vocis* [empty sound].

The "*humanum*" in the word points to *humanitas*, the essence of man; the "-ism" indicates that the essence of man is meant to be taken essentially. This is the sense that the word "humanism" has as such. To restore a sense to it can only mean to redefine the meaning of the word. That requires that we first experience the essence of man more primordially; but it also demands that we show to what extent this essence in its own way becomes fateful. The essence of man lies in ek-sistence. That is what is essentially—that is, from Being itself—at issue here, insofar as Being appropriates man as ek-sisting for guardianship over the truth of Being into this truth itself. "Humanism" now means, in case we decide to retain the word, that the essence of man is essential for the truth of Being, specifically in such a way that what matters is not man simply as such. So we are thinking a curious kind of "humanism." The word results in a name that is a *lucus a non lucendo* [literally, a grove where no light penetrates].

Should we still keep the name "humanism" for a "humanism" that contradicts all previous humanism—although it in no way advocates the inhuman? And keep it just so that by sharing in the use of the name we might perhaps swim in the predominant currents, stifled in metaphysical subjectivism and submerged in oblivion of Being? Or should thinking, by means of open resistance to "humanism," risk a shock that could for the first time cause perplexity concerning the *humanitas* of *homo humanus* and its basis? In this way it could awaken a reflection—if the world-historical moment did not itself already compel such a reflection—that thinks not only about man but also about the "nature" of man, not only about his nature but even more primordially about the dimension in which the essence of man, determined by Being itself, is at home. Should we not rather suffer a little while longer those inevitable misinterpretations to which the path of thinking in the element of Being and time has hitherto been exposed and let them slowly dissipate? These misinterpretations are natural reinterpretations of what was read, or simply mirrorings of what one believes he knows already before he reads. They all betray the same structure and the same foundation.

Because we are speaking against "humanism" people fear a defense of the inhuman and a glorification of barbaric brutality. For what is more "logical" than that for somebody who negates humanism nothing remains but the affirmation of inhumanity?

Because we are speaking against "logic" people believe we are demanding that the rigor of thinking be renounced and in its place the arbitrariness of drives and feelings be installed and thus that "irrationalism" be proclaimed as true. For what is more "logical" than that whoever speaks against the logical is defending the alogical?

Because we are speaking against "values" people are horrified at a philosophy that ostensibly dares to despise humanity's best qualities. For what is more "logical" than that a thinking that denies values must necessarily pronounce everything valueless?

Because we say that the Being of man consists in "being-in-the-world" people find that man is downgraded to a merely terrestrial being, whereupon philosophy sinks into positivism. For what is more "logical" than that whoever asserts the worldliness of human being holds only this life as valid, denies the beyond, and renounces all "Transcendence"?

Because we refer to the word of Nietzsche on the "death of God" people regard such a gesture as atheism. For what is more "logical" than that whoever has experienced the death of God is godless?

Because in all the respects mentioned we everywhere speak against all that humanity deems high and holy our philosophy teaches an irresponsible and destructive "nihilism." For what is more "logical" than that whoever roundly denies what is truly in being puts himself on the side of nonbeing and thus professes the pure nothing as the meaning of reality?

What is going on here? People hear talk about "humanism," "logic," "values," "world," and "God." They hear something about opposition to these. They recognize and accept these things as positive. But with hearsay—in a way that is not strictly deliberate—they immediately assume that what speaks against something is automatically its negation and that this is "negative" in the sense of destructive. And somewhere in *Being and Time* there is explicit talk of "the phenomenological destructuring." With the assistance of logic and *ratio*—so often invoked—people come to believe that whatever is not positive is negative and thus that it seeks to degrade reason—and therefore deserves to be branded as depravity. We are so filled with "logic" that anything that disturbs the habitual somnolence of prevailing opinion is automatically registered as a despicable contradiction. We pitch everything that does not stay close to the familiar and beloved positive into the previously excavated pit of pure negation, which negates everything, ends in nothing, and so consummates nihilism. Following this logical course we let everything expire in a nihilism we invented for ourselves with the aid of logic.

But does the "against" which a thinking advances against ordinary opinion necessarily point toward pure negation and the negative? This happens—and then, to be sure, happens inevitably and conclusively, that is, without a clear prospect of anything else—only when one posits in advance what is meant by the "positive" and on this basis makes an absolute and absolutely negative decision about the range of possible opposition to it. Concealed in such a procedure is the refusal to subject to reflection this presupposed "positive" in which one believes oneself saved, together with its position and opposition. By continually appealing to the logical one conjures up the illusion that one is entering straightforwardly into thinking when in fact one has disavowed it.

It ought to be somewhat clearer now that opposition to "humanism" in no way implies a defense of the inhuman but rather opens other vistas.

"Logic" understands thinking to be the representation of beings in their Being, which representation proposes to itself in the generality of the concept. But how is it with meditation on Being itself, that is, with the thinking that thinks the truth of Being? This thinking alone reaches the primordial essence of *logos*, which was already obfuscated and lost in Plato and in Aristotle, the founder of "logic." To think against "logic" does not mean to break a lance for the illogical but simply to trace in thought the *logos* and its essence, which appeared in the dawn of thinking, that is, to exert ourselves for the first time in preparing for such reflection. Of what value are even far-reaching systems of logic to us if, without really knowing what they are doing, they recoil before the task of simply inquiring into the essence of *logos?* If we wished to bandy about objections, which is of course fruitless, we could say with more right: irrationalism, as a denial of *ratio*, rules unnoticed and uncontested in the defense of "logic," which believes it can eschew meditation on *logos* and on the essence of *ratio*, which has its ground in *logos*.

To think against "values" is not to maintain that everything interpreted as "a value"—"culture," "art," "science," "human dignity," "world," and "God"—is valueless. Rather, it is important finally to realize that precisely through the characterization of something as "a value" what is so valued is robbed of its worth. That is to say, by the assessment of something as a value what is valued is admitted only as an object for man's estimation. But what a thing is in its Being is not exhausted by its being an object, particularly when objectivity takes the form of value. Every valuing, even where it values positively, is a subjectivizing. It does not let beings: be. Rather, valuing lets beings: be valid—solely as the objects of its doing. The bizarre effort to prove the objectivity of values does not know what it is doing. When one proclaims "God" the altogether "highest value," this is a degradation of God's essence. Here as elsewhere thinking in values is the greatest blasphemy imaginable against Being. To think against values therefore does not mean to beat the drum for the valuelessness and nullity of beings. It means rather to bring the clearing of the truth of Being before thinking, as against subjectivizing beings into mere objects.

The reference to "being-in-the-world" as the basic trait of the *humanitas* of *homo humanus* does not assert that man is merely a "worldly" creature understood in a Christian sense, thus a creature turned away from God and so cut loose from "Transcendence." What is really meant by this word would be more clearly called "the transcendent." The transcendent is supersensible being. This is considered the highest being in the sense of the first cause of all beings. God is thought as this first cause. However, in the name "being-in-the-world," "world" does not in any way imply earthly as opposed to heavenly being, nor the "worldly" as opposed to the "spiritual." For us "world" does not at all signify beings or any realm of beings but the openness of Being. Man is, and is man, insofar as he is the ek-sisting one. He stands out into the openness of Being. Being itself, which as the throw has projected the essence of man into "care," is as this openness. Thrown in such fashion, man stands "in" the openness of Being. "World" is the clearing of Being into which man stands out on the basis of his thrown essence. "Being-in-the-world" designates the essence of ek-sistence with regard to the cleared dimension out of which the "ek-" of ek-sistence essentially unfolds. Thought in terms of ek-sistence, "world" is in a certain sense precisely "the beyond" within existence and for it. Man is never first and foremost man on the hither side of the world, as a "subject," whether this is taken as "I" or "We." Nor is he ever simply a mere subject which always simultaneously is related to objects, so that his essence lies in the subject-object relation. Rather, before all this, man in his essence is ek-sistent into the openness of Being, into the open region that clears the "between" within which a "relation" of subject to object can "be."

The statement that the essence of man consists in being-in-the-world likewise contains no decision about whether man in a theologico-metaphysical sense is merely a this-worldly or an other-worldly creature.

With the existential determination of the essence of man, therefore, nothing is decided about the "existence of God" or his "nonbeing," no more than about the possibility or impossibility of gods. Thus it is not only rash but also an error in procedure to maintain that the interpretation of the essence of man from the relation of his essence to the truth of Being is atheism. And what is more, this arbitrary classification betrays a lack of careful reading. No one bothers to notice that in my essay "On the Essence of Ground" the following appears: "Through the ontological interpretation of Dasein as being-in-the-world no decision, whether positive or negative, is made concerning a possible being toward God. It is, however, the case that through an illumination of transcendence we first achieve an *adequate concept of Dasein*, with respect to which it can now be asked how the relationship of Dasein to God is ontologically ordered."[5] If we think about this remark too quickly, as is usually the case, we will declare that such a philosophy does not decide either for or against the existence of God. It remains stalled in indifference. Thus it is unconcerned with the religious question. Such indifferentism ultimately falls prey to nihilism.

But does the foregoing observation teach indifferentism? Why then are particular words in the note italicized—and not just random ones? For no other reason than to indicate that the thinking that thinks from the question concerning the truth of Being questions more primordially than metaphysics can. Only from the truth of Being can the essence of the holy be thought. Only from the essence of the holy is the essence of divinity to be thought. Only in the light of the essence of divinity can it be thought or said what the word "God" is to signify. Or should we not first be able to hear and understand all these words carefully if we are to be permitted as men, that is, as ek-sistent creatures, to experience a relation of God to man? How can man at the present stage of world history ask at all seriously and rigorously whether the god nears or withdraws, when he has above all neglected to think into the dimension in which alone that question can be asked? But this is the dimension

5. Martin Heidegger, *Vom Wesen des Grundes*, p. 28 n. 1.

of the holy, which indeed remains closed as a dimension if the open region of Being is not cleared and in its clearing is near man. Perhaps what is distinctive about this world-epoch consists in the closure of the dimension of the hale [*des Heilen*]. Perhaps that is the sole malignancy [*Unheil*].

But with this reference the thinking that points toward the truth of Being as what is to be thought has in no way decided in favor of theism. It can be theistic as little as atheistic. Not, however, because of an indifferent attitude, but out of respect for the boundaries that have been set for thinking as such, indeed set by what gives itself to thinking as what is to be thought, by the truth of Being. Insofar as thinking limits itself to its task it directs man at the present moment of the world's destiny into the primordial dimension of his historical abode. When thinking of this kind speaks the truth of Being it has entrusted itself to what is more essential than all values and all types of beings. Thinking does not overcome metaphysics by climbing still higher, surmounting it, transcending it somehow or other; thinking overcomes metaphysics by climbing back down into the nearness of the nearest. The descent, particularly where man has strayed into subjectivity, is more arduous and more dangerous than the ascent. The descent leads to the poverty of the ek-sistence of *homo humanus*. In ek-sistence the region of *homo animalis*, of metaphysics, is abandoned. The dominance of that region is the mediate and deeply rooted basis for the blindness and arbitrariness of what is called "biologism," but also of what is known under the heading "pragmatism." To think the truth of Being at the same time means to think the humanity of *homo humanus*. What counts is *humanitas* in the service of the truth of Being, but without humanism in the metaphysical sense.

But if *humanitas* must be viewed as so essential to the thinking of Being, must not "ontology" therefore be supplemented by "ethics"? Is not that effort entirely essential which you express in the sentence, "*Ce que je cherche à faire, depuis longtemps déjà, c'est préciser le rapport de l'ontologie avec une éthique possible*" ["What I have been trying to do for a long time now is to determine precisely the relation of ontology to a possible ethics"]?

Soon after *Being and Time* appeared a young friend asked me, "When are you going to write an ethics?" Where the essence of man is thought so essentially, i.e., solely from the question concerning the truth of Being, but still without elevating man to the center of beings, a longing necessarily awakens for a peremptory directive and for rules that say how man, experienced from ek-sistence toward Being, ought to live in a fitting manner. The desire for an ethics presses ever more ardently for fulfillment as the obvious no less than the hidden perplexity of man soars to immeasurable heights. The greatest care must be fostered upon the ethical bond at a time when technological man, delivered over to mass society, can be kept reliably on call only by gathering and ordering all his plans and activities in a way that corresponds to technology.

Who can disregard our predicament? Should we not safeguard and secure the existing bonds even if they hold human beings together ever so tenuously and merely for the present? Certainly. But does this need ever release thought from the task of thinking what still remains principally to be thought and, as Being, prior to all beings, is their guarantor and their truth? Even further, can thinking refuse to think Being after the latter has lain hidden so long in oblivion but at the same time has made itself known in the present moment of world history by the uprooting of all beings?

Before we attempt to determine more precisely the relationship between "ontology" and "ethics" we must ask what "ontology" and "ethics" themselves are. It becomes necessary to ponder whether what can be designated by both terms still remains near and proper to what is assigned to thinking, which as such has to think above all the truth of Being.

Of course if both "ontology" and "ethics," along with all thinking in terms of disciplines, become untenable, and if our thinking therewith becomes more disciplined, how then do matters stand with the question about the relation between these two philosophical disciplines?

Along with "logic" and "physics," "ethics" appeared for the first time in the school of Plato. These disciplines arose at a time when thinking was becoming "philosophy," philosophy *epistēmē* (science), and science itself a matter for schools and academic pursuits. In the course of a philosophy so understood, science waxed and thinking waned. Thinkers prior to this period knew neither a "logic" nor an "ethics" nor "physics." Yet their thinking was neither illogical nor immoral. But they did think *physis* in a depth and breadth that no subsequent "physics" was ever again able to attain. The tragedies of Sophocles—provided such a comparison is at all permissible—preserve the *ēthos* in their sagas more primordially than Aristotle's lectures on "ethics." A saying of Heraclitus which consists of only three words says something so simply that from it the essence of the *ēthos* immediately comes to light.

The saying of Heraclitus (Fragment 119) goes: *ēthos anthrōpōi daimōn*. This is usually translated, "A man's character is his daimon." This translation thinks in a modern way, not a Greek one. *Ēthos* means abode, dwelling place. The word names the open region in which man dwells. The open region of his abode allows what pertains to man's essence, and what in thus arriving resides in nearness to him, to appear. The abode of man contains and preserves the advent of what belongs to man in his essence. According to Heraclitus's phrase this is *daimōn*, the god. The fragment says: Man dwells, insofar as he is man, in the nearness of god. A story that Aristotle reports (*De partibus animalium*, I, 5, 645a 17ff.) agrees with this fragment of Heraclitus.

> The story is told of something Heraclitus said to some strangers who wanted to come visit him. Having arrived, they saw him warming himself at a stove. Surprised, they stood there in consternation—above all because he encouraged them, the astounded ones, and called for them to come in, with the words, "For here too the gods are present."

The story certainly speaks for itself, but we may stress a few aspects.

The group of foreign visitors, in their importunate curiosity about the thinker, are disappointed and perplexed by their first glimpse of his abode. They believe they should meet the thinker in circumstances which, contrary to the ordinary round of human life, everywhere bear traces of the exceptional and rare and so of the exciting. The group hopes that in their visit to the thinker they will find things that will provide material for entertaining conversation—at least for a while. The foreigners who wish to visit the thinker expect to catch sight of him perchance at that very moment when, sunk in profound meditation, he is thinking. The visitors want this "experience" not in order to be overwhelmed by thinking but simply so they can say they saw and heard someone everybody says is a thinker.

Instead of this the sightseers find Heraclitus by a stove. That is surely a common and insignificant place. True enough, bread is baked here. But Heraclitus is not even busy baking at the stove. He stands there merely to warm himself. In this altogether everyday place he betrays the whole poverty of his life. The vision of a shivering thinker offers little of interest. At this disappointing spectacle even the curious lose their desire to come any closer. What are they supposed to do here? Such an everyday and unexciting occurrence—somebody who is chilled warming himself at a stove—anyone can find any time at home. So why look up a thinker? The visitors are on the verge of going away again. Heraclitus reads the frustrated curiosity in their faces. He knows that for the crowd the failure of an expected sensation to materialize is enough to make those who have just arrived leave. He therefore encourages them. He invites them explicitly to come in with the words, *Einai gar kai entautha theous*, "Here too the gods come to presence."

This phrase places the abode (*ēthos*) of the thinker and his deed in another light. Whether the visitors understood this phrase at once—or at all—and then saw everything differently in this other light the story does not say. But the story was told and has come down to us today because what it reports derives from and characterizes the atmosphere surrounding this thinker. *Kai entautha*, "even here," at the stove, in that ordinary place where every thing and every condition, each deed and thought is intimate and commonplace, that is, familiar [*geheuer*], "even there" in the sphere of the familiar, *einai theous*, it is the case that "the gods come to presence."

Heraclitus himself says, *ēthos anthrōpōi daimōn*, "The (familiar) abode for man is the open region for the presencing of god (the unfamiliar one)."

If the name "ethics," in keeping with the basic meaning of the word *ēthos*, should now say that "ethics" ponders the abode of man, then that thinking which thinks the truth of Being as the primordial element of man, as one who ek-sists, is in itself the original ethics. However, this thinking is not ethics in the first instance, because it is ontology. For ontology always thinks solely the being (*on*) in its Being. But as long as the truth of Being is not thought all ontology remains without its foundation. Therefore the thinking that in *Being and Time* tries to advance thought in a preliminary way into the truth of Being characterizes itself as "fundamental ontology." [See *Being and Time*, sections 3 and 4, above.] It strives to reach back into the essential ground from which thought concerning the truth of Being emerges. By initiating another inquiry this thinking is already removed from the "ontology" of metaphysics (even that of Kant). "Ontology" itself, however, whether transcendental or precritical, is subject to criticism, not because it thinks the Being of beings and thereby reduces Being to a concept, but because it does not think the truth of Being and so fails to recognize that there is a thinking more rigorous than the conceptual. In the poverty of its first breakthrough, the thinking that tries to advance thought into the truth of Being brings only a small part of that wholly other dimension to language. This language even falsifies itself, for it does not yet succeed in retaining the essential help of phenomenological seeing while dispensing with the inappropriate concern with "science" and "research." But in order to make the attempt at thinking recognizable and at the same time understandable for existing philosophy, it could at first be expressed only within the horizon of that existing philosophy and its use of current terms.

In the meantime I have learned to see that these very terms were bound to lead immediately and inevitably into error. For the terms and the conceptual language corresponding to them were not rethought by readers from the matter particularly to be thought; rather, the matter was conceived according to the established terminology in its customary meaning. The thinking that inquires into the truth of Being and so defines man's essential abode from Being and toward Being is neither ethics nor ontology. Thus the question about the relation of each to the other no longer has any basis in this sphere. Nonetheless, your question, thought in a more original way, retains a meaning and an essential importance.

For it must be asked: If the thinking that ponders the truth of Being defines the essence of *humanitas* as ek-sistence from the latter's belongingness to Being, then does thinking remain only a theoretical representation of Being and of man; or can we obtain from such knowledge directives that can be readily applied to our active lives?

The answer is that such thinking is neither theoretical nor practical. It comes to pass before this distinction. Such thinking is, insofar as it is, recollection of Being and nothing else. Belonging to Being, because thrown by Being into the preservation of its truth and claimed for such preservation, it thinks Being. Such thinking has no result. It has no effect. It satisfies its essence in that it is. But it is by saying its matter. Historically, only one saying [*Sage*] belongs to the matter of thinking, the one that is in each case appropriate to its matter. Its material relevance is essentially higher than the validity of the sciences, because it is freer. For it lets Being—be.

Thinking builds upon the house of Being, the house in which the jointure of Being fatefully enjoins the essence of man to dwell in the truth of Being. This dwelling is the essence of "being-in-the-world." The reference in *Being and Time* (p. 54) to "being-in" as "dwelling" is no etymological game.* The same reference in the 1936 essay on Hölderlin's verse, "Full of merit, yet poetically, man dwells on this earth," is no adornment of a thinking that rescues itself from science by means of poetry. The talk about the house of Being is no transfer of the image "house" to Being. But one day we will, by thinking the essence of Being in a way appropriate to its matter, more readily be able to think what "house" and "to dwell" are.

And yet thinking never creates the house of Being. Thinking conducts historical ek-sistence, that is, the *humanitas* of *homo humanus*, into the realm of the upsurgence of healing [*des Heilens*].

With healing, evil appears all the more in the clearing of Being. The essence of evil does not consist in the mere baseness of human action, but rather in the malice of rage. Both of these, however, healing and the raging, can essentially occur only in Being, insofar as Being itself is what is contested. In it is concealed the essential provenance of nihilation. What nihilates illuminates itself as the negative. This can be addressed in the "no." The "not" in no way arises from the no-saying of negation. Every "no" that does not mistake itself as willful assertion of the positing power of subjectivity, but rather remains a letting-be of ek-sistence, answers to the claim of the nihilation illumined. Every "no" is simply the affirmation of the "not." Every affirmation consists in acknowledgment. Acknowledgment lets that toward which it goes come toward it. It is believed that nihilation is nowhere to be found in beings themselves. This is correct as long as one seeks nihilation as some kind of being, as an existing quality in beings. But in so seeking, one is not seeking nihilation. Neither is Being any existing quality that allows itself to be fixed among beings. And yet Being is more in

*Citing an analysis of the word "in" by Jacob Grimm, Heidegger relates "being-in" to *innan*, *wohnen*, inhabit, reside, or dwell. To be *in* the world means to dwell and be at home there, i.e., to be familiar with meaningful structures that articulate people and things. On the meaning of *dwelling*, see Reading VIII.—Ed.

being than any being. Because nihilation occurs essentially in Being itself we can never discern it as a being among beings. Reference to this impossibility never in any way proves that the origin of the not is no-saying. This proof appears to carry only if one posits beings as what is objective for subjectivity. From this alternative it follows that every "not," because it never appears as something objective, must inevitably be the product of a subjective act. But whether no-saying first posits the "not" as something merely thought, or whether nihilation first requires the "no" as what is to be said in the letting-be of beings—this can never be decided at all by a subjective reflection of a thinking already posited as subjectivity. In such a reflection we have not yet reached the dimension where the question can be appropriately formulated. It remains to ask, granting that thinking belongs to ek-sistence, whether every "yes" and "no" are not themselves already dependent upon Being. As these dependents, they can never first posit the very thing to which they themselves belong.

Nihilation unfolds essentially in Being itself, and not at all in the existence of man—so far as this is thought as the subjectivity of the *ego cogito*. Dasein in no way nihilates as a human subject who carries out nihilation in the sense of denial; rather, Da-sein nihilates inasmuch as it belongs to the essence of Being as that essence in which man ek-sists. Being nihilates—as Being. Therefore the "not" appears in the absolute Idealism of Hegel and Schelling as the negativity of negation in the essence of Being. But there Being is thought in the sense of absolute actuality as unconditioned will that wills itself and does so as the will of knowledge and of love. In this willing Being as will to power is still concealed. But just why the negativity of absolute subjectivity is "dialectical," and why nihilation comes to the fore through this dialectic but at the same time is veiled in its essence, cannot be discussed here.

The nihilating in Being is the essence of what I call the nothing. Hence, because it thinks Being, thinking thinks the nothing.

To healing Being first grants ascent into grace; to raging its compulsion to malignancy.

Only so far as man, ek-sisting into the truth of Being, belongs to Being can there come from Being itself the assignment of those directives that must become law and rule for man. In Greek, to assign is *nemein*. *Nomos* is not only law but more originally the assignment contained in the dispensation of Being. Only the assignment is capable of dispatching man into Being. Only such dispatching is capable of supporting and obligating. Otherwise all law remains merely something fabricated by human reason. More essential than instituting rules is that man find the way to his abode in the truth of Being. This abode first yields the experience of something we can hold on to. The truth of Being offers a hold for all conduct. "Hold" in our language means protective heed. Being is the protective heed that holds man in his ek-sistent essence to the truth of such protective heed—in such a way that it houses ek-sistence in language. Thus language is at once the house of Being and the home of human beings. Only because language is the home of the essence of man can historical mankind and human beings not be at home in their language, so that for them language becomes a mere container for their sundry preoccupations.

But now in what relation does the thinking of Being stand to theoretical and practical behavior? It exceeds all contemplation because it cares for the light in which a seeing, as *theoria*, can first live and move. Thinking attends to the clearing of Being in that it puts its saying of Being into language as the home of ek-sistence. Thus thinking is a deed. But a deed that also surpasses all *praxis*. Thinking towers above action and production, not through the grandeur of its achievement and not as a consequence of its effect, but through the humbleness of its inconsequential accomplishment.

For thinking in its saying merely brings the unspoken word of Being to language.

The usage "bring to language" employed here is now to be taken quite literally. Being comes, clearing itself, to language. It is perpetually under way to language. Such arriving in its turn brings ek-sisting thought to language in a saying. Thus language itself is raised into the clearing of Being. Language *is* only in this mysterious and yet for us always pervasive way. To the extent that language which has thus been brought fully into its essence is historical, Being is entrusted to recollection. Ek-sistence thoughtfully dwells in the house of Being. In all this it is as if nothing at all happens through thoughtful saying.

But just now an example of the inconspicuous deed of thinking manifested itself. For to the extent that we expressly think the usage "bring to language," which was granted to language, think only that and nothing further, to the extent that we retain this thought in the heedfulness of saying as what in the future continually has to be thought, we have brought something of the essential unfolding of Being itself to language.

What is strange in the thinking of Being is its simplicity. Precisely this keeps us from it. For we look for thinking—which has its world-historical prestige under the name "philosophy"—in the form of the unusual, which is accessible only to initiates. At the same time we conceive of thinking on the model of scientific knowledge and its research projects. We measure deeds by the impressive and successful achievements of *praxis*. But the deed of thinking is neither theoretical nor practical, nor is it the conjunction of these two forms of behavior.

Through its simple essence, the thinking of Being makes itself unrecognizable to us. But if we become acquainted with the unusual character of the simple, then another plight immediately befalls us. The suspicion arises that such thinking of Being falls prey to arbitrariness; for it cannot cling to beings. Whence does thinking take its measure? What law governs its deed?

Here the third question of your letter must be entertained: *Comment sauver l'élément d'aventure que comporte toute recherche sans faire de la philosophie une simple aventurière?* [How can we preserve the element of adventure that all research contains without simply turning philosophy into an adventuress?] I shall mention poetry now only in passing. It is confronted by the same question, and in the same manner, as thinking. But Aristotle's words in the *Poetics*, although they have scarcely been pondered, are still valid—that poetic composition is truer than exploration of beings.

But thinking is an *adventure* not only as a search and an inquiry into the unthought. Thinking, in its essence as thinking of Being, is claimed by Being. Thinking is related to Being as what arrives (*l'avenant**). Thinking as such is bound to the advent of Being, to Being as advent. Being has already been dispatched to thinking. Being *is* as the destiny of thinking. But destiny is in itself historical. Its history has already come to language in the saying of thinkers.

To bring to language ever and again this advent of Being that remains, and in its remaining waits for man, is the sole matter of thinking. For this reason essential thinkers always say the Same. But that does not mean the identical. Of course they say it only to one who undertakes to think back on them. Whenever thinking, in historical recollection, attends to the destiny of Being, it has already bound itself to what is fitting for it, in accord with its destiny. To flee into the identical is not dangerous. To risk discord in order to say the Same is the danger. Ambiguity threatens, and mere quarreling.

The fittingness of the saying of Being, as of the destiny of truth, is the first law of thinking—not the rules of logic, which can become rules only on the basis of the law of Being. To attend to the fittingness of thoughtful saying does not only imply, however, that we contemplate at every turn *what* is to be said of Being and *how* it is to be said. It is equally essential to ponder *whether* what is to be thought is to be said—to what extent, at what moment of the history of Being, in what sort of dialogue with this history, and on the

**L'avenant* (cf. the English *advenient*) is most often used as an adverbial phrase, *à l'avenant*, to be in accord, conformity, or relation to something. It is related to *l'aventure*, the arrival of some unforeseen challenge, and *l'avenir*, the future, literally, what is to come. Thinking is in relation to Being insofar as Being advenes or arrives. Being as arrival of presencing is the "adventure" toward which Heidegger's thought is on the way.—ED.

basis of what claim, it ought to be said. The threefold thing mentioned in an earlier letter is determined in its cohesion by the law of the fittingness of thought on the history of Being: rigor of meditation, carefulness in saying, frugality with words.

It is time to break the habit of overestimating philosophy and of thereby asking too much of it. What is needed in the present world crisis is less philosophy, but more attentiveness in thinking; less literature, but more cultivation of the letter.

The thinking that is to come is no longer philosophy, because it thinks more originally than metaphysics—a name identical to philosophy. However, the thinking that is to come can no longer, as Hegel demanded, set aside the name "love of wisdom" and become wisdom itself in the form of absolute knowledge. Thinking is on the descent to the poverty of its provisional essence. Thinking gathers language into simple saying. In this way language is the language of Being, as clouds are the clouds of the sky. With its saying, thinking lays inconspicuous furrows in language. They are still more inconspicuous than the furrows that the farmer, slow of step, draws through the field.

# VI

# MODERN SCIENCE, METAPHYSICS, AND MATHEMATICS
## (from *What Is a Thing?*)

*The oldest of the old follows behind us in our thinking, and yet it comes to meet us.*

In *Being and Time* (p. 50, above) Heidegger remarked that the level of advance in the sciences may be gauged by their readiness to undergo a crisis in fundamental concepts. Such a crisis brewed in Western Europe between the fifteenth and seventeenth centuries, an age historians dub "The Scientific Revolution." In a 1936 lecture course entitled "Basic Questions of Metaphysics," Heidegger had occasion to comment on this crisis, preparatory to an analysis of Kant's *Critique of Pure Reason*. One of those basic metaphysical questions came to be the title of the published lectures, *What Is a Thing?* (Recall that the question of the "thing" served as the point of departure for Heidegger's inquiry into the work of art; the same question recurs later, in Reading VIII, "Building Dwelling Thinking.")

How does modern science differ from ancient and medieval science? Popular belief asserts that early science fiddled with "concepts" while modern science faces "facts." Historians of science have been busy for a long time discouraging this facile interpretation. Science—ancient, medieval, and modern—measures, experiments, and works with concepts, in order to learn about things. But modern science diverges from its antecedents by its *manner* of measuring, experimenting, and conceptualizing. Its manner is prescribed by what Heidegger calls the *mathematical projection*. "Mathematical" does not mean merely what pertains to numbers, but the way something is learned. More specifically, it means what the learner brings to things when he or she learns. "Projection" here means the fundamental presuppositions and expectations science entertains with respect to the "thingness" of things. Heidegger contrasts Aristotle's projection in *On the Heavens* to that of Newton in the *Principia mathematica* and Galileo in his *Discourses*. Especially in the latter work the "mathematical" element crystallizes, reflecting what Galileo wants to learn from things and so anticipates in them. Modern science proves to be grounded in the will to axiomatic knowledge of unshakably certain propositions. It projects a universally valid ground-plan or blueprint for all things in a way that is, in Heidegger's words, "neither arbitrary nor readily comprehensible." The essence of beings as a whole, what makes things be what they are, becomes accessible only to a kind of thinking entirely conformable to this ground-plan. Mathematics in the narrower sense is a response to, and by no means the ground of, this will to axiomatic knowledge. Neither the controversy between formalistic and intuitionistic mathematics in our own time, nor the emendations of classical physics in contemporary quantum mechanics, relativity theory, and thermodynamics, alter the basic structure of the modern mathematical projection.

The change from ancient and medieval to modern science is one from "bodies" to "mass," "places" to "position," "motions" to "inertia," "tendencies" to "force." "Things" become aggregates of calculable mass located on the grid of space-time and ultimately a play of forces issuing in (partly) discernible and (variably) predictable jumps across that grid; even when new discoveries require a change in its design, the grid remains a transparency laminated on the "things." Although it is undeniably "successful" the transparency remains mysterious: Newton's First Law applies to all things but no thing can behave the way it presupposes, and when Galileo searches for an inclined plane that all the real things in the world can roll along without resistance he can find it only in his head. Transparencies cloud. But to put them to question is not to deny their efficacy and to turn "anti-science."

Not only physics but also metaphysics shares in the mathematical projection. Descartes's quest for a self-grounding, hence absolutely certain, foundation for "first philosophy" or ontology makes the mathematical projection that of modern philosophy as a whole. Thus "Modern Science, Metaphysics, and Mathematics" is invaluable for our understanding of Heidegger's own project, which aims to raise the question of Being through a kind of thought different from that of Descartes, Kant, Hegel, and Husserl. His opposition to their preoccupation with "method" and "certainty" (see Reading XI) stems from Heidegger's perception of the historical cast of both physics and metaphysics in the mathematical projection. His resistance to the totalizing tendency of that projection in positivism and in the compulsive pursuit of technological mastery is perhaps best grasped through careful study of the following material. It accomplishes essential steps in the destructuring of the history of ontology and therefore at least partly fulfills the intentions of the second major part—the unpublished part—of *Being and Time*.

# MODERN SCIENCE, METAPHYSICS, AND MATHEMATICS

## A. The Characteristics of Modern Science in Contrast to Ancient and Medieval Science

One commonly characterizes modern science in contradistinction to medieval science by saying that modern science starts from facts while the medieval started from general speculative propositions and concepts. This is true in a certain respect. But it is equally undeniable that the medieval and ancient sciences also observed the facts and that modern science also works with universal propositions and concepts. This went so far that Galileo, one of the founders of modern science, suffered the same reproach that he and his disciples actually made against Scholastic science. They said it was "abstract," that is, that it proceeded with general propositions and principles. Yet in an even more distinct and conscious way the same was the case with Galileo. The contrast between the ancient and the modern attitude toward science cannot therefore be established by saying, there concepts and principles, and here facts. Both ancient and modern science have to do with both facts and concepts. However, the way the facts are conceived and how the concepts are established are decisive.

---

This selection appears in Martin Heidegger, *What Is a Thing?* translated by W. B. Barton, Jr., and Vera Deutsch, with an analysis by Eugene T. Gendlin (Chicago: Henry Regnery Co., 1967), pp. 66–108. Reprinted by permission, with minor changes and some deletions. Original edition: Martin Heidegger, *Die Frage nach dem Ding* (Tübingen: Max Niemeyer Verlag, 1962), pp. 50–83.

The greatness and superiority of natural science during the sixteenth and seventeenth centuries rests in the fact that all the scientists were philosophers. They understood that there are no mere facts, but that a fact is only what it is in the light of the fundamental conception, and always depends upon how far that conception reaches. The characteristic of positivism—which is where we have been for decades, today more than ever—by way of contrast is that it thinks it can manage sufficiently with facts, or other and new facts, while concepts are merely expedients that one somehow needs but should not get too involved with, since that would be philosophy. Furthermore, the comedy—or rather the tragedy—of the present situation of science is that one thinks to overcome positivism through positivism. To be sure, this attitude prevails only where average and supplemental work is done. Where genuine and discovering research is done the situation is no different from that of three hundred years ago. That age also had its indolence, just as, conversely, the present leaders of atomic physics, Niels Bohr and Heisenberg, think in a thoroughly philosophical way, and only therefore create new ways of posing questions and, above all, hold out in the questionable.

Hence it remains basically inadequate to try to distinguish modern from medieval science by calling it the science of facts. Further, the difference between the old and the new science is often seen in the fact that the latter experiments and proves "experimentally" its cognitions. But the experiment or test to get information concerning the behavior of things through a definite ordering of things and events was also already familiar in ancient times and in the Middle Ages. This kind of experience lies at the basis of all contact with things in the crafts and in the use of tools. Here too what matters is not the experiment as such in the wide sense of testing through observation but the manner of setting up the test and the intent with which it is undertaken and in which it is grounded. The manner of experimentation is presumably connected with the kind of conceptual determination of the facts and way of applying concepts, i.e., with the kind of preconception about things.

Besides these two constantly cited characteristics of modern science, science of facts and experimental research, one also usually meets a third. This third affirms that modern science is a calculating and measuring investigation. That is true. However, it is also true of ancient science, which also worked with measurement and number. Again it is a question of how and in what sense calculating and measuring are applied and carried out, and what importance they have for the determination of the objects themselves.

With these three characteristics of modern science, that it is a factual, experimental, measuring science, we still miss the fundamental characteristic of modern science. The fundamental feature must consist in what rules and determines the basic movement of science itself. This characteristic is the manner of working with the things and the metaphysical projection of the thingness of the things. How are we to conceive this fundamental feature?

We attain this fundamental feature of modern science for which we are searching by saying that modern science is *mathematical*. From Kant comes the oft-quoted but still little understood sentence, "However, I maintain that in any particular doctrine of nature only so much *genuine* science can be found as there is mathematics to be found in it" (Preface to *Metaphysical Beginning Principles of Natural Science*).

The decisive question is: What do "mathematics" and "mathematical" mean here? It seems as though we can take the answer to this question only from mathematics itself. This is a mistake, because mathematics itself is only a particular formation of the mathematical. . . .

### B. The Mathematical, *Mathēsis*

How do we explain the mathematical if not by mathematics? In such questions we do well to keep to the word itself. Of course, the issue is not always there where the word occurs. But with the Greeks, from whom the word stems, we may safely make this assumption. In its formation the word "mathematical" stems from the

Greek expression *ta mathēmata*, which means what can be learned and thus, at the same time, what can be taught; *manthanein* means to learn, *mathēsis* the teaching, and this in a twofold sense. First, it means studying and learning; then it means the doctrine taught. To teach and to learn are here intended in a wide and at the same time essential sense, and not in the later narrow and trite sense of schools and scholars. However, this distinction is not sufficient to grasp the proper sense of the "mathematical." To do this we must inquire in what further connection the Greeks employ the mathematical and from what they distinguish it.

We experience what the mathematical properly is when we inquire *under what* the Greeks classify the mathematical and *against what* they distinguish it within this classification. The Greeks identify the mathematical, *ta mathēmata*, in connection with the following determinations:

1. *Ta physica:* the things insofar as they originate and come forth from themselves.
2. *Ta poioumena:* the things insofar as they are produced by the human hand and subsist as such.
3. *Ta chrēmata:* the things insofar as they are in use and therefore stand at our constant disposal—they may be either *physica*, rocks and so on, or *poioumena*, something specially made.
4. *Ta pragmata:* the things insofar as we have to do with them at all, whether we work on them, use them, transform them, *or* only look at and examine them, *pragmata* being related to *praxis:* here *praxis* is taken in a truly wide sense, neither in the narrow meaning of practical use (*chrēsthai*) nor in the sense of *praxis* as ethical action; *praxis* is all doing, pursuing, and sustaining, which also includes *poiēsis*; and finally,
5. *Ta mathēmata:* According to the characterization running through these last four, we must also say here of *mathēmata:* the things insofar as they . . . but the question is: In what respect?

. . . We are long used to thinking of numbers when we think of the mathematical. The mathematical and numbers are obviously connected. But the question remains: Is this connection due to the fact that the mathematical is numerical in character, or, on the contrary, is the numerical something mathematical? The second is the case. But insofar as numbers are in this way connected with the mathematical the question still remains: Why precisely are numbers something mathematical? What is the mathematical itself, that something like numbers must be conceived as mathematical and are primarily presented as the mathematical? *Mathēsis* means learning; *mathēmata*, what is learnable. In accord with what has been said, this designation is intended of things insofar as they are learnable. Learning is a kind of grasping and appropriating. But not every taking is a learning. We can take a thing, for instance, a rock, take it with us and put it in a rock collection. We can do the same with plants. It says in our cookbooks that one "takes," i.e., uses. To take means in some way to take possession of a thing and have disposal over it. Now, what kind of taking is learning? *Mathēmata*—things, insofar as we learn them. . . .

The *mathēmata* are the things insofar as we take cognizance of them as what we already know them to be in advance, the body as the bodily, the plant-like of the plant, the animal-like of the animal, the thingness of the thing, and so on. This genuine learning is therefore an extremely peculiar taking, a taking where one who takes only takes what one basically already has. Teaching corresponds to *this* learning. Teaching is a giving, an offering; but what is offered in teaching is not the learnable, for the student is merely instructed to take for himself what he already has. If the student only takes over something that is offered he does not learn. He comes to learn only when he experiences what he takes as something he himself really already has. True learning occurs only where the taking of what one already has is a *self-giving* and is experienced as such. Teaching therefore does not mean anything else than to let the others learn, that is, to bring one another to learning. Teaching is more difficult than learning; for only he who can truly learn—and only as long as he can do it—can truly teach. The genuine teacher differs from the pupil only in that he can learn better and that he more genuinely wants to learn. In all teaching, the teacher learns the most.

The most difficult learning is coming to know actually and to the very foundations what we already know. Such learning, with which we are here solely concerned, demands dwelling continually on what appears to be nearest to us, for instance, on the question of what a thing is. We steadfastly ask the *same* question—which in terms of utility is obviously useless—of what a thing is, what tools are, what man is, what a work of art is, what the state and the world are.

In ancient times there was a famous Greek scholar who traveled everywhere lecturing. Such people were called Sophists. This famous Sophist, returning to Athens once from a lecture tour in Asia Minor, met Socrates on the street. It was Socrates' habit to hang around on the street and talk with people, for example, with a cobbler about what a shoe is. Socrates had no other topic than what the things are. "Are you still standing there," condescendingly asked the much-traveled Sophist of Socrates, "and still saying the same thing about the same thing?" "Yes," answered Socrates, "that I am. But you who are so extremely smart, you *never* say the same thing about the same thing."

The *mathēmata*, the mathematical, is that "about" things which we really already know. Therefore we do not first get it out of things, but, in a certain way, we bring it already with us. From this we can now understand why, for instance, number is something mathematical. We see three chairs and say that there are three. What "three" is the three chairs do not tell us, nor three apples, three cats, nor any other three things. Rather, we can count three things only if we already know "three." In thus grasping the number three as such, we only expressly recognize something which, in some way, we already have. This recognition is genuine learning. The number is something in the proper sense learnable, a *mathēma*, i.e., something mathematical. Things do not help us to grasp "three" as such, i.e., threeness. "Three"—what exactly is it? It is the number in the natural series of numbers that stands in third place. In "third"? It is only the third number because it is the three. And "place"—where do places come from? "Three" is not the third number, but the first number. "One" isn't really the first number. For instance, we have before us one loaf of bread and one knife, this one and, in addition, another one. When we take both together we say, "both of these," the one and the other, but we do not say, "these two," or 1 + 1. Only when we add a cup to the bread and the knife do we say "all." Now we take them as a sum, i.e., as a whole and so and so many. Only when we perceive it from the third is the former one the first, the former other the second, so that one and two arise, and "and" becomes "plus," and there arises the possibility of places and of a series. What we now take cognizance of is not drawn from any of the things. We take what we ourselves somehow already have. What must be understood as mathematical is what we can learn in this way.

We take cognizance of all this and learn it without regard for the things. Numbers are the most familiar form of the mathematical because, in our usual dealing with things, when we calculate or count, numbers are the closest to that which we recognize in things without deriving it from them. For this reason numbers are the most familiar form of the mathematical. In this way, this most familiar mathematical becomes mathematics. But the essence of the mathematical does not lie in number, as purely delimiting the pure "how much," but vice versa. Because number has such a nature, therefore, it belongs to the learnable in the sense of *mathēsis*.

Our expression "the mathematical" always has two meanings. It means, first, what can be learned in the manner we have indicated, and only in that way, and, second, the manner of learning and the process itself. The mathematical is that evident aspect of things within which we are always already moving and according to which we experience them as things at all, and as such things. The mathematical is this fundamental position we take toward things by which we take up things as already given to us, and as they must and should be given. The mathematical is thus the fundamental presupposition of the knowledge of things.

Therefore, Plato put over the entrance to his Academy the words: *Ageometrētos mēdeis eisito!* "Let no one who has not grasped the mathematical enter here!"* These words do not mean that one must be educated in only one subject—"geometry"—but that one must grasp that the fundamental condition for the proper possibility of knowing is knowledge of the fundamental presuppositions of all knowledge and the position we take based on such knowledge. A knowledge which does not build its foundation knowledgeably, and thereby notes its limits, is not knowledge but mere opinion. The mathematical, in the original sense of learning what one already knows, is the fundamental presupposition of "academic" work. This saying over the Academy thus contains nothing more than a hard condition and a clear circumscription of work. Both have had the consequence that we today, after two thousand years, are still not through with this academic work and never will be as long as we take ourselves seriously.

This brief reflection on the essence of the mathematical was brought about by our maintaining that the basic character of modern science is the mathematical. After what has been said, this cannot mean that this science employs mathematics. Our inquiry showed that, *in consequence* of this basic character of science, mathematics in the narrower sense first had to come into play.

Therefore, we must now show in what sense the foundation of modern thought and knowledge is essentially mathematical. With this intention we shall try to set forth an essential step of modern science in its main outline. This will make clear what the mathematical consists of and how it unfolds its essence, but also how it becomes established in a certain direction.

*Elias Philosophus, sixth century A.D. Neoplatonist, in *Aristotelis Categorias Commentaria* (*Commentaria in Aristotelem Graeca*), A. Busse, ed. (Berlin, 1900), 118.18.—Tr.

## C. The Mathematical Character of Modern Natural Science; Newton's First Law of Motion

Modern thought does not appear all at once. Its beginnings stir during the later Scholasticism of the fifteenth century; the sixteenth century brings sudden advances as well as setbacks; but it is only during the seventeenth century that the decisive clarifications and foundations are accomplished. This entire happening finds its first systematic and creative culmination in the English mathematician and physicist Newton, in his major work, *Philosophiae Naturalis Principia Mathematica*, 1686–87. In the title, "philosophy" indicates general science (compare "*Philosophia experimentalis*"); "*principia*" indicates first principles, the beginning ones, i.e., the *very first* principles. But these starting principles by no means deal with an introduction for beginners.

This work was not only a culmination of preceding efforts, but at the same time the foundation for the succeeding natural science. It has both promoted and limited the development of natural science. When we talk about classical physics today, we mean the form of knowledge, questioning, and evidence as Newton established it. When Kant speaks of "science," he means Newton's physics.

This work is preceded by a short section entitled "*Definitiones.*" These are definitions of *quantitas materiae, quantitas motus*, force, and, above all, *vis centripeta*. Then there follows an additional *scholium* that contains the series of famous conceptions of absolute and relative time, absolute and relative space, and finally, of absolute and relative motion. Then follows a section with the title "*Axiomata, sive leges motus*" ("Principles or Laws of Motion"). This contains the proper content of the work. It is divided into three volumes. The first two deal with the motion of bodies, *de motu corporum*, the third with the system of the world, *de mundi systemate*.

Here we shall merely take a look at the first principle, i.e., that Law of Motion which Newton sets at the apex of his work. . . . "Every body continues in its state of rest, or uniform motion in a straight line, unless it is compelled to change that state by force impressed upon it."* This is called the principle of inertia (*lex inertiae*).

The second edition of this work was published in 1713, while Newton was still alive. It included an extended preface by Cotes, then professor at Cambridge. In it Cotes says about this basic principle: "It is a law of nature universally received by all philosophers."

Students of physics do not puzzle over this law today and have not for a long time. If we mention it at all and know anything about it, that and to what extent it is a fundamental principle, we consider it self-evident. And yet, one hundred years before Newton at the apex of his physics put this law in this form, it was still unknown. It was not even Newton himself who discovered it, but Galileo; the latter, however, applied it only in his last works and did not even express it as such. Only the Genoese Professor Baliani articulated this discovered law in general terms. Descartes then took it into his *Principia Philosophiae* and tried to ground it metaphysically. With Leibniz it plays the role of a metaphysical law (C.I. Gerhardt, *Die philosophischen Schriften von G. W. Leibniz* [Berlin, 1875–1890], IV, 518, *contra* Bayle).

This law, however, was not at all self-evident even in the seventeenth century. During the preceding fifteen hundred years it was not only unknown, but nature and beings in general were experienced in such a way that it would have been senseless. In its discovery and its establishment as the fundamental law lies a revolution that belongs to the greatest in human thought, and which first provides the ground for the turning from the Ptolemaic to the Copernican conception of the universe. To be sure, the law of inertia and its definition already had their predecessors in ancient times. Certain fundamental principles of Democritus (460–370 B.C.) tend in this direction. It has also been shown that Galileo and his

*Isaac Newton, *Mathematical Principles of Natural Philosophy and His System of the World*, Andrew Motte, trans., 1729; revised translation, Florian Cajori (Berkeley: University of California Press, 1946), p. 13—Tr.

age (partly directly and partly indirectly) knew of the thought of Democritus. But, as is always the case, that which can already be found in the older philosophers is seen only when one has newly thought it out for oneself. . . . After people understood Democritus with the help of Galileo they could reproach the latter for not really reporting anything new. All great insights and discoveries are not only usually thought by several people at the same time, they must also be rethought in that unique effort to truly say the same thing about the same thing.

## D. The Difference Between the Greek Experience of Nature and That of Modern Times

### 1. *The experience of nature in Aristotle and Newton*

How does the aforementioned fundamental law relate to the earlier conception of nature? The idea of the universe (world) which reigned in the West up to the seventeenth century was determined by Platonic and Aristotelian philosophy. Scientific conceptual thought especially was guided by those fundamental representations, concepts, and principles which Aristotle had set forth in his lectures on physics and the heavens (*De caelo*), and which were taken over by the medieval Scholastics.

We must, therefore, briefly go into the fundamental conceptions of Aristotle in order to evaluate the significance of the revolution articulated in Newton's First Law. But we must first liberate ourselves from a prejudice which was partly nourished by modern science's sharp criticism of Aristotle: that his propositions were merely concepts he thought up, which lacked any support in the things themselves. This might be true of later medieval Scholasticism, which often in a purely dialectical way was concerned with a foundationless analysis of concepts. It is certainly not true of Aristotle himself. Moreover, Aristotle fought in his time precisely to make thought, inquiry, and assertion always a *legein homologoumena tois phainomenois*, "saying what corresponds to that which shows itself in beings" (*De caelo*, III, 7, 306a 6).

In the same place Aristotle expressly says: *telos de tēs men poiētikēs epistēmēs to ergon, tēs de physikēs to phainomenon aei kuriōs kata tēn aisthēsin*. ["And that issue, which in the case of productive knowledge is the product, in the knowledge of nature is the unimpeachable evidence of the senses as to each fact."*] We have heard (p. 274, above) that the Greeks characterize the thing as *physica* and *poioumena*, such as occurs from out of itself, or such as is produced. Corresponding to this, there are two different kinds of knowledge (*epistēmē*), knowledge of what occurs from out of itself and knowledge of what is produced. Corresponding to this, the *telos* of knowledge, that is, that whereby this knowledge comes to an end point, where it stops, *what it genuinely holds to*, is different. Therefore the above sentence states, "That at which productive knowledge comes to a halt, where from the beginning it takes hold, is the *work* to be produced. That, however, in which the knowledge of 'nature' takes hold is *to phainomenon*, what shows itself in that which occurs out of itself. This is always predominant and is the standard, especially for perception, that is, for mere 'taking-in-and-up'" (in contradistinction to making and concerning oneself busily with creation of things). What Aristotle here expresses as a basic principle of scientific method differs in no way from the principles of modern science. Newton writes (*Principia*, Bk. III, *Regulae* IV): ". . . In experimental philosophy we are to look upon propositions inferred by general induction from phenomena as accurate or very nearly true, notwithstanding contrary hypotheses that may be imagined, till such times as other phenomena occur, by which they may either be made more accurate, or liable to exceptions."

But despite this similar basic attitude toward procedure, the basic position of Aristotle is essentially different from that of Newton. For

*De caelo*, III, 7, 306a 16–17. The translation is taken from *The Works of Aristotle*, W. D. Ross, ed. and trans., 11 vols. (Oxford: Clarendon Press, 1931).—Tr.

*what* is actually apprehended as appearing and *how* it is interpreted are not alike.

2. *The doctrine of motion in Aristotle*

Nevertheless, they share from the start the experience that beings, in the general sense of nature—earth, sky, and stars—are in motion or at rest. Rest means only a special case of motion. It is everywhere a question of the motion of bodies. But how motion and bodies are to be conceived and what relation they have to each other is not established and not self-evident. From the general and indefinite experience that things change, come into existence and pass away, thus are in motion, it is a long way to an insight into the essence of motion and into the manner of its belonging to things. The ancient Greek conception of the earth is of a disc around which floats Okeanos. The sky overarches it and turns around it. Later, Plato, Aristotle, and Eudoxus—though each differently—present the earth as a ball, but still as a center of everything.

We restrict ourselves to the presentation of the Aristotelian conception, which later became widely dominant, and this only sufficiently to show the contrast that expresses itself in the first axiom of Newton.

First, we ask in general what, according to Aristotle, is the essence of a thing in nature? The answer is: *ta physica sōmata* are *kath' auta kinēta kata topon*. "Those bodies which belong to 'nature' and constitute it are, in themselves, movable with respect to location." Motion, in general, is *metabolē*, the alteration of something into something else. Motion in this wide sense includes, for instance, turning pale or blushing. But it is also an alteration when a body is transferred from one place to another. This being transported, altered, or conveyed is expressed in Greek as *phora*. *Kinēsis kata topon* means in Greek what constitutes the proper motion of Newtonian bodies. In this motion lies a definite relation to place. The motion of bodies, however, is *kath' auta*, according to them, themselves. That is to say, how a body moves, i.e., how it relates to place and to which place it relates—all this has its basis in the body itself. This basis is *archē*, which has a double meaning: that from which something emerges, and that which governs over what emerges in this way. The body is *archē kinēseōs*. What an *archē kinēseōs* in this manner is, is *physis*, the original mode of emergence, which, however, remains limited solely to pure movement in space. Herein appears an essential transformation of the concept of *physis*. The body moves according to its nature. A moving body, which is itself an *archē kinēseōs*, is a natural body. The purely earthly body moves downward, the purely fiery body—as every blazing flame demonstrates—moves upward. Why? Because the earthly has its place below, the fiery, above. Each body has *its* place *according to its kind*, and it strives toward that place. Around the earth is water, around this, the air, and around this, fire—the four elements. When a body moves toward its place this motion accords with nature, *kata physin*. A rock falls down to the earth. However, if a rock is thrown upward by a sling, this motion is essentially against the nature of the rock, *para physin*. All motions against nature are *biai*, violent.

The kind of motion and place of the body are determined according to the nature of the body. Earth is the center for all characterization and evaluation of motion. The rock that falls moves toward this center, *epi to meson*. The fire that rises, *apo tou mesou*, moves away from the center. In both cases the motion is *kinēsis eutheia*, in a straight line. But the stars and the entire heavens move around the center, *peri to meson*. This motion is *kyklōi*. Circular motion and motion in a straight line are the simple movements, *haplai*. Of these two, circular motion is first, i.e., is the higher, and thus, of the highest order. For *proteron to teleion tou atelous*, the complete precedes the incomplete. The motion of bodies accords with their place. In circular motion the body has its place in the motion itself; for this reason such motion is perpetual and truly in being. In rectilinear motion the place lies only in one direction, away from another place, so that motion comes to an end over there. Besides these two forms of simple motion there are mixtures of both, *miktē*. The purest motion, in the sense of change of place, is circular motion; it contains, as it were, its place in itself. A body that so moves itself, moves itself completely. This is true of all celestial bodies. Compared to this, earthly motion is always in a straight line, or mixed, or violent, but always incomplete.

There is an essential difference between the motion of celestial bodies and earthly bodies. The domains of these motions are different. How a body moves depends upon its species and the place to which it belongs. The *where* determines the *how* of its Being, for Being means *presence* [*Anwesenheit*]. Because it moves in a circle, that is, moves completely and permanently in the simplest motion, the moon does not fall earthward. This circular motion is in itself completely independent of anything outside itself—for instance, from the earth as center. But, by way of contrast, to anticipate, in modern thought circular motion is understood only in such a way that a perpetual attracting force from the center is necessary for its formation and preservation. With Aristotle, however, this "force," *dynamis*, the capacity for its motion, lies in the *nature* of the body itself. The kind of motion of the body and its relation to its place depend upon the nature of the body. The velocity of natural motion increases the nearer the body comes to its place; that is, increase and decrease of velocity and cessation of motion depend upon the nature of the body. A motion contrary to nature, i.e., violent motion, has its cause in the force that affects it. However, according to its motion, the body, driven forcibly, must withdraw from this power, and since the body itself does not bring with it any basis *for* this violent motion, its motion must necessarily become slower and finally stop (cf. *De caelo*, I, 8, 277b 6; I, 2, 269b 9).

This corresponds distinctly to the common conception: a motion imparted to a body continues for a certain time and then ceases, passing over into a state of rest. Therefore we must look for the causes of the continuation or endurance of the motion. According to Aristotle the basis for natural motion lies in the nature of the body itself, in its essence, in its most proper Being. A later Scholastic proposition is in accord with this: *Operari* (*agere*) *sequitur esse*, "The kind of motion follows from the kind of Being."

3. *Newton's doctrine of motion*

How do Aristotle's observation of nature and concept of motion as we have described them relate to the modern ones, which got an essential foundation in the first axiom of Newton? We shall try to present in order a few main distinctions. For this purpose we give the axiom an abridged form: "Every body left to itself moves uniformly in a straight line." *Corpus omne, quod a viribus impressis non cogitur, uniformiter in directum movetur.* We shall discuss what is new in eight points.

1. Newton's axiom begins with *corpus omne*, "every body." That means that the distinction between earthly and celestial bodies has become obsolete. The universe is no longer divided into two well-separated realms, the one beneath the stars, the other the realm of the stars themselves. All natural bodies are essentially of the same kind. The upper realm is not a superior one.

2. In accord with this, the priority of circular motion over motion in a straight line also disappears. And although now, on the contrary, motion in a straight line becomes decisive, still this does not lead to a division of bodies and of different domains according to their kind of motion.

3. Accordingly, the distinguishing of certain places also disappears. Each body can in principle be in any place. The concept of place itself is changed: place no longer is where the body belongs according to its inner nature, but only a position in relation to other positions. (Cf. points 5 and 7.) *Phora* and change of place in the modern sense are not the same.

With respect to the causation and determination of motion, one does not ask for the cause of the continuity of motion and therefore for its perpetual occurrence, but the reverse: being in motion is presupposed, and one asks for the causes of a change in the kind of motion presupposed as uniform and in a straight line. The circularity of the moon's motion does not cause its uniform perpetual motion around the earth. Precisely the reverse. It is this motion for whose cause we must search. According to the law of inertia, the body of the moon should move from every point of its circular orbit in a straight line, i.e., in the form of a tangent. Since the moon does not do so, the question—based upon the presupposition of the law of inertia—arises: Why does the moon decline from the line of a tangent? Why does it move, as the Greeks put it, in a circle? The circular movement is now not cause but, on the contrary, precisely what requires a reason. (We know that Newton arrived at a new answer when he proposed that the force according to which bodies fall to the ground is also the one according to which the celestial bodies remain in their orbits: gravity. Newton compared the centripetal declination of the moon from the tangent of its orbit during a fraction of time with the linear distance that a falling body achieves at the surface of earth in an equal time. At this point we see immediately the elimination of the distinction already mentioned between earthly and celestial motions and thus between bodies.)

4. Motions themselves are not determined according to different natures, capacities, and forces, or the elements of the body, but, in reverse, the essence of force is determined by the fundamental law of motion: every body, left to itself, moves uniformly in a straight line. According to this, a force is that whose impact results in a declination from rectilinear, uniform motion. "An impressed force is an action exerted upon a body, in order to change its state, either of rest, or of uniform motion in a right line" (*Principia*, Def. IV). This new determination of force leads at the same time to a new determination of mass.

5. Corresponding to the change of the concept of place, motion is seen only as a change of position and relative position, as distances between places. Therefore, the determination of motion develops into one regarding distances, stretches of the measurable, of

the so and so large. Motion is determined as the amount of motion, and, similarly, mass is determined as weight.

6. Therefore, the difference between natural and against nature, i.e., violent, is also eliminated; the *bia*, violence, is as force only a measure of the change of motion and is no longer special in kind. Impact, for instance, is only a particular form of impressed force, along with pressure and centripetality.

7. Therefore, the concept of nature in general changes. Nature is no longer the *inner* principle out of which the motion of the body follows; rather, nature is the mode of the variety of the changing relative positions of bodies, the manner in which they are present in space and time, which themselves are domains of possible positional orders and determinations of order and have no special traits anywhere.

8. Thereby the manner of questioning nature also changes and, in a certain respect, becomes the reverse.

We cannot set forth here the full implications of the revolution of inquiry into nature. It should have become clear only that, and how, the application of the First Law of Motion implies all the essential changes. All these changes are linked together and uniformly based on the new basic position expressed in the First Law and which we call mathematical.

### E. The Essence of the Mathematical Project [*Entwurf*]* (Galileo's Experimental with Free Fall)

For us, for the moment, the question concerns the application of the First Law, more precisely, the question in what sense the mathematical becomes decisive in it.

How about this law? It speaks of a body, *corpus quod a viribus impressis non cogitur*, a body which is left to itself. Where do we

*Perhaps the best insight as to what Heidegger means by *Entwurf* is Kant's use of the word in the *Critique of Pure Reason*. "When Galileo experimented with balls

find it? There is no such body. There is also no experiment that could ever bring such a body to direct perception. But modern science, in contrast to the mere dialectical, poetic conception of medieval Scholasticism and science, is supposed to be based upon experience. Instead, it has such a law at its apex. This law speaks of a thing that does not exist. It demands a fundamental representation of things that contradict the ordinary.

The mathematical is based on such a claim, i.e., the application of a determination of the thing which is not experientially derived from the thing and yet lies at the base of every determination of the things, making them possible and making room for them. Such a fundamental conception of things is neither arbitrary nor self-evident. Therefore, it required a long controversy to bring it into power. It required a change in the mode of approach to things along with the achievement of a new manner of thought. We can accurately follow the history of this battle. Let us cite *one* example from it. In the Aristotelian view, bodies move according to their nature, the heavy ones downward, the light ones upward. When both fall, heavy ones fall faster than light ones, since the latter have the urge to move upward. It becomes a decisive insight of Galileo that all bodies fall equally fast, and that the differences in the time of fall derive only from the resistance of the air, not from the different inner natures of the bodies or from their own corresponding relation to their particular place. Galileo did his experiment at the leaning tower in the town of Pisa, where he was professor of

---

whose weight he himself had already predetermined, when Torricelli caused the air to carry a weight which he had calculated beforehand to be equal to that of a definite column of water, or, at a later time, when Stahl converted metal into lime and this again into metal by withdrawing something and then adding it, a light broke in on all investigators of nature. They learned that reason only gains insight into what it produces itself according to its own project [*was sie selbst nach ihrem Entwurfe hervorbringt*]; that it must go before with principles of judgment according to constant laws, and constrain nature to reply to its questions, not content merely to follow her leading-strings" (B XIII).—Tr.

mathematics, in order to prove his statement. In it bodies of different weights did not arrive at precisely the same time after having fallen from the tower, but the difference in time was slight. In spite of these differences and therefore really *against* the evidence of experience, Galileo upheld his proposition. The witnesses to this experiment, however, became really perplexed by the experiment and Galileo's upholding his view. They persisted the more obstinately in their former view. By reason of this experiment the opposition toward Galileo increased to such an extent that he had to give up his professorship and leave Pisa.

Both Galileo and his opponents saw the same "fact." But they interpreted the same fact differently and made the same happening visible to themselves in different ways. Indeed, what appeared for them as the essential fact and truth was something different. Both thought something along with the same appearance but they thought something different, not only about the single case, but fundamentally, regarding the essence of a body and the nature of its motion. What Galileo thought in advance about motion was the determination that the motion of every body is uniform and rectilinear, when every obstacle is excluded, but that it also changes uniformly when an equal force affects it. In his *Discorsi*, which appeared in 1638, Galileo said: "I think of a body thrown on a horizontal plane and every obstacle excluded. This results in what has been given a detailed account in another place, that the motion of the body over this plane would be uniform and perpetual if the plane were extended infinitely."

In this proposition, which may be considered the antecedent of the First Law of Newton, what we have been looking for is clearly expressed. Galileo says: *Mobile . . . mente concipio omni secluso impedimento*, "I think in my mind of something movable that is left entirely to itself." This "to think in the mind" is that giving oneself a cognition about a determination of things. It is a procedure of going ahead in advance, which Plato once characterized regarding *mathēsis* in the following way: *analabōn autos ex autou tēn epistēmēn* (*Meno* 85d), "bringing up and taking up—above and beyond the other—taking the knowledge itself from out of himself."

There is a prior grasping together in this *mente concipere* of what should be uniformly determinative of each body as such, i.e., for being bodily. All bodies are alike. No motion is special. Every place is like every other, each moment like any other. Every force becomes determinable only by the change of motion which it causes—this change in motion being understood as a change of place. All determinations of bodies have one basic blueprint, according to which the natural process is nothing but the space-time determination of the motion of points of mass. This fundamental design of nature at the same time circumscribes its realm as everywhere uniform.

Now, if we summarize at a glance all that has been said, we can grasp the essence of the mathematical more sharply. Up to now we have stated only its general characteristic, that it is a taking cognizance of something, what it takes being something it gives to itself from itself, thereby giving to itself what it already has. We now summarize the fuller essential determination of the mathematical in a few separate points:

1. The mathematical is, as *mente concipere*, a project of thingness which, as it were, skips over the things. The project first opens a domain where things—i.e., facts—show themselves.

2. In this projection is posited that which things are taken as, what and how they are to be evaluated beforehand. Such evaluation and taking-for is called in Greek *axioō*. The anticipating determinations and assertions in the project are *axiōmata*. Newton therefore entitles the section in which he presents the fundamental determinations about things as moved *Axiōmata, sive leges motus* [The Axioms or Laws of Motion]. The project is axiomatic. Insofar as every science and cognition is expressed in propositions, the cognition that is taken and posited in the mathematical project is of such a kind as to set things upon their foundation in advance. The axioms are *fundamental* propositions.

3. As axiomatic, the mathematical project is the anticipation of the essence of things, of bodies; thus the basic blueprint of the structure of every thing and its relation to every other thing is sketched in advance.

4. This basic plan at the same time provides the measure for laying out the realm which in the future will encompass all things of that sort. Now nature is no longer an inner capacity of a body, determining its form of motion and its place. Nature is now the realm of the uniform space-time context of motion, which is outlined in the axiomatic project and in which alone bodies can be bodies as a part of it and anchored in it.

5. The realm of nature, axiomatically determined in outline by this project, now also requires for the bodies and corpuscles within it a *mode of access* appropriate to the axiomatically predetermined objects. The mode of questioning and the cognitive determination of nature are now no longer ruled by traditional opinions and concepts. Bodies have no concealed qualities, powers, and capacities. Natural bodies are now only what they *show* themselves as, within this projected realm. Things now show themselves only in the relations of places and time points and in the measures of mass and working forces. How they show themselves is prefigured in the project. Therefore, the project also determines the mode of taking in and studying what shows itself, experience, the *experiri*. However, because inquiry is now predetermined by the outline of the project, a line of questioning can be instituted in such a way that it poses conditions in advance to which nature must answer in one way or another. Upon the basis of the mathematical, the *experientia* becomes the modern experiment. Modern science is experimental because of the mathematical project. The experimenting urge to the facts is a necessary consequence of the preceding mathematical skipping of all facts. But where this skipping ceases or becomes weak, mere facts as such are collected, and positivism arises.

6. Because the project establishes a uniformity of all bodies according to relations of space, time, and motion, it also makes pos-

sible and requires a universal uniform measure as an essential determinant of things, i.e., numerical measurement. The mathematical project of Newtonian bodies leads to the development of a certain "mathematics" in the narrow sense. The new form of modern science did not arise because mathematics became an essential determinant. Rather, that mathematics, and a particular kind of mathematics, could come into play and had to come into play is a *consequence* of the mathematical project. The founding of analytical geometry by Descartes, the founding of the infinitesimal calculus by Newton, the simultaneous founding of the differential calculus by Leibniz—all these novelties, this mathematical in a narrower sense, first became possible and above all necessary on the grounds of the basically mathematical character of the thinking.

We would certainly fall into great error if we were to think that with this characterization of the reversal from ancient to modern natural science, and with this sharpened essential outline of the mathematical, we had already gained a picture of the actual science itself.

What we have been able to cite is only the fundamental outline along which there unfolds the entire realm of posing questions and experiments, establishing laws, and disclosing new regions of beings. Within this fundamental mathematical position the questions about the nature of space and time, motion and force, body and matter remain open. These questions now receive a new sharpness; for instance, the question whether motion is sufficiently formulated by the designation "change of location." Regarding the concept of force, the question arises whether it is sufficient to represent force only as a cause that is effective from the outside. Concerning the basic law of motion, the law of inertia, the question arises whether this law is not to be subordinated under a more general one, i.e., the law of the conservation of energy, which is now determined in accordance with its *expenditure* and *consumption*, as *work*—names for new basic representations that now enter into the study of nature and betray a notable accord with economics, with the "calculation" of success. All this develops within and according to the fundamental mathematical position. What remains questionable in all this is a closer determination of the relation of the mathematical in the sense of mathematics to the intuitive experience of the given things and to these things themselves. Up to this hour such questions have been open. Their questionability is concealed by the results and the progress of scientific work. One of these burning questions concerns the justification and limits of mathematical formalism in contrast to the demand for an immediate return to intuitively given nature.

If we have grasped some of what has been said up till now, then it is understandable that the question cannot be decided by way of an either/or, either formalism or immediate intuitive determination of things; for the nature and direction of the mathematical project participate in deciding their possible relation to the intuitively experienced, and vice versa. Behind this question concerning the relation of mathematical formalism to the intuition of nature stands the fundamental question of the justification and limits of the mathematical in general, within a fundamental position we take toward beings as a whole. But in this regard the delineation of the mathematical has gained an importance for us.

## F. The Metaphysical Meaning of the Mathematical

To reach our goal, the understanding of the mathematical we have gained by now is not sufficient. To be sure, we shall now no longer conceive of it as a generalization of the procedure of a particular mathematical discipline, but rather the particular discipline as a special form developing from the mathematical. But this mathematical must, in turn, be grasped from causes that lie even deeper. We have said that it is a fundamental trait of modern thought. Every sort of thought, however, is always only the execution and consequence of a mode of historical Dasein, of the fundamental position taken toward Being and toward the way in which beings are manifest as such, i.e., toward truth. . . .

1. *The principles: new freedom, self-binding, and self-grounding*

We inquire, therefore, about the metaphysical meaning of the mathematical in order to evaluate its importance for modern metaphysics. We divide the question into two subordinate ones: (1) What new fundamental position of Dasein shows itself in this rise of the dominance of the mathematical? (2) How does the mathematical, according to its own inner direction, drive toward an ascent to a metaphysical determination of Dasein?

The second question is the more important for us. We shall answer the first one only in the merest outline.

Up to the distinct emergence of the mathematical as a fundamental characteristic of thought, the authoritative truth was considered that of Church and faith. The means for the proper knowledge of beings were obtained by way of the interpretation of the sources of revelation, the writ and the tradition of the Church. Whatever more experience and knowledge had been won adjusted itself (as if by itself) to this frame. For basically there was no worldly knowledge. The so-called natural knowledge not based upon any revelation therefore did not have its own form of intelligibility or grounds for itself, let alone from out of itself. Thus, what is decisive for the history of science is not that all truth of natural knowledge was measured by the supernatural. Rather, it is that this natural knowledge, disregarding this criterion, arrived at no independent foundation and character out of itself. For adoption of the logic of Aristotelian syllogism cannot be reckoned such.

In the essence of the mathematical, as the project we delineated, lies a specific will to a new formation and self-grounding of the form of knowledge as such. The detachment from revelation as the first source for truth and the rejection of tradition as the authoritative means of knowledge—all these rejections are only negative consequences of the mathematical project. He who dared to project the mathematical project put himself as the projector of this project upon a base which is first projected only in the project. There is not only a liberation in the mathematical project, but also a new experience and formation of freedom itself, i.e., a binding with obligations that are self-imposed. In the mathematical project develops an obligation to principles demanded by the mathematical itself. According to this inner drive, a liberation to a new freedom, the mathematical strives out of itself to establish its own essence as the ground of itself and thus of all knowledge.

Therewith we come to the second question: How does the mathematical, according to its own inner drive, move toward an ascent to a metaphysical determination of Dasein? We can abridge this question as follows: In what way does modern metaphysics arise out of the spirit of the mathematical? It is already obvious from the form of the question that mathematics could not become the standard of philosophy, as if mathematical methods were only appropriately generalized and then transferred to philosophy.

Rather, modern natural science, modern mathematics, and modern metaphysics sprang from the same root of the mathematical in the wider sense. Because metaphysics, of these three, reaches farthest—to beings in totality—and because at the same time it also reaches deepest, toward the Being of beings as such, therefore it is precisely metaphysics that must dig down to the bedrock of its mathematical base and ground. . . .

### 2. *Descartes:* Cogito Sum; *"I" as a special subject*

Modern philosophy is usually considered to have begun with Descartes (1596–1650), who lived a generation after Galileo. Contrary to the attempts which appear from time to time to have modern philosophy begin with Meister Eckhart or in the time between Eckhart and Descartes, we must adhere to the usual beginning. The only question is how one understands Descartes's philosophy. It is no accident that the philosophical formation of the mathematical foundation of modern Dasein is primarily achieved in France, England, and Holland any more than it is accidental that Leibniz received his decisive inspiration from there, especially during his sojourn in Paris from 1672–76. Only because he passed through that world and truly appraised its greatness in greater reflection was he in a position to lay the first foundation for its overcoming.

The following is the usual image of Descartes and his philosophy: During the Middle Ages philosophy stood—if it stood independently at all—under the exclusive domination of theology, and gradually degenerated into a mere analysis of concepts and elucidations of traditional opinions and propositions. It petrified into an academic knowledge which no longer concerned man and was unable to illuminate reality as a whole. Then Descartes appeared and liberated philosophy from this disgraceful position. He began by doubting everything, but this doubt finally did run into something that could no longer be doubted, for, inasmuch as the doubter doubts, he cannot doubt that he is present and must be present in order to doubt at all. As I doubt I must admit that "I am." The "I," accordingly, is the indubitable. As the doubter, Descartes forced men into doubt in this way; he led them to think of themselves, of their "I." Thus the "I," human subjectivity, came to be declared the center of thought. From here originated the I-viewpoint of modern times and its subjectivism. Philosophy itself, however, was thus brought to the insight that doubting must stand at the beginning of philosophy: reflection upon knowledge itself and its possibility. A theory of knowledge had to be erected before a theory of the world. From then on epistemology is the foundation of philosophy, and that distinguishes modern from medieval philosophy. Since then, the attempts to renew Scholasticism also strive to demonstrate the epistemology in their system, or to add it where it is missing, in order to make it usable for modern times. Accordingly, Plato and Aristotle are reinterpreted as epistemologists.

This story of Descartes, who came and doubted and so became a subjectivist, thus grounding epistemology, does give the usual picture; but at best it is only a bad novel, and anything but a story in which the movement of Being becomes visible.

The main work of Descartes carries the title *Meditationes de prima philosophia* (1641). *Prima philosophia*—this is the *protē philosophia* of Aristotle, the question concerning the Being of beings, in the form of the question concerning the thingness of things. *Meditationes de metaphysica*—nothing about theory of knowledge. The sentence or proposition constitutes the guide for the question about the Being of beings (for the categories). (The essential historical-metaphysical basis for the priority of *certainty*, which first made the acceptance and metaphysical development of the mathematical possible—Christianity and the *certainty of salvation*, the security of the individual as such—will not be considered here.)*

In the Middle Ages, the doctrine of Aristotle was taken over in a very special way. In later Scholasticism, through the Spanish philosophical schools, especially through the Jesuit, Suárez, the "medieval" Aristotle went through an extended interpretation. Descartes received his first and fundamental philosophical education from the Jesuits at La Flèche. The title of his main work expresses both his argument with this tradition and his will to take up anew the question about the Being of beings, the thingness of the thing, "substance."

But all this happened in the midst of a period in which, for a century, mathematics had already been emerging more and more as the foundation of thought and was pressing toward clarity. It was a time which, in accordance with this free projection of the world, embarked on a new assault upon reality. There is nothing of skepticism here, nothing of the I-viewpoint and subjectivity—but just the contrary. Therefore, it is the passion of the new thought and inquiry to bring to clarification and display in its innermost essence

*See Martin Heidegger, *Nietzsche*, two vols. (Pfullingen: G. Neske Verlag, 1961), II, 141–48 and ff.; in English translation, *Nietzsche, vol. IV: Nihilism*, trans. Frank A. Capuzzi, ed. D. F. Krell (San Francisco: Harper & Row, 1982), sections 15–16. —ED.

the at first dark, unclear, and often misinterpreted fundamental position, which has progressed only by fits and starts. But this means that the mathematical wills to ground itself in the sense of its own inner requirements. It expressly intends to explicate itself as the standard of *all* thought and to establish the rules which thereby arise. Descartes substantially participates in this work of reflection upon the fundamental meaning of the mathematical. Because this reflection concerned the totality of beings and the knowledge of it, it had to become a reflection on metaphysics. This simultaneous advance in the direction of a foundation of mathematics and of a reflection on metaphysics above all characterizes his fundamental philosophical position. We can pursue this clearly in an unfinished early work which did not appear in print until fifty years after Descartes's death (1701). This work is called *Regulae ad directionem ingenii.*

(1) *Regulae:* basic and guiding propositions in which mathematics submits itself to its own essence; (2) *ad directionem ingenii:* laying the foundation of the mathematical in order that it, as a whole, becomes the measure of the inquiring mind. In the enunciation of something subject to rules as well as with regard to the inner free determination of the mind, the basic mathematical-metaphysical character is already expressed in the title. Here, by way of a reflection upon the essence of mathematics, Descartes grasps the idea of a *scientia universalis,* to which everything must be directed and ordered as the one authoritative science. Descartes expressly emphasizes that it is not a question of *mathematica vulgaris* but of mathesis universalis.

We cannot here present the inner construction and the main content of this unfinished work. In it the modern concept of science is coined. Only one who has really thought through this relentlessly sober volume long enough, down to its remotest and coldest corner, fulfills the prerequisite for getting an inkling of what is going on in modern science. In order to convey a notion of the intention and attitude of this work, we shall quote only three of the twenty-one rules, namely, the third, fourth, and fifth. Out of these the basic character of modern thought leaps before our eyes.

Regula III: "Concerning the objects before us, we should pursue the questions, not what others have thought, nor what we ourselves conjecture, but what we can clearly and insightfully intuit, or deduce with steps of certainty, for in no other way is knowledge arrived at."*

Regula IV: "Method is necessary for discovering the truth of nature."

This rule does not intend the platitude that a science must also have its method, but it wants to say that the procedure, i.e., how in general we are to pursue things (*methodos*), decides in advance what truth we shall seek out in the things.

Method is not one piece of equipment of science among others but the primary component out of which is first determined what can become object and how it becomes object.

Regula V: "Method consists entirely in the order and arrangement of that upon which the sharp vision of the mind must be directed in order to discover some truth. But we will follow such a method only if we lead complex and obscure propositions back step by step to the simpler ones and then try to ascend by the same steps from the insight of the very simplest propositions to the knowledge of all the others."

What remains decisive is how this reflection on the mathematical affects the argument with traditional metaphysics (*prima philosophia*), and how, starting from there, the further destiny and form of modern philosophy is determined.

To the essence of the mathematical as a projection belongs the axiomatical, the beginning of basic principles upon which everything further is based in insightful order. If mathematics, in the

*Descartes, *Rules for the Direction of the Mind*, F. P. Lafleur, trans. (Indianapolis: Library of Liberal Arts, 1961), p. 8.—Tr.

sense of a *mathesis universalis,* is to ground and form the whole of knowledge, then it requires the formulation of special axioms.

(1) They must be absolutely first, intuitively evident in and of themselves, i.e., absolutely certain. This certainty participates in deciding their truth. (2) The highest axioms, as mathematical, must establish in advance, concerning the whole of beings, what is in being and what Being means, from where and how the thingness of things is determined. According to tradition this happens along the guidelines of the proposition. But up till now, the proposition had been taken only as what offered itself, as it were, of itself. The simple proposition about the simply present things contains and retains what the things are. Like the things, the proposition too is simply at hand and is the container of Being.

However, there can be no pregiven things for a basically mathematical position. The proposition cannot be an arbitrary one. The proposition, and precisely it, must itself be based on its foundation. It must be a basic principle—*the* basic principle absolutely. One must therefore find such a principle of all positing, i.e., a proposition in which that about which it says something, the *subjectum* (*hypokeimenon*), is not just taken from somewhere else. That underlying subject must as such first emerge for itself in this original proposition and be established. Only in this way is the *subjectum* a *fundamentum absolutum*, purely posited from the proposition as such, a basis and ground established in the mathematical; only in this way is a *fundamentum absolutum* at the same time *inconcussum*, and thus indubitable and absolutely certain. Because the mathematical now sets itself up as the principle of all knowledge, all knowledge up to now must necessarily be put into question, regardless of whether it is tenable or not.

Descartes does not doubt because he is a skeptic; rather, he must become a doubter because he posits the mathematical as the absolute ground and seeks for all knowledge a foundation that will be in accord with it. It is a question not only of finding a fundamental law for the realm of nature, but finding the very first and highest basic principle for the Being of beings in general. This absolutely mathematical principle cannot have anything in front of it and cannot allow what might be given to it beforehand. If anything is given at all, it is only the *proposition* in general *as such*, i.e., the positing, the position, in the sense of a thinking that asserts. The positing, the proposition, has only itself as that which can be posited. Only where thinking thinks itself is it absolutely mathematical, i.e., a taking cognizance of that which we already have. Insofar as thinking and positing directs itself toward itself, it finds the following: *whatever* may be asserted, and in whatever sense, this asserting and thinking is always an "*I* think." Thinking *is* always an "*I* think," *ego cogito*. Therein lies: I am, *sum*. *Cogito, sum*—this is the highest certainty lying immediately in the proposition as such. In "I posit" the "I" as the positer is co- and pre-posited as that which is already present, as the being. The Being of beings is determined out of the "I am" as the certainty of the positing.

The formula which the proposition sometimes has, "*Cogito ergo sum*," suggests the misunderstanding that it is here a question of inference. That is not the case and cannot be so, because this conclusion would have to have as its major premise: *Id quod cogitat, est*; and the minor premise: *cogito*; conclusion: *ergo sum*. However, the major premise would be only a formal generalization of what lies in the proposition: "*cogito—sum*." Descartes himself emphasizes that no inference is present. The *sum* is not a consequence of the thinking, but vice versa; it is the ground of thinking, the *fundamentum*. In the essence of positing lies the proposition: I posit. That is a proposition which does not depend upon something given beforehand, but only gives to itself what lies within it. In it lies "I posit": I am the one who posits and thinks. This proposition has the peculiarity of first positing that about which it makes an assertion, the *subjectum*. What it posits in this case is the "*I*." The I is the *subjectum* of the very first principle. The I is therefore a special something which underlies [*Zugrundeliegendes*]—*hypokeimenon*, *subjectum*—the *subjectum* of the positing as such. Hence it came about that ever since then the "I" has especially been called the *subjectum*, "subject." The character of the ego, as what is especially already present before one, remains unnoticed. Instead, the subjectivity of the subject is determined by the "I-ness" [*Ichheit*] of the "I think." That the "I" comes to be defined as that which is already present for representation (the "objective" in today's sense) is not because of any I-viewpoint or any subjectivistic doubt, but because of the essential predominance and the definitely directed radicalization of the mathematical and the axiomatic.

This "I," which has been raised to be the special *subjectum* on the basis of the mathematical, is in its meaning nothing "subjective" at all, in the sense of an incidental quality of just this particular human being. This "subject" designated in the "I think," this I, is subjectivistic only when its essence is no longer understood, i.e., is not unfolded from its origin considered in terms of its mode of Being.

Until Descartes, every thing at hand for itself was a "subject"; but now the "I" becomes the special subject, that with regard to which all the remaining things first determine themselves as such. Because—mathematically—they first receive their thingness only through the founding relation to the highest principle and its "subject" (I), they are essentially such as stand as something else in relation to the "subject," which lie over against it as *objectum*. The things themselves become "objects."

The word *objectum* now passes through a corresponding change of meaning. For up to then the word *objectum* denoted what one cast before oneself in mere fantasy: I imagine a golden mountain. This thus-represented—an *objectum* in the language of the Middle Ages—is, according to the usage of language today, merely something "subjective"; for "a golden mountain" does not exist "objectively" in the meaning of the changed linguistic use. This reversal of the meanings of the words *subjectum* and *objectum* is no mere affair of usage; it is a radical change of Dasein, that is to say, of the clearing of the Being of beings on the basis of the predominance of the *mathematical. It is a stretch of the way of actual history necessarily hidden from the usual view*, a history that always concerns the openness of Being—or nothing at all.

3. *Reason as the highest ground: the principle of the I; the principle of contradiction*

The I, as "I think," is the ground upon which hereafter all certainty and truth are based. But thought, assertion, *logos*, is at the same time the guideline for the determinations of Being, the categories. These are found by the guideline of the "I think," in viewing the "I." By virtue of this fundamental significance for the foundation of all knowledge, the "I" thus becomes the accentuated and essential definition of man. Up to Descartes's time, and even later, man was conceived as the *animal rationale*, as a rational living being. With this peculiar emphasis on the I, that is, with the "I think," the determination of the rational and of reason now takes on a distinct priority. For thinking is the fundamental act of reason. With the *cogito—sum*, reason now becomes *explicitly* posited according to its own demand as the first ground of all knowledge and the guideline of the determination of the things.

Already in Aristotle the assertion, the *logos*, was the guideline for the determination of the categories, i.e., of the Being of beings. However, the locus of this guideline—human reason, reason in general—was not characterized as the subjectivity of the subject. But now reason is expressly set forth as the "I think" in the highest principle as guideline and court of appeal for all determinations of Being. The highest principle is the "I" principle: *cogito—sum*. It is the fundamental axiom of all knowledge; but it is not the only fundamental axiom, simply because in this I-principle itself there is included and posited with this one, and thereby with every proposition, yet another. When we say "*cogito—sum*," we express what lies in the *subjectum* (*ego*). If the assertion is to be an assertion, it must always posit what lies in the *subjectum*. What is posited and spoken in the predicate may not and cannot speak against the subject. The *kataphasis* must always be such that it avoids the *antiphasis*, i.e., saying in the sense of speaking against, of contradiction. In the proposition as proposition, and accordingly in the highest principle as I-principle, there is co-posited as equally as valid the principle of the avoidance of contradiction (briefly: the principle of contradiction).

Since the mathematical as the axiomatic project posits itself as the authoritative principle of knowledge, the positing is thereby established as the thinking, as the "*I* think," the I-principle. "I think" signifies that I avoid contradiction and follow the principle of contradiction.

The I-principle and the principle of contradiction spring from the essence of thinking itself, and in such a way that one looks only to the essence of the "I think" and what lies in it and in it alone. The "I think" is reason, is its fundamental act; what is drawn solely from the "I think" is gained solely out of reason itself. Reason so comprehended is purely itself, pure reason.

These principles, which in accord with the fundamental mathematical feature of thinking spring solely from reason, become the principles of knowledge proper, i.e., philosophy in the primary sense, metaphysics. The principles of mere reason are the axioms of pure reason. Pure reason, *logos* so understood, the proposition in this form, becomes the guideline and standard of metaphysics, i.e., the court of appeal for the determination of the Being of beings, the thingness of things. The question about the thing is now anchored in pure reason, i.e., in the mathematical unfolding of its principles.

In the title "pure reason" lies the *logos* of Aristotle, and in the "pure" a certain special formation of the mathematical.

# VII

# THE QUESTION CONCERNING TECHNOLOGY

. . . *Thinking holds to the coming of what has been, and is remembrance.*

It is a question raised on all sides and always with a sense of urgency. On it hinges nothing less than the survival of the species man and the planet earth. Yet the question concerning technology is usually posed within a purely technical framework as one to be debated solely by technicians. Technological problems, we say, require technological solutions which no layman can fashion or fathom. Just as there are "technical philosophical" questions which none but the philosopher can answer, so are there "technical technological" problems that the philosopher had best let alone. Surely technology and philosophy are as far apart as any two fields could possibly be.

Historians and social scientists define "modern technology" as the application of power machinery to production. They locate its beginnings in eighteenth-century England, where large coal deposits provide a source of energy for the production of steam, which in turn propels machinery in textile and other mills. But already at this relatively primitive stage of development the nexus of events becomes so complicated that nobody can neatly separate cause from effect or even establish the customary hierarchy of causes. Everything is jumbled together into inscrutable "factors"—revolutionary discoveries in the natural sciences, detection and extraction of energy resources, invention of mechanical devices and chemical processes, availability of investment capital, improved means of transportation and communication, land enclosures, mechanization of agriculture, concentration of unskilled labor, a happy combination of this-worldly and otherworldly incentives—and the age of modern technology is off and running before anyone can catch their breath and raise a question.

On December 1, 1949, Heidegger delivered four lectures to the Bremen Club under the general title "Insight into What Is." Each lecture had its own title: "The Thing," "The Enframing," "The Danger," "The Turning." Heidegger read the first two to the Bavarian Academy of Fine Arts, "The Thing" on June 6, 1950, and "The Enframing," completely revised as "The Question Concerning Technology," on November 18, 1953.

In the last-named lecture, here printed complete, Heidegger poses the question of the *essence* of technology. He asserts that it is nothing technological and suggests that purely technical modes of thought and discussion do not suit it. For the essence of technology is ultimately a way of revealing the totality of beings. As a way of revealing it is pervasive and fundamental in our time, so much so that we cannot "opt for" technology or "opt out" of it. The advent of technology—and it is this historic, essential unfolding or provenance that Heidegger means by "essence"—is something destined or sent our way long before the eighteenth century. One of Heidegger's most daring theses is that the essence of technology is prior to, and by no means a consequence of, the Scientific Revolution.

However, to insist that technology belongs to the destiny of the West in no way implies that it does not menace. On the contrary, the question concerning the essence of technology confronts the supreme danger, which is that this one way of revealing beings may overwhelm man and beings and all other possible ways of revealing. Such danger is impacted in the essence of technology, which is an ordering of, or setting-upon, both nature and man, a defiant challenging of beings that aims at total and exclusive mastery. The technological framework is inherently expansionist and can reveal only by reduction. Its attempt to enclose all beings in a particular claim—utter availability and sheer manipulability—Heidegger calls *Ge-stell,* "enframing."

As the essence of technology, enframing would be absolute. It would reduce man and beings to a sort of "standing reserve" or stockpile in service to, and on call for, technological purposes. But enframing cannot overpower or even reveal its own historic, essential unfolding, nor indeed the advent, endurance, and departure of beings. Behind all the confident and even arrogant manipulations of the technological will to power something remains mysterious about technology that only a thoughtful recollection can appreciate—though indeed it cannot explain (and so enframe) what is transpiring all over the globe.

This mysterious coming to presence and withdrawal into absence that includes technology and that technology would but cannot entirely master relates the essence of technology to what Heidegger speaks of in his treatise on the essence of truth: the presencing of beings in unconcealment (see Reading III). Finally, Heidegger asks whether the kind of revealing of beings that occurs in the work of *art* (see Reading IV) can rescue human beings for the role they must play—whether as technician or philosopher—in the safeguarding of Being. Indeed, the work of art now comes to be more prominent in Heidegger's thought than ever: whereas in 1935 "the deed that founds the political state" participates in the revelation of beings, in 1953 the political is in total eclipse. Not the political but the poetical appears as the saving power; not *praxis* but *poiēsis* may enable us to confront the essential unfolding of technology.

Yet the suppression of the political and of *praxis* by *poiēsis* and the work of art ought to disturb us. If thinking perdures beyond or beneath the distinction between theory and practice (see Reading V), does it also remain untouched by the apparent split between *poiēsis* and *praxis?*

# THE QUESTION CONCERNING TECHNOLOGY

In what follows we shall be *questioning* concerning technology. Questioning builds a way. We would be advised, therefore, above all to pay heed to the way, and not to fix our attention on isolated sentences and topics. The way is one of thinking. All ways of thinking, more or less perceptibly, lead through language in a manner that is extraordinary. We shall be questioning concerning *technology*, and in so doing we should like to prepare a free relationship to it. The relationship will be free if it opens our human existence to the essence of technology. When we can respond to this essence, we shall be able to experience the technological within its own bounds.

Technology is not equivalent to the essence of technology. When we are seeking the essence of "tree," we have to become aware that what pervades every tree, as tree, is not itself a tree that can be encountered among all the other trees.

Likewise, the essence of technology is by no means anything technological. Thus we shall never experience our relationship to the essence of technology so long as we merely represent and pursue the technological, put up with it, or evade it. Everywhere we remain unfree and chained to technology, whether we passionately affirm or deny it. But we are delivered over to it in the worst possible

This essay appears in Martin Heidegger, *The Question Concerning Technology and Other Essays*, translated by William Lovitt, (New York: Harper & Row, 1977). I have altered the translation slightly here. The German text appears in Martin Heidegger, *Vorträge und Aufsätze* (Pfullingen: Günther Neske Verlag, 1954), pp. 13–44, and in the same publisher's "Opuscula" series under the title, *Die Technik und die Kehre* (1962), pp. 5–36.

way when we regard it as something neutral; for this conception of it, to which today we particularly like to pay homage, makes us utterly blind to the essence of technology.

According to ancient doctrine, the essence of a thing is considered to be *what* the thing is. We ask the question concerning technology when we ask what it is. Everyone knows the two statements that answer our question. One says: Technology is a means to an end. The other says: Technology is a human activity. The two definitions of technology belong together. For to posit ends and procure and utilize the means to them is a human activity. The manufacture and utilization of equipment, tools, and machines, the manufactured and used things themselves, and the needs and ends that they serve, all belong to what technology is. The whole complex of these contrivances is technology. Technology itself is a contrivance—in Latin, an *instrumentum*.

The current conception of technology, according to which it is a means and a human activity, can therefore be called the instrumental and anthropological definition of technology.

Who would ever deny that it is correct? It is in obvious conformity with what we are envisaging when we talk about technology. The instrumental definition of technology is indeed so uncannily correct that it even holds for modern technology, of which, in other respects, we maintain with some justification that it is, in contrast to the older handicraft technology, something completely different and therefore new. Even the power plant with its turbines and generators is a man-made means to an end established by man. Even the jet aircraft and the high-frequency apparatus are means to ends. A radar station is of course less simple than a weather vane. To be sure, the construction of a high-frequency apparatus requires the interlocking of various processes of technical-industrial production. And certainly a sawmill in a secluded valley of the Black Forest is a primitive means compared with the hydroelectric plant on the Rhine River.

But this much remains correct: Modern technology too is a means to an end. This is why the instrumental conception of technology conditions every attempt to bring man into the right relation to technology. Everything depends on our manipulating technology in the proper manner as a means. We will, as we say, "get" technology "intelligently in hand." We will master it. The will to mastery becomes all the more urgent the more technology threatens to slip from human control.

But suppose now that technology were no mere means: how would it stand with the will to master it? Yet we said, did we not, that the instrumental definition of technology is correct? To be sure. The correct always fixes upon something pertinent in whatever is under consideration. However, in order to be correct, this fixing by no means needs to uncover the thing in question in its essence. Only at the point where such an uncovering happens does the true propriate. For that reason the merely correct is not yet the true. Only the true brings us into a free relationship with that which concerns us from its essence. Accordingly, the correct instrumental definition of technology still does not show us technology's essence. In order that we may arrive at this, or at least come close to it, we must seek the true by way of the correct. We must ask: What is the instrumental itself? Within what do such things as means and end belong? A means is that whereby something is effected and thus attained. Whatever has an effect as its consequence is called a cause. But not only that by means of which something else is effected is a cause. The end that determines the kind of means to be used may also be considered a cause. Wherever ends are pursued and means are employed, wherever instrumentality reigns, there reigns causality.

For centuries philosophy has taught that there are four causes: (1) the *causa materialis*, the material, the matter out of which, for example, a silver chalice is made; (2) the *causa formalis*, the form, the shape into which the material enters; (3) the *causa finalis*, the end, for example, the sacrificial rite in relation to which the required chalice is determined as to its form and matter; (4) the *causa efficiens*, which brings about the effect that is the finished, actual chalice, in this instance, the silversmith. What technology is, when represented as a means, discloses itself when we trace instrumentality back to fourfold causality.

But suppose that causality, for its part, is veiled in darkness with respect to what it is? Certainly for centuries we have acted as though the doctrine of the four causes had fallen from heaven as a truth as clear as daylight. But it might be that the time has come to ask: Why are there only four causes? In relation to the aforementioned four, what does "cause" really mean? From whence does it come that the causal character of the four causes is so unifiedly determined that they belong together?

So long as we do not allow ourselves to go into these questions, causality, and with it instrumentality, and with this the accepted definition of technology, remain obscure and groundless.

For a long time we have been accustomed to representing cause as that which brings something about. In this connection, to bring about means to obtain results, effects. The *causa efficiens*, but one among the four causes, sets the standard for all causality. This goes so far that we no longer even count the *causa finalis*, telic finality, as causality. *Causa*, *casus*, belongs to the verb *cadere*, to fall, and means that which brings it about that something turns out as a result in such and such a way. The doctrine of the four causes goes back to Aristotle. But everything that later ages seek in Greek thought under the conception and rubric "causality" in the realm of Greek thought and for Greek thought *per se* has simply nothing at all to do with bringing about and effecting. What we call cause [*Ursache*] and the Romans call *causa* is called *aition* by the Greeks, that to which something else is indebted [*das, was ein anderes verschuldet*]. The four causes are the ways, all belonging at once to each other, of being responsible for something else. An example can clarify this.

Silver is that out of which the silver chalice is made. As this matter (*hyle*), it is co-responsible for the chalice. The chalice is indebted to, i.e., owes thanks to, the silver for that of which it consists. But the sacrificial vessel is indebted not only to the silver. As a chalice, that which is indebted to the silver appears in the aspect of a chalice, and not in that of a brooch or a ring. Thus the sacred vessel is at the same time indebted to the aspect (*eidos*) of chaliceness. Both the silver into which the aspect is admitted as chalice and the aspect in which the silver appears are in their respective ways co-responsible for the sacrificial vessel.

But there remains yet a third something that is above all responsible for the sacrificial vessel. It is that which in advance confines the chalice within the realm of consecration and bestowal. Through this the chalice is circumscribed as sacrificial vessel. Circumscribing gives bounds to the thing. With the bounds the thing does not stop; rather, from within them it begins to be what after production it will be. That which gives bounds, that which completes, in this sense is called in Greek *telos*, which is all too often translated as "aim" and "purpose," and so misinterpreted. The *telos* is responsible for what as matter and what as aspect are together co-responsible for the sacrificial vessel.

Finally, there is a fourth participant in the responsibility for the finished sacrificial vessel's lying before us ready for use, i.e., the silversmith—but not at all because he, in working, brings about the finished sacrificial chalice as if it were the effect of a making; the silversmith is not a *causa efficiens*.

The Aristotelian doctrine neither knows the cause that is named by this term, nor uses a Greek word that would correspond to it.

The silversmith considers carefully and gathers together the three aforementioned ways of being responsible and indebted. To consider carefully [*überlegen*] is in Greek *legein*, *logos*. *Legein* is rooted in *apophainesthai*, to bring forward into appearance. The silversmith is co-responsible as that from which the sacred vessel's being brought forth and subsistence take and retain their first departure.

The three previously mentioned ways of being responsible owe thanks to the pondering of the silversmith for the "that" and the "how" of their coming into appearance and into play for the production of the sacrificial vessel.

Thus four ways of owing hold sway in the sacrificial vessel that lies ready before us. They differ from one another, yet they belong together. What unites them from the beginning? In what does this playing in unison of the four ways of being responsible play? What is the source of the unity of the four causes? What, after all, does this owing and being responsible mean, thought as the Greeks thought it?

Today we are too easily inclined either to understand being responsible and being indebted moralistically as a lapse, or else to construe them in terms of effecting. In either case we bar from ourselves the way to the primal meaning of that which is later called causality. So long as this way is not opened up to us we shall also fail to see what instrumentality, which is based on causality, properly is.

In order to guard against such misinterpretations of being responsible and being indebted, let us clarify the four ways of being responsible in terms of that for which they are responsible. According to our example, they are responsible for the silver chalice's lying ready before us as a sacrificial vessel. Lying before and lying ready (*hypokeisthai*) characterize the presencing of something that is present. The four ways of being responsible bring something into appearance. They let it come forth into presencing [*Anwesen*]. They set it free to that place and so start it on its way, namely, into its complete arrival. The principal characteristic of being responsible is this starting something on its way into arrival. It is in the sense of such a starting something on its way into arrival that being responsible is an occasioning or an inducing to go forward [*Ver-anlassen*]. On the basis of a look at what the Greeks experienced in being responsible, in *aitia*, we now give this verb "to occasion" a more inclusive meaning, so that it now is the name for the essence of causality thought as the Greeks thought it. The common and narrower meaning of "occasion," in contrast, is nothing more than a colliding and releasing; it means a kind of secondary cause within the whole of causality.

But in what, then, does the playing in unison of the four ways of occasioning play? These let what is not yet present arrive into presencing. Accordingly, they are unifiedly governed by a bringing that brings what presences into appearance. Plato tells us what this bringing is in a sentence from the *Symposium* (205b): *hē gar toi ek tou mē ontos eis to on ionti hotōioun aitia pasa esti poiēsis.* "Every occasion for whatever passes beyond the nonpresent and goes forward into presencing is *poiēsis*, bringing-forth [*Her-vor-bringen*]."

It is of utmost importance that we think bringing-forth in its full scope and at the same time in the sense in which the Greeks thought it. Not only handicraft manufacture, not only artistic and poetical bringing into appearance and concrete imagery, is a bringing-forth, *poiēsis*. *Physis*, also, the arising of something from out of itself, is a bringing-forth, *poiēsis*. *Physis* is indeed *poiēsis* in the highest sense. For what presences by means of *physis* has the irruption belonging to bringing-forth, e.g., the bursting of a blossom into bloom, in itself (*en heautōi*). In contrast, what is brought forth by the artisan or the artist, e.g., the silver chalice, has the irruption belonging to bringing-forth, not in itself, but in another (*en allōi*), in the craftsman or artist.

The modes of occasioning, the four causes, are at play, then, within bringing-forth. Through bringing-forth the growing things of nature as well as whatever is completed through the crafts and the arts come at any given time to their appearance.

But how does bringing-forth happen, be it in nature or in handicraft and art? What is the bringing-forth in which the fourfold way of occasioning plays? Occasioning has to do with the presencing [*Anwesen*] of that which at any given time comes to appearance in bringing-forth. Bringing-forth brings out of concealment into unconcealment. Bringing-forth propriates only insofar as something concealed comes into unconcealment. This coming rests and moves freely within what we call revealing [*das Entbergen*]. The Greeks have the word *alētheia* for revealing. The Romans translate this with *veritas*. We say "truth" and usually understand it as correctness of representation.

But where have we strayed to? We are questioning concerning technology, and we have arrived now at *alētheia*, at revealing. What has the essence of technology to do with revealing? The answer: everything. For every bringing-forth is grounded in revealing. Bringing-forth, indeed, gathers within itself the four modes of occasioning—causality—and rules them throughout. Within its domain belong end and means as well as instrumentality. Instrumentality is considered to be the fundamental characteristic of technology. If we inquire step by step into what technology, represented as means, actually is, then we shall arrive at revealing. The possibility of all productive manufacturing lies in revealing.

Technology is therefore no mere means. Technology is a way of revealing. If we give heed to this, then another whole realm for the essence of technology will open itself up to us. It is the realm of revealing, i.e., of truth.

This prospect strikes us as strange. Indeed, it should do so, as persistently as possible and with so much urgency that we will finally take seriously the simple question of what the name "technology" means. The word stems from the Greek. *Technikon* means that which belongs to *technē*. We must observe two things with respect to the meaning of this word. One is that *technē* is the name not only for the activities and skills of the craftsman but also for the arts of the mind and the fine arts. *Technē* belongs to bringing-forth, to *poiēsis*; it is something poetic.

The other thing that we should observe with regard to *technē* is even more important. From earliest times until Plato the word *technē* is linked with the word *epistēmē*. Both words are terms for knowing in the widest sense. They mean to be entirely at home in something, to understand and be expert in it. Such knowing provides an opening up. As an opening up it is a revealing. Aristotle, in a discussion of special importance (*Nicomachean Ethics*, Bk. VI, chaps. 3 and 4), distinguishes between *epistēmē* and *technē* and indeed with respect to what and how they reveal. *Technē* is a mode of *alētheuein*. It reveals whatever does not bring itself forth and does not yet lie here before us, whatever can look and turn out now one way and now another. Whoever builds a house or a ship or forges a sacrificial chalice reveals what is to be brought forth, according to the terms of the four modes of occasioning. This revealing gathers together in advance the aspect and the matter of ship or house, with a view to the finished thing envisaged as completed, and from this gathering determines the manner of its construction. Thus what is decisive in *technē* does not at all lie in making and manipulating, nor in the using of means, but rather in the revealing mentioned before. It is as revealing, and not as manufacturing, that *technē* is a bringing-forth.

Thus the clue to what the word *technē* means and to how the Greeks defined it leads us into the same context that opened itself to us when we pursued the question of what instrumentality as such in truth might be.

Technology is a mode of revealing. Technology comes to presence in the realm where revealing and unconcealment take place, where *alētheia*, truth, happens.

In opposition to this definition of the essential domain of technology, one can object that it indeed holds for Greek thought and that at best it might apply to the techniques of the handicraftsman, but that it simply does not fit modern machine-powered technology. And it is precisely the latter and it alone that is the disturbing thing, that moves us to ask the question concerning technology *per se*. It is said that modern technology is something incomparably different from all earlier technologies because it is based on modern physics as an exact science. Meanwhile, we have come to understand more clearly that the reverse holds true as well: modern physics, as experimental, is dependent upon technical apparatus and upon progress in the building of apparatus. The establishing of this mutual relationship between technology and physics is correct. But it remains a merely historiological establishing of facts and says nothing about that in which this mutual relationship is grounded. The decisive question still remains: Of what essence is modern technology that it thinks of putting exact science to use?

What is modern technology? It too is a revealing. Only when we allow our attention to rest on this fundamental characteristic does that which is new in modern technology show itself to us.

And yet, the revealing that holds sway throughout modern technology does not unfold into a bringing-forth in the sense of *poiēsis*. The revealing that rules in modern technology is a challenging [*Herausfordern*], which puts to nature the unreasonable demand that it supply energy which can be extracted and stored as such. But does this not hold true for the old windmill as well? No. Its sails do indeed turn in the wind; they are left entirely to the wind's blowing. But the windmill does not unlock energy from the air currents in order to store it.

In contrast, a tract of land is challenged in the hauling out of coal and ore. The earth now reveals itself as a coal mining district, the soil as a mineral deposit. The field that the peasant formerly cultivated and set in order appears differently than it did when to set in order still meant to take care of and maintain. The work of the peasant does not challenge the soil of the field. In sowing grain it places seed in the keeping of the forces of growth and watches over its increase. But meanwhile even the cultivation of the field has come under the grip of another kind of setting-in-order, which *sets upon* nature. It sets upon it in the sense of challenging it. Agriculture is now the mechanized food industry. Air is now set upon to yield nitrogen, the earth to yield ore, ore to yield uranium, for example; uranium is set upon to yield atomic energy, which can be unleashed either for destructive or for peaceful purposes.

This setting-upon that challenges the energies of nature is an expediting, and in two ways. It expedites in that it unlocks and exposes. Yet that expediting is always itself directed from the beginning toward furthering something else, i.e., toward driving on to the maximum yield at the minimum expense. The coal that has been hauled out in some mining district has not been produced in order that it may simply be at hand somewhere or other. It is being stored; that is, it is on call, ready to deliver the sun's warmth that is stored in it. The sun's warmth is challenged forth for heat, which in turn is ordered to deliver steam whose pressure turns the wheels that keep a factory running.

The hydroelectric plant is set into the current of the Rhine. It sets the Rhine to supplying its hydraulic pressure, which then sets the turbines turning. This turning sets those machines in motion whose thrust sets going the electric current for which the long-distance power station and its network of cables are set up to dispatch electricity. In the context of the interlocking processes pertaining to the orderly disposition of electrical energy, even the Rhine itself appears to be something at our command. The hydroelectric plant is not built into the Rhine River as was the old wooden bridge that joined bank with bank for hundreds of years. Rather, the river is dammed up into the power plant. What the river is now, namely, a water-power supplier, derives from the essence of the power station. In order that we may even remotely consider the monstrousness that reigns here, let us ponder for a moment the contrast that is spoken by the two titles: "The Rhine," as dammed up into the *power* works, and "The Rhine," as uttered by the *art*-work, in Hölderlin's hymn by that name. But, it will be replied, the Rhine is still a river in the landscape, is it not? Perhaps. But how? In no other way than as an object on call for inspection by a tour group ordered there by the vacation industry.

The revealing that rules throughout modern technology has the character of a setting-upon, in the sense of a challenging-forth.

Such challenging happens in that the energy concealed in nature is unlocked, what is unlocked is transformed, what is transformed is stored up, what is stored up is in turn distributed, and what is distributed is switched about ever anew. Unlocking, transforming, storing, distributing, and switching about are ways of revealing. But the revealing never simply comes to an end. Neither does it run off into the indeterminate. The revealing reveals to itself its own manifoldly interlocking paths, through regulating their course. This regulating itself is, for its part, everywhere secured. Regulating and securing even become the chief characteristics of the revealing that challenges.

What kind of unconcealment is it, then, that is peculiar to that which results from this setting-upon that challenges? Everywhere everything is ordered to stand by, to be immediately on hand, indeed to stand there just so that it may be on call for a further ordering. Whatever is ordered about in this way has its own standing. We call it the standing-reserve [*Bestand*]. The word expresses here something more, and something more essential, than mere "stock." The word "standing-reserve" assumes the rank of an inclusive rubric. It designates nothing less than the way in which everything presences that is wrought upon by the revealing that challenges. Whatever stands by in the sense of standing-reserve no longer stands over against us as object.

Yet an airliner that stands on the runway is surely an object. Certainly. We can represent the machine so. But then it conceals itself as to what and how it is. Revealed, it stands on the taxi strip only as standing-reserve, inasmuch as it is ordered to insure the possibility of transportation. For this it must be in its whole structure and in every one of its constituent parts itself on call for duty, i.e., ready for takeoff. (Here it would be appropriate to discuss Hegel's definition of the machine as an autonomous tool. When applied to the tools of the craftsman, his characterization is correct. Characterized in this way, however, the machine is not thought at all from the essence of technology within which it belongs. Seen in terms of the standing-reserve, the machine is completely nonautonomous, for it has its standing only on the basis of the ordering of the orderable.)

The fact that now, wherever we try to point to modern technology as the revealing that challenges, the words "setting-upon," "ordering," "standing-reserve," obtrude and accumulate in a dry, monotonous, and therefore oppressive way—this fact has its basis in what is now coming to utterance.

Who accomplishes the challenging setting-upon through which what we call the actual is revealed as standing-reserve? Obviously, man. To what extent is man capable of such a revealing? Man can indeed conceive, fashion, and carry through this or that in one way or another. But man does not have control over unconcealment itself, in which at any given time the actual shows itself or withdraws. The fact that it has been showing itself in the light of Ideas ever since the time of Plato, Plato did not bring about. The thinker only responded to what addressed itself to him.

Only to the extent that man for his part is already challenged to exploit the energies of nature can this revealing that orders happen. If man is challenged, ordered, to do this, then does not man himself belong even more originally than nature within the standing-reserve? The current talk about human resources, about the supply of patients for a clinic, gives evidence of this. The forester who measures the felled timber in the woods and who to all appearances walks the forest path in the same way his grandfather did is today ordered by the industry that produces commercial woods, whether he knows it or not. He is made subordinate to the orderability of cellulose, which for its part is challenged forth by the need for paper, which is then delivered to newspapers and illustrated magazines. The latter, in their turn, set public opinion to swallowing what is printed, so that a set configuration of opinion becomes available on demand. Yet precisely because man is challenged more originally than are the energies of nature, i.e., into the process of ordering, he never is transformed into mere standing-reserve. Since man drives technology forward, he takes part in ordering as a way of revealing. But the unconcealment itself, within which ordering unfolds, is never a human handiwork, any more than is the realm man traverses every time he as a subject relates to an object.

Where and how does this revealing happen if it is no mere handiwork of man? We need not look far. We need only apprehend in an unbiased way that which has already claimed man so decisively that he can only be man at any given time as the one so claimed. Wherever man opens his eyes and ears, unlocks his heart, and gives himself over to meditating and striving, shaping and working, entreating and thanking, he finds himself everywhere already brought into the unconcealed. The unconcealment of the unconcealed has already propriated whenever it calls man forth into the modes of revealing allotted to him. When man, in his way, from within unconcealment reveals that which presences, he merely responds to the call of unconcealment, even when he contradicts it. Thus when man, investigating, observing, pursues nature as an area of his own conceiving, he has already been claimed by a way of revealing that challenges him to approach nature as an object of research, until even the object disappears into the objectlessness of standing-reserve.

Modern technology, as a revealing that orders, is thus no mere human doing. Therefore we must take the challenging that sets upon man to order the actual as standing-reserve in accordance with the way it shows itself. That challenging gathers man into ordering. This gathering concentrates man upon ordering the actual as standing-reserve.

That which primordially unfolds the mountains into mountain ranges and pervades them in their folded contiguity is the gathering that we call *Gebirg* [mountain chain].

That original gathering from which unfold the ways in which we have feelings of one kind or another we name *Gemüt* [disposition].

We now name the challenging claim that gathers man with a view to ordering the self-revealing as standing-reserve: *Ge-stell* [enframing].

We dare to use this word in a sense that has been thoroughly unfamiliar up to now.

According to ordinary usage, the word *Gestell* [frame] means some kind of apparatus, e.g., a bookrack. *Gestell* is also the name for a skeleton. And the employment of the word *Gestell* [enframing] that is now required of us seems equally eerie, not to speak of the arbitrariness with which words of a mature language are so misused. Can anything be more strange? Surely not. Yet this strangeness is an old custom of thought. And indeed thinkers follow this custom precisely at the point where it is a matter of thinking that which is highest. We, late born, are no longer in a position to appreciate the significance of Plato's daring to use the word *eidos* for that which in everything and in each particular thing endures as present. For *eidos*, in the common speech, meant the outward aspect [*Ansicht*] that a visible thing offers to the physical eye. Plato exacts of this word, however, something utterly extraordinary: that it name what precisely is not and never will be perceivable with physical eyes. But even this is by no means the full extent of what is extraordinary here. For *idea* names not only the nonsensuous aspect of what is physically visible. Aspect (*idea*) names and also is that which constitutes the essence in the audible, the tasteable, the tactile, in everything that is in any way accessible. Compared with the demands that Plato makes on language and thought in this and in other instances, the use of the word *Gestell* as the name for the essence of modern technology, which we are venturing, is almost harmless. Even so, the usage now required remains something exacting and is open to misinterpretation.

Enframing means the gathering together of the setting-upon that sets upon man, i.e., challenges him forth, to reveal the actual, in the mode of ordering, as standing-reserve. Enframing means the way of revealing that holds sway in the essence of modern technology and that is itself nothing technological. On the other hand, all those things that are so familiar to us and are standard parts of assembly, such as rods, pistons, and chassis, belong to the technological. The assembly itself, however, together with the aforementioned stockparts, fall within the sphere of technological activity. Such activity always merely responds to the challenge of enframing, but it never comprises enframing itself or brings it about.

The word *stellen* [to set] in the name *Ge-stell* [enframing] does not only mean challenging. At the same time it should preserve the suggestion of another *Stellen* from which it stems, namely that producing and presenting [*Her-und Dar-stellen*], which, in the sense of *poiēsis*, lets what presences come forth into unconcealment. This producing that brings forth, e.g., erecting a statue in the temple precinct, and the ordering that challenges now under consideration are indeed fundamentally different, and yet they remain related in their essence. Both are ways of revealing, of *alētheia*. In enframing, the unconcealment propriates in conformity with which the work of modern technology reveals the actual as standing-reserve. This work is therefore neither only a human activity nor a mere means within such activity. The merely instrumental, merely anthropological definition of technology is therefore in principle untenable. And it may not be rounded out by being referred back to some metaphysical or religious explanation that undergirds it.

It remains true nonetheless that man in the technological age is, in a particularly striking way, challenged forth into revealing. Such revealing concerns nature, above all, as the chief storehouse of the standing energy reserve. Accordingly, man's ordering attitude and behavior display themselves first in the rise of modern physics as an exact science. Modern science's way of representing pursues and entraps nature as a calculable coherence of forces. Modern physics is not experimental physics because it applies apparatus to the questioning of nature. The reverse is true. Because physics, indeed already as pure theory, sets nature up to exhibit itself as a coherence of forces calculable in advance, it orders its experiments precisely for the purpose of asking whether and how nature reports itself when set up in this way.

But, after all, mathematical science arose almost two centuries before technology. How, then, could it have already been set upon by modern technology and placed in its service? The facts testify to the contrary. Surely technology got under way only when it could be supported by exact physical science. Reckoned chronologically, this is correct. Thought historically, it does not hit upon the truth.

The modern physical theory of nature prepares the way not simply for technology but for the essence of modern technology. For such gathering-together, which challenges man to reveal by way of ordering, already holds sway in physics. But in it that gathering does not yet come expressly to the fore. Modern physics is the herald of enframing, a herald whose provenance is still unknown. The essence of modern technology has for a long time been concealed, even where power machinery has been invented, where electrical technology is in full swing, and where atomic technology is well under way.

All coming to presence, not only modern technology, keeps itself everywhere concealed to the last. Nevertheless, it remains, with respect to its holding sway, that which precedes all: the earliest. The Greek thinkers already knew of this when they said: That which is earlier with regard to its rise into dominance becomes manifest to us men only later. That which is primally early shows itself only ultimately to men. Therefore, in the realm of thinking, a painstaking effort to think through still more primally what was primally thought is not the absurd wish to revive what is past, but rather the sober readiness to be astounded before the coming of the dawn.

Chronologically speaking, modern physical science begins in the seventeenth century. In contrast, machine-power technology develops only in the second half of the eighteenth century. But modern technology, which for chronological reckoning is the later, is, from the point of view of the essence holding sway within it, historically earlier.

If modern physics must resign itself ever increasingly to the fact that its realm of representation remains inscrutable and incapable of being visualized, this resignation is not dictated by any committee of researchers. It is challenged forth by the rule of enframing, which demands that nature be orderable as standing-reserve. Hence physics, in its retreat from the kind of representation that turns only to objects, which has been the sole standard until recently, will never be able to renounce this one thing: that nature report itself in some way or other that is identifiable through calculation and that it remain orderable as a system of information. This system is then determined by a causality that has changed once again. Causality now displays neither the character of the occasioning that brings forth nor the nature of the *causa efficiens*, let alone that of the *causa formalis*. It seems as though causality is shrinking into a reporting—a reporting challenged forth—of standing-reserves that must be guaranteed either simultaneously or in sequence. To this shrinking would correspond the process of growing resignation that Heisenberg's lecture depicts in so impressive a manner.[1]

Because the essence of modern technology lies in enframing, modern technology must employ exact physical science. Through its so doing the deceptive appearance arises that modern technology is applied physical science. This illusion can maintain itself precisely insofar as neither the essential provenance of modern science nor indeed the essence of modern technology is adequately sought in our questioning.

We are questioning concerning technology in order to bring to light our relationship to its essence. The essence of modern technology shows itself in what we call enframing. But simply to point to this is still in no way to answer the question concerning technology, if to answer means to respond, in the sense of correspond, to the essence of what is being asked about.

Where do we find ourselves if now we think one step further regarding what enframing itself actually is? It is nothing technolog-

1. W. Heisenberg, "Das Naturbild in der heutigen Physik," in *Die Künste im technischen Zeitalter* (Munich, 1954), pp. 43ff. [See also W. Heisenberg, *Physics and Philosophy: The Revolution in Modern Science* (New York: Harper & Row, 1958).—Ed.]

ical, nothing on the order of a machine. It is the way in which the actual reveals itself as standing-reserve. Again we ask: Does such revealing happen somewhere beyond all human doing? No. But neither does it happen exclusively *in* man, or definitively *through* man.

Enframing is the gathering together which belongs to that setting-upon which challenges man and puts him in position to reveal the actual, in the mode of ordering, as standing-reserve. As the one who is challenged forth in this way, man stands within the essential realm of enframing. He can never take up a relationship to it only subsequently. Thus the question as to how we are to arrive at a relationship to the essence of technology, asked in this way, always comes too late. But never too late comes the question as to whether we actually experience ourselves as the ones whose activities everywhere, public and private, are challenged forth by enframing. Above all, never too late comes the question as to whether and how we actually admit ourselves into that wherein enframing itself essentially unfolds.

The essence of modern technology starts man upon the way of that revealing through which the actual everywhere, more or less distinctly, becomes standing-reserve. "To start upon a way" means "to send" in our ordinary language. We shall call the sending that gathers [*versammelnde Schicken*], that first starts man upon a way of revealing, *destining* [*Geschick*]. It is from this destining that the essence of all history [*Geschichte*] is determined. History is neither simply the object of written chronicle nor merely the process of human activity. That activity first becomes history as something destined.[2] And it is only the destining into objectifying representation that makes the historical accessible as an object for historiography, i.e., for a science, and on this basis makes possible the current equating of the historical with that which is chronicled.

2. See "On the Essence of Truth" (1930), first edition 1943, pp. 16ff. [Cf. above, p. 126ff.—Ed.]

Enframing, as a challenging-forth into ordering, sends into a way of revealing. Enframing is an ordaining of destining, as is every way of revealing. Bringing-forth, *poiēsis*, is also a destining in this sense.

Always the unconcealment of that which is goes upon a way of revealing. Always the destining of revealing holds complete sway over men. But that destining is never a fate that compels. For man becomes truly free only insofar as he belongs to the realm of destining and so becomes one who listens, though not one who simply obeys.

The essence of freedom is *originally* not connected with the will or even with the causality of human willing.

Freedom governs the free space in the sense of the cleared, that is to say, the revealed. To the occurrence of revealing, i.e., of truth, freedom stands in the closest and most intimate kinship. All revealing belongs within a harboring and a concealing. But that which frees—the mystery—is concealed and always concealing itself. All revealing comes out of the free, goes into the free, and brings into the free. The freedom of the free consists neither in unfettered arbitrariness nor in the constraint of mere laws. Freedom is that which conceals in a way that opens to light, in whose clearing shimmers the veil that hides the essential occurrence of all truth and lets the veil appear as what veils. Freedom is the realm of the destining that at any given time starts a revealing on its way.

The essence of modern technology lies in enframing. Enframing belongs within the destining of revealing. These sentences express something different from the talk that we hear more frequently, to the effect that technology is the fate of our age, where "fate" means the inevitableness of an unalterable course.

But when we consider the essence of technology we experience enframing as a destining of revealing. In this way we are already sojourning within the free space of destining, a destining that in no way confines us to a stultified compulsion to push on blindly with technology or, what comes to the same, to rebel helplessly against it and curse it as the work of the devil. Quite to the contrary, when we once open ourselves expressly to the *essence* of technology we find ourselves unexpectedly taken into a freeing claim.

The essence of technology lies in enframing. Its holding sway belongs within destining. Since destining at any given time starts man on a way of revealing, man, thus under way, is continually approaching the brink of the possibility of pursuing and promulgating nothing but what is revealed in ordering, and of deriving all his standards on this basis. Through this the other possibility is blocked—that man might rather be admitted sooner and ever more primally to the essence of what is unconcealed and to its unconcealment, in order that he might experience as his essence the requisite belonging to revealing.

Placed between these possibilities, man is endangered by destining. The destining of revealing is as such, in every one of its modes, and therefore necessarily, *danger.*

In whatever way the destining of revealing may hold sway, the unconcealment in which everything that is shows itself at any given time harbors the danger that man may misconstrue the unconcealed and misinterpret it. Thus where everything that presences exhibits itself in the light of a cause-effect coherence, even God, for representational thinking, can lose all that is exalted and holy, the mysteriousness of his distance. In the light of causality, God can sink to the level of a cause, of *causa efficiens*. He then becomes even in theology the God of the philosophers, namely, of those who define the unconcealed and the concealed in terms of the causality of making, without ever considering the essential provenance of this causality.

In a similar way the unconcealment in accordance with which nature presents itself as a calculable complex of the effects of forces can indeed permit correct determinations; but precisely through these successes the danger may remain that in the midst of all that is correct the true will withdraw.

The destining of revealing is in itself not just any danger, but *the* danger.

Yet when destining reigns in the mode of enframing, it is the supreme danger. This danger attests itself to us in two ways. As soon as what is unconcealed no longer concerns man even as object, but exclusively as standing-reserve, and man in the midst of objectlessness is nothing but the orderer of the standing-reserve, then he comes to the very brink of a precipitous fall; that is, he comes to the point where he himself will have to be taken as standing-reserve. Meanwhile, man, precisely as the one so threatened, exalts himself and postures as lord of the earth. In this way the illusion comes to prevail that everything man encounters exists only insofar as it is his construct. This illusion gives rise in turn to one final delusion: it seems as though man everywhere and always encounters only himself. Heisenberg has with complete correctness pointed out that the actual must present itself to contemporary man in this way.[3] *In truth, however, precisely nowhere does man today any longer encounter himself, i.e., his essence.* Man stands so decisively in subservience to on the challenging-forth of enframing that he does not grasp enframing as a claim, that he fails to see himself as the one spoken to, and hence also fails in every way to hear in what respect he ek-sists, in terms of his essence, in a realm where he is addressed, so that he *can never* encounter only himself.

But enframing does not simply endanger man in his relationship to himself and to everything that is. As a destining, it banishes man into the kind of revealing that is an ordering. Where this ordering holds sway, it drives out every other possibility of revealing. Above all, enframing conceals that revealing which, in the sense of *poiēsis*, lets what presences come forth into appearance. As compared with that other revealing, the setting-upon that challenges forth thrusts man into a relation to whatever is that is at once antithetical and rigorously ordered. Where enframing holds sway, regulating and securing of the standing-reserve mark all revealing. They no longer

3. "Das Naturbild," pp. 60ff.

even let their own fundamental characteristic appear, namely, this revealing as such.

Thus the challenging-enframing not only conceals a former way of revealing (bringing-forth) but also conceals revealing itself and with it that wherein unconcealment, i.e., truth, propriates.

Enframing blocks the shining-forth and holding sway of truth. The destining that sends into ordering is consequently the extreme danger. What is dangerous is not technology. Technology is not demonic; but its essence is mysterious. The essence of technology, as a destining of revealing, is the danger. The transformed meaning of the word "enframing" will perhaps become somewhat more familiar to us now if we think enframing in the sense of destining and danger.

The threat to man does not come in the first instance from the potentially lethal machines and apparatus of technology. The actual threat has already afflicted man in his essence. The rule of enframing threatens man with the possibility that it could be denied to him to enter into a more original revealing and hence to experience the call of a more primal truth.

Thus where enframing reigns, there is *danger* in the highest sense.

> But where danger is, grows
> The saving power also.

Let us think carefully about these words of Hölderlin.* What does it mean to "save"? Usually we think that it means only to seize hold of a thing threatened by ruin in order to secure it in its former continuance. But the verb "to save" says more. "To save" is to fetch something home into its essence, in order to bring the essence for the first time into its proper appearing. If the essence of technology, enframing, is the extreme danger, if there is truth in Hölderlin's

*From "Patmos." Cf. *Friedrich Hölderlin Poems and Fragments*, trans. Michael Hamburger (Ann Arbor: The University of Michigan Press, 1966), pp. 462–63.—Ed.

words, then the rule of enframing cannot exhaust itself solely in blocking all lighting-up of every revealing, all appearing of truth. Rather, precisely the essence of technology must harbor in itself the growth of the saving power. But in that case, might not an adequate look into what enframing is, as a destining of revealing, bring the upsurgence of the saving power into appearance?

In what respect does the saving power grow also there where the danger is? Where something grows, there it takes root, from thence it thrives. Both happen concealedly and quietly and in their own time. But according to the words of the poet we have no right whatsoever to expect that there where the danger is we should be able to lay hold of the saving power immediately and without preparation. Therefore we must consider now, in advance, in what respect the saving power does most profoundly take root and thence thrive even where the extreme danger lies—in the holding sway of enframing. In order to consider this it is necessary, as a last step upon our way, to look with yet clearer eyes into the danger. Accordingly, we must once more question concerning technology. For we have said that in technology's essence roots and thrives the saving power.

But how shall we behold the saving power in the essence of technology so long as we do not consider in what sense of "essence" it is that enframing properly is the essence of technology?

Thus far we have understood "essence" in its current meaning. In the academic language of philosophy "essence" means *what* something is; in Latin, *quid*. *Quidditas*, whatness, provides the answer to the question concerning essence. For example, what pertains to all kinds of trees—oaks, beeches, birches, firs—is the same "treeness." Under this inclusive genus—the "universal"—fall all actual and possible trees. Is then the essence of technology, enframing, the common genus for everything technological? If this were the case then the steam turbine, the radio transmitter, and the cyclotron would each be an enframing. But the word "enframing" does not mean here a tool or any kind of apparatus. Still less does it mean the general concept of such resources. The machines and apparatus are no more cases and kinds of enframing than are the man at the switchboard and the engineer in the drafting room. Each of these in its own way indeed belongs as stockpart, available resource, or executor, within enframing; but enframing is never the essence of technology in the sense of a genus. Enframing is a way of revealing that is a destining, namely, the way that challenges forth. The revealing that brings forth (*poiēsis*) is also a way that has the character of destining. But these ways are not kinds that, arrayed beside one another, fall under the concept of revealing. Revealing is that destining which, ever suddenly and inexplicably to all thinking, apportions itself into the revealing that brings forth and the revealing that challenges, and which allots itself to man. The revealing that challenges has its origin as a destining in bringing-forth. But at the same time enframing, in a way characteristic of a destining, blocks *poiēsis*.

Thus enframing, as a destining of revealing, is indeed the essence of technology, but never in the sense of genus and *essentia*. If we pay heed to this, something astounding strikes us: it is technology itself that makes the demand on us to think in another way what is usually understood by "essence." But in what way?

If we speak of the "essence of a house" and the "essence of a state" we do not mean a generic type; rather we mean the ways in which house and state hold sway, administer themselves, develop, and decay—the way they "essentially unfold" [*wesen*]. Johann Peter Hebel in a poem, "Ghost on Kanderer Street," for which Goethe had a special fondness, uses the old word *die Weserei*. It means the city hall, inasmuch as there the life of the community gathers and village existence is constantly in play, i.e., essentially unfolds. It is from the verb *wesen* that the noun is derived. *Wesen* understood as a verb is the same as *währen* [to last or endure], not only in terms of meaning, but also in terms of the phonetic formation of the word. Socrates and Plato already think the essence of something as what it is that unfolds essentially, in the sense of what endures. But they think what endures is what remains permanently (*aei on*). And they find what endures permanently in what persists throughout all that happens, in what remains. That which remains they discover, in turn, in the aspect (*eidos*, *idea*), for example, the Idea "house."

The Idea "house" displays what anything is that is fashioned as a house. Particular, real, and possible houses, in contrast, are changing and transitory derivatives of the Idea and thus belong to what does not endure.

But it can never in any way be established that enduring is based solely on what Plato thinks as *idea* and Aristotle thinks as *to ti ēn einai* (that which any particular thing has always been), or what metaphysics in its most varied interpretations thinks as *essentia*.

All unfolding endures. But is enduring only permanent enduring? Does the essence of technology endure in the sense of the permanent enduring of an Idea that hovers over everything technological, thus making it seem that by technology we mean some mythological abstraction? The way in which technology unfolds lets itself be seen only on the basis of that permanent enduring in which enframing propriates as a destining of revealing. Goethe once uses the mysterious word *fortgewähren* [to grant continuously] in place of *fortwähren* [to endure continuously].[4] He hears *währen* [to endure] and *gewähren* [to grant] here in one unarticulated accord. And if we now ponder more carefully than we did before what it is that properly endures and perhaps alone endures, we may venture to say: *Only what is granted endures. What endures primally out of the earliest beginning is what grants.*

As the essencing of technology, enframing is what endures. Does enframing hold sway at all in the sense of granting? No doubt the question seems a horrendous blunder. For according to everything that has been said, enframing is rather a destining that gathers together into the revealing that challenges forth. Challenging is any-

4. "Die Wahlverwandtschaften," pt. 2, chap. 10, in the novel *Die wunderlichen Nachbarskinder.*

thing but a granting. So it seems, so long as we do not notice that the challenging-forth into the ordering of the actual as standing-reserve remains a destining that starts man upon a way of revealing. As this destining, the essential unfolding of technology gives man entry into something which, of himself, he can neither invent nor in any way make. For there is no such thing as a man who exists singly and solely on his own.

But if this destining, enframing, is the extreme danger, not only for man's essential unfolding, but for all revealing as such, should this destining still be called a granting? Yes, most emphatically, if in this destining the saving power is said to grow. Every destining of revealing propriates from a granting and as such a granting. For it is granting that first conveys to man that share in revealing that the propriative event of revealing needs. So needed and used, man is given to belong to the propriative event of truth. The granting that sends one way or another into revealing is as such the saving power. For the saving power lets man see and enter into the highest dignity of his essence. This dignity lies in keeping watch over the unconcealment—and with it, from the first, the concealment—of all essential unfolding on this earth. It is precisely in enframing, which threatens to sweep man away into ordering as the ostensibly sole way of revealing, and so thrusts man into the danger of the surrender of his free essence—it is precisely in this extreme danger that the innermost indestructible belongingness of man within granting may come to light, provided that we, for our part, begin to pay heed to the essence of technology.

Thus the essential unfolding of technology harbors in itself what we least suspect, the possible rise of the saving power.

Everything, then, depends upon this: that we ponder this rising and that, recollecting, we watch over it. How can this happen? Above all through our catching sight of the essential unfolding in technology, instead of merely gaping at the technological. So long as we represent technology as an instrument, we remain transfixed in the will to master it. We press on past the essence of technology.

When, however, we ask how the instrumental unfolds essentially as a kind of causality, then we experience this essential unfolding as the destining of a revealing.

When we consider, finally, that the essential unfolding of the essence of technology propriates in the granting that needs and uses man so that he may share in revealing, then the following becomes clear:

The essence of technology is in a lofty sense ambiguous. Such ambiguity points to the mystery of all revealing, i.e., of truth.

On the one hand, enframing challenges forth into the frenziedness of ordering that blocks every view into the propriative event of revealing and so radically endangers the relation to the essence of truth.

On the other hand, enframing propriates for its part in the granting that lets man endure—as yet inexperienced, but perhaps more experienced in the future—that he may be the one who is needed and used for the safekeeping of the essence of truth. Thus the rising of the saving power appears.

The irresistibility of ordering and the restraint of the saving power draw past each other like the paths of two stars in the course of the heavens. But precisely this, their passing by, is the hidden side of their nearness.

When we look into the ambiguous essence of technology, we behold the constellation, the stellar course of the mystery.

The question concerning technology is the question concerning the constellation in which revealing and concealing, in which the essential unfolding of truth propriates.

But what help is it to us to look into the constellation of truth? We look into the danger and see the growth of the saving power.

Through this we are not yet saved. But we are thereupon summoned to hope in the growing light of the saving power. How can this happen? Here and now and in little things, that we may foster the saving power in its increase. This includes holding always before our eyes the extreme danger.

The essential unfolding of technology threatens revealing, threatens it with the possibility that all revealing will be consumed in ordering and that everything will present itself only in the unconcealment of standing-reserve. Human activity can never directly counter this danger. Human achievement alone can never banish it. But human reflection can ponder the fact that all saving power must be of a higher essence than what is endangered, though at the same time kindred to it.

But might there not perhaps be a more primally granted revealing that could bring the saving power into its first shining-forth in the midst of the danger that in the technological age rather conceals than shows itself?

There was a time when it was not technology alone that bore the name *technē*. Once the revealing that brings forth truth into the splendor of radiant appearance was also called *technē*.

There was a time when the bringing-forth of the true into the beautiful was called *technē*. The *poiēsis* of the fine arts was also called *technē*.

At the outset of the destining of the West, in Greece, the arts soared to the supreme height of the revealing granted them. They illuminated the presence [*Gegenwart*] of the gods and the dialogue of divine and human destinings. And art was called simply *technē*. It was a single, manifold revealing. It was pious, *promos*, i.e., yielding to the holding sway and the safekeeping of truth.

The arts were not derived from the artistic. Artworks were not enjoyed aesthetically. Art was not a sector of cultural activity.

What was art—perhaps only for that brief but magnificent age? Why did art bear the modest name *technē?* Because it was a revealing that brought forth and made present, and therefore belonged within *poiēsis*. It was finally that revealing which holds complete sway in all the fine arts, in poetry, and in everything poetical that obtained *poiēsis* as its proper name.

The same poet from whom we heard the words

> But where danger is, grows
> The saving power also . . .

says to us:

> . . . poetically man dwells on this earth.

The poetical brings the true into the splendor of what Plato in the *Phaedrus* calls *to ekphanestaton*, that which shines forth most purely. The poetical thoroughly pervades every art, every revealing of essential unfolding into the beautiful.

Could it be that the fine arts are called to poetic revealing? Could it be that revealing lays claim to the arts most primally, so that they for their part may expressly foster the growth of the saving power, may awaken and found anew our vision of, and trust in, that which grants?

Whether art may be granted this highest possibility of its essence in the midst of the extreme danger, no one can tell. Yet we can be astounded. Before what? Before this other possibility: that the frenziedness of technology may entrench itself everywhere to such an extent that someday, throughout everything technological, the essence of technology may unfold essentially in the propriative event of truth.

Because the essence of technology is nothing technological, essential reflection upon technology and decisive confrontation with it must happen in a realm that is, on the one hand, akin to the essence of technology and, on the other, fundamentally different from it.

Such a realm is art. But certainly only if reflection upon art, for its part, does not shut its eyes to the constellation of truth, concerning which we are *questioning*.

Thus questioning, we bear witness to the crisis that in our sheer preoccupation with technology we do not yet experience the essential unfolding of technology, that in our sheer aesthetic-mindedness we no longer guard and preserve the essential unfolding of art. Yet the more questioningly we ponder the essence of technology, the more mysterious the essence of art becomes.

The closer we come to the danger, the more brightly do the ways into the saving power begin to shine and the more questioning we become. For questioning is the piety of thought.

# VIII

# BUILDING DWELLING THINKING

*As soon as we have the thing before our eyes, and in our hearts an ear for the word, thinking prospers.*

Not much more than a year before his death, Rainer Maria Rilke began a poem with the following lines:

> Jetzt wär es Zeit, daß Götter träten
> aus bewohnten Dingen. . . .
>
> (*Insel ed.,* II, 185)
>
> Now it is time that gods emerge
> from things by which we dwell.. . . .

To the thing as technological component and as scientific object Heidegger opposes the thing as the place where the truth of Being, disclosedness, happens. In the work of art such disclosedness is compellingly experienced—perhaps most of all in the work of poetry. In poetry we are less disposed to manipulate things or reduce them to our own technical-scientific, quantitative frames of reference; we are encouraged rather to let things be what they are and show their many-sidedness.

Heidegger presented the lecture "Building Dwelling Thinking" ("*Bauen Wohnen Denken*") to the Darmstadt Symposium on *Man and Space* on August 5, 1951. It belongs to a group of three lectures composed in the early 1950s that unravel new though not wholly unfamiliar strands of the question of Being. These lectures, "Building Dwelling Thinking," "The Thing," and "Poetically Man Dwells," are dominated less by scholarly, technical-philosophical language than by figures of myth and poetry. In them Heidegger seeks further insight into that "saving power" that begins to surge in meditation on the essence of technology, a new way of envisaging man's position with regard to things. In the present piece, here printed complete, the primary issue is the relation of "building" to "dwelling" and the kind of "thinking" that results from attention to that relation.

For modern metaphysics *Denken* is representation of objects and assertion of propositions by a subject. The axiomatic proposition and founding representation is *cogito sum,* I think, I am, *ich denke, ich bin. Bin,* like the English *be,* stems from the Indo-Germanic *bheu,* as does the Latin *fui* (I have been) and the Greek *phuō* (I come to light, grow, engender). But these words also give rise to the German word *bauen,* to build. The Cartesian *ich bin,* floating in the unextended realm of the *res cogitans* and representing all extended things out of itself, is now required to build on the earth—and that means to dwell, since the original meaning of *bauen* is *wohnen,* to settle a piece of land, work it by farming, mining, or viniculture, and build a home on it. (Also the English verb *to be* originally has the sense of place-dwelling.) In short, to think about building and dwelling appears to advance thought on the meaning of "Being." There is an essential connection between the present essay and the earlier remarks on "dwelling" in the "Letter on Humanism" and *Being and Time* (see pp. 260 and 54, above).

To be sure, there are differences in such an advance from what has gone before. Instead of artworks we now hear of "everyday" things in familiar locations, such as bridges and houses. Instead of the strife of world and earth we hear of something even more alien to our customary ways of looking at things. For here Heidegger sees the thing as the concrescence of what he calls the fourfold (*das Geviert*) of earth, sky, mortals, and divinities. No introductory word of ours can explain what Heidegger means by this fourfold. We can only point back to the essays on the work of art, technology, and modern science and metaphysics, and elsewhere to the poetry of Rilke and Hölderlin and the archetypes of mythology, for possible comparisons and contrasts. At the risk of making what is strange in Heidegger's essay even more foreign, we add the following brief remarks on *bauen* and *wohnen,* building and dwelling.

*Wohnen* means to reside or stay, to dwell at peace, to be content; it is related to words that mean to grow accustomed to, or feel at home in, a place. It is also tied to the German word for "delight," *Wonne.* For Heidegger to *dwell* signifies the way "we human beings *are* on the earth." Man's Being rests in his capacity to cultivate and safeguard the earth, to protect it from thoughtless exploitation and to defend it against the calumnies of the metaphysical tradition. *Bauen* in its origins reflects *phuein,* the coming to light of things that grow in time from the earth skyward. Sky suggests divinities that epiphanize and depart and in departing gesture toward mortals who delight in the earth. In the unfathomable depths of this delight, at the source of man's being at home on the earth, occurs what Heidegger elsewhere has called "being held out into the nothing," which preserves the unconcealment and secures the concealment at play in Being. "Being" originally names the unified presencing of the fourfold of earth, sky, divinities, and mortals—in the things. To open thinking to this onefold presencing in things is indeed to persevere in the question of Being.

# BUILDING DWELLING THINKING

In what follows we shall try to think about dwelling and building. This thinking about building does not presume to discover architectural ideas, let alone to give rules for building. This venture in thought does not view building as an art or as a technique of construction; rather, it traces building back into that domain to which everything that *is* belongs. We ask:

1. What is it to dwell?
2. How does building belong to dwelling?

## I

We attain to dwelling, so it seems, only by means of building. The latter, building, has the former, dwelling, as its goal. Still, not every building is a dwelling. Bridges and hangars, stadiums and power stations are buildings but not dwellings; railway stations and highways, dams and market halls are built, but they are not dwelling places. Even so, these buildings are in the domain of our dwelling. That domain extends over these buildings and so is not limited to the dwelling place. The truck driver is at home on the highway, but he does not have his lodgings there; the working woman is at home in the spinning mill, but does not have her dwelling place there; the chief engineer is at home in the power station, but he does not

Martin Heidegger, "Building Dwelling Thinking," appears in Martin Heidegger, *Poetry, Language, Thought*, translated by Albert Hofstadter (New York: Harper & Row, 1971), pp. 145–61. The German text appears in Martin Heidegger, *Vorträge und Aufsätze* (Pfullingen: Günther Neske Verlag, 1954), pp. 145–62.

dwell there. These buildings house man. He inhabits them and yet does not dwell in them, if to dwell means solely to have our lodgings in them. In today's housing shortage even this much is reassuring and to the good; residential buildings do indeed provide lodgings; today's houses may even be well planned, easy to keep, attractively cheap, open to air, light, and sun, but—do the houses in themselves hold any guarantee that *dwelling* occurs in them? Yet those buildings that are not dwelling places remain in turn determined by dwelling insofar as they serve man's dwelling. Thus dwelling would in any case be the end that presides over all building. Dwelling and building are related as end and means. However, as long as this is all we have in mind, we take dwelling and building as two separate activities, an idea that has something correct in it. Yet at the same time by the means-end schema we block our view of the essential relations. For building is not merely a means and a way toward dwelling—to build is in itself already to dwell. Who tells us this? Who gives us a standard at all by which we can take the measure of the essence of dwelling and building?

It is language that tells us about the essence of a thing, provided that we respect language's own essence. In the meantime, to be sure, there rages round the earth an unbridled yet clever talking, writing, and broadcasting of spoken words. Man acts as though *he* were the shaper and master of language, while in fact *language* remains the master of man. Perhaps it is before all else man's subversion of *this* relation of dominance that drives his essential being into alienation. That we retain a concern for care in speaking is all to the good, but it is of no help to us as long as language still serves us even then only as a means of expression. Among all the appeals that we human beings, on our part, can help to be voiced, language is the highest and everywhere the first.

Now, what does *bauen*, to build, mean? The Old High German word for building, *buan*, means to dwell. This signifies to remain, to stay in a place. The proper meaning of the verb *bauen*, namely, to dwell, has been lost to us. But a covert trace of it has been preserved in the German word *Nachbar,* neighbor. The *Nachbar* is the *Nachgebur,* the *Nachgebauer,* the near-dweller, he who dwells nearby. The verbs *buri, büren, beuren, beuron,* all signify dwelling, the place of dwelling. Now, to be sure, the old word *buan* not only tells us that *bauen,* to build, is really to dwell; it also gives us a clue as to how we have to think about the dwelling it signifies. When we speak of dwelling we usually think of an activity that man performs alongside many other activities. We work here and dwell there. We do not merely dwell—that would be virtual inactivity—we practice a profession, we do business, we travel and find shelter on the way, now here, now there. *Bauen* originally means to dwell. Where the word *bauen* still speaks in its original sense it also says *how far* the essence of dwelling reaches. That is, *bauen, buan, bhu, beo* are our word *bin* in the versions: *ich bin,* I am, *du bist,* you are, the imperative form *bis,* be. What then does *ich bin* mean? The old word *bauen,* to which the *bin* belongs, answers: *ich bin, du bist* mean I dwell, you dwell. The way in which you are and I am, the manner in which we humans *are* on the earth, is *buan,* dwelling. To be a human being means to be on the earth as a mortal. It means to dwell. The old word *bauen,* which says that man *is* insofar as he *dwells,* this word *bauen,* however, *also* means at the same time to cherish and protect, to preserve and care for, specifically to till the soil, to cultivate the vine. Such building only takes care—it tends the growth that ripens into fruit of its own accord. Building in the sense of preserving and nurturing is not making anything. Shipbuilding and temple-building, on the other hand, do in a certain way make their own works. Here building, in contrast with cultivating, is a constructing. Both modes of building—building as cultivating, Latin *colere, cultura,* and building as the raising up of edifices, *aedificare*—are comprised within genuine building, that is, dwelling. Building as dwelling, that is, as being on the earth, however, remains for man's everyday experience that which is from the outset "habitual"—we inhabit it, as our language says so beautifully: it is the *Gewohnte.* For this reason it recedes behind the manifold ways in which dwelling is accomplished, the activities of cultivation and construction. These activities later claim the name of *bauen*, building, and with it the matter of building, exclusively for themselves. The proper sense of *bauen*, namely dwelling, falls into oblivion.

At first sight this event looks as though it were no more than a change of meaning of mere terms. In truth, however, something decisive is concealed in it; namely, dwelling is not experienced as man's Being; dwelling is never thought of as the basic character of human being.

That language in a way retracts the proper meaning of the word *bauen*, which is dwelling, is evidence of the original one of these meanings; for with the essential words of language, what they genuinely say easily falls into oblivion in favor of foreground meanings. Man has hardly yet pondered the mystery of this process. Language withdraws from man its simple and high speech. But its primal call does not thereby become incapable of speech; it merely falls silent. Man, though, fails to heed this silence.

But if we listen to what language says in the word *bauen* we hear three things:

1. Building is really dwelling.
2. Dwelling is the manner in which mortals are on the earth.
3. Building as dwelling unfolds into the building that cultivates growing things and the building that erects buildings.

If we give thought to this threefold fact, we obtain a clue and note the following: as long as we do not bear in mind that all building is in itself a dwelling, we cannot even adequately *ask*, let alone properly decide, what the building of buildings might be in its essence. We do not dwell because we have built, but we build and have built because we dwell, that is, because we are *dwellers*. But in what does the essence of dwelling consist? Let us listen once more to what language says to us. The Old Saxon *wuon*, the Gothic *wunian*, like the old word *bauen*, mean to remain, to stay in a place.

But the Gothic *wunian* says more distinctly how this remaining is experienced. *Wunian* means to be at peace, to be brought to peace, to remain in peace. The word for peace, *Friede*, means the free, das *Frye*; and *fry* means preserved from harm and danger, preserved *from* something, safeguarded. To free actually means to spare. The sparing itself consists not only in the fact that we do not harm the one whom we spare. Real sparing is something *positive* and takes place when we leave something beforehand in its own essence, when we return it specifically to its essential being, when we "free" it in the proper sense of the word into a preserve of peace. To dwell, to be set at peace, means to remain at peace within the free, the preserve, the free sphere that safeguards each thing in its essence. *The fundamental character of dwelling is this sparing.* It pervades dwelling in its whole range. That range reveals itself to us as soon as we recall that human being consists in dwelling and, indeed, dwelling in the sense of the stay of mortals on the earth.

But "on the earth" already means "under the sky." Both of these *also* mean "remaining before the divinities" and include a "belonging to men's being with one another." By a *primal* oneness the four—earth and sky, divinities and mortals—belong together in one.

Earth is the serving bearer, blossoming and fruiting, spreading out in rock and water, rising up into plant and animal. When we say earth, we are already thinking of the other three along with it, but we give no thought to the simple oneness of the four.

The sky is the vaulting path of the sun, the course of the changing moon, the wandering glitter of the stars, the year's seasons and their changes, the light and dusk of day, the gloom and glow of night, the clemency and inclemency of the weather, the drifting clouds and blue depth of the ether. When we say sky, we are already thinking of the other three along with it, but we give no thought to the simple oneness of the four.

The divinities are the beckoning messengers of the godhead. Out of the holy sway of the godhead, the god appears in his presence or withdraws into his concealment. When we speak of the divinities, we are already thinking of the other three along with them, but we give no thought to the simple oneness of the four.

The mortals are the human beings. They are called mortals because they can die. To die means to be capable of death *as* death. Only man dies, and indeed continually, as long as he remains on earth, under the sky, before the divinities. When we speak of mortals, we are already thinking of the other three along with them, but we give no thought to the simple oneness of the four.

This simple oneness of the four we call *the fourfold*. Mortals *are* in the fourfold by *dwelling*. But the basic character of dwelling is safeguarding. Mortals dwell in the way they safeguard the fourfold in its essential unfolding. Accordingly, the safeguarding that dwells is fourfold.

Mortals dwell in that they save the earth—taking the word in the old sense still known to Lessing. Saving does not only snatch something from a danger. To save properly means to set something free into its own essence. To save the earth is more than to exploit it or even wear it out. Saving the earth does not master the earth and does not subjugate it, which is merely one step from boundless spoliation.

Mortals dwell in that they receive the sky as sky. They leave to the sun and the moon their journey, to the stars their courses, to the seasons their blessing and their inclemency; they do not turn night into day nor day into a harassed unrest.

Mortals dwell in that they await the divinities as divinities. In hope they hold up to the divinities what is unhoped for. They wait for intimations of their coming and do not mistake the signs of their absence. They do not make their gods for themselves and do not worship idols. In the very depth of misfortune they wait for the weal that has been withdrawn.

Mortals dwell in that they initiate their own essential being—their being capable of death as death—into the use and practice of this capacity, so that there may be a good death. To initiate mortals into the essence of death in no way means to make death, as the empty nothing, the goal. Nor does it mean to darken dwelling by blindly staring toward the end.

In saving the earth, in receiving the sky, in awaiting the divinities, in initiating mortals, dwelling propriates as the fourfold preservation of the fourfold. To spare and preserve means to take under our care, to look after the fourfold in its essence. What we take under our care must be kept safe. But if dwelling preserves the fourfold, where does it keep the fourfold's essence? How do mortals make their dwelling such a preserving? Mortals would never be capable of it if dwelling were merely a staying on earth under the sky, before the divinities, among mortals. Rather, dwelling itself is always a staying with things. Dwelling, as preserving, keeps the fourfold in that with which mortals stay: in things.

Staying with things, however, is not merely something attached to this fourfold preservation as a fifth something. On the contrary: staying with things is the only way in which the fourfold stay within the fourfold is accomplished at any time in simple unity. Dwelling preserves the fourfold by bringing the essence of the fourfold into things. But things themselves secure the fourfold *only when* they themselves *as* things are let be in their essence. How does this happen? In this way, that mortals nurse and nurture the things that grow, and specially construct things that do not grow. Cultivating and construction are building in the narrower sense. *Dwelling*, inasmuch as it keeps the fourfold in things, is, as this keeping, a *building*. With this, we are on our way to the second question.

## II

In what way does building belong to dwelling?

The answer to this question will clarify for us what building, understood by way of the essence of dwelling, really is. We limit ourselves to building in the sense of constructing things and inquire:

what is a built thing? A bridge may serve as an example for our reflections.

The bridge swings over the stream "with ease and power." It does not just connect banks that are already there. The banks emerge as banks only as the bridge crosses the stream. The bridge expressly causes them to lie across from each other. One side is set off against the other by the bridge. Nor do the banks stretch along the stream as indifferent border strips of the dry land. With the banks, the bridge brings to the stream the one and the other expanse of the landscape lying behind them. It brings stream and bank and land into each other's neighborhood. The bridge *gathers* the earth as landscape around the stream. Thus it guides and attends the stream through the meadows. Resting upright in the stream's bed, the bridge-piers bear the swing of the arches that leave the stream's waters to run their course. The waters may wander on quiet and gay, the sky's floods from storm or thaw may shoot past the piers in torrential waves—the bridge is ready for the sky's weather and its fickle nature. Even where the bridge covers the stream, it holds its flow up to the sky by taking it for a moment under the vaulted gateway and then setting it free once more.

The bridge lets the stream run its course and at the same time grants mortals their way, so that they may come and go from shore to shore. Bridges initiate in many ways. The city bridge leads from the precincts of the castle to the cathedral square; the river bridge near the country town brings wagons and horse teams to the surrounding villages. The old stone bridge's humble brook crossing gives to the harvest wagon its passage from the fields into the village and carries the lumber cart from the field path to the road. The highway bridge is tied into the network of long-distance traffic, paced and calculated for maximum yield. Always and ever differently the bridge initiates the lingering and hastening ways of men to and fro, so that they may get to other banks and in the end, as mortals, to the other side. Now in a high arch, now in a low, the bridge vaults over glen and stream—whether mortals keep in mind this vaulting of the bridge's course or forget that they, always themselves on their way to the last bridge, are actually striving to surmount all that is common and unsound in them in order to bring themselves before the haleness of the divinities. The bridge *gathers*, as a passage that crosses, before the divinities—whether we explicitly think of, and visibly *give thanks for,* their presence, as in the figure of the saint of the bridge, or whether that divine presence is obstructed or even pushed wholly aside.

The bridge *gathers* to itself in *its own* way earth and sky, divinities and mortals.

Gathering [*Versammlung*], by an ancient word of our language, is called *thing*. The bridge is a thing—and, indeed, it is such *as* the gathering of the fourfold which we have described. To be sure, people think of the bridge as primarily and properly *merely* a bridge; after that, and occasionally, it might possibly express much else besides; and as such an expression it would then become a symbol, for instance a symbol of those things we mentioned before. But the bridge, if it is a true bridge, is never first of all a mere bridge and then afterward a symbol. And just as little is the bridge in the first place exclusively a symbol, in the sense that it expresses something that strictly speaking does not belong to it. If we take the bridge strictly as such, it never appears as an expression. The bridge is a thing and *only that*. Only? As this thing it gathers the fourfold.

Our thinking has of course long been accustomed to *understate* the essence of the thing. The consequence, in the course of Western thought, has been that the thing is represented as an unknown X to which perceptible properties are attached. From this point of view, everything *that already belongs to the gathering essence of this thing* does, of course, appear as something that is afterward read into it. Yet the bridge would never be a mere bridge if it were not a thing.

To be sure, the bridge is a thing of its *own* kind; for it gathers the fourfold in *such* a way that it allows a *site* for it. But only something *that is itself a locale* can make space for a site. The locale is not already there before the bridge is. Before the bridge stands, there are of course many spots along the stream that can be occupied by something. One of them proves to be a locale, and does so *because of the bridge*. Thus the bridge does not first come to a locale to stand in it; rather, a locale comes into existence only by virtue of the bridge. The bridge is a thing; it gathers the fourfold, but in such a way that it allows a site for the fourfold. By this site are determined the places and paths by which a space is provided for.

Only things that are locales in this manner allow for spaces. What the word for space, *Raum*, designates is said by its ancient meaning. *Raum*, *Rum*, means a place that is freed for settlement and lodging. A space is something that has been made room for, something that has been freed, namely, within a boundary, Greek *peras*. A boundary is not that at which something stops but, as the Greeks recognized, the boundary is that from which something *begins its essential unfolding*. That is why the concept is that of *horismos*, that is, the horizon, the boundary. Space is in essence that for which room has been made, that which is let into its bounds. That for which room is made is always granted and hence is joined, that is, gathered, by virtue of a locale, that is, by such a thing as the bridge. *Accordingly, spaces receive their essential being from locales and not from "space."*

Things which, as locales, allow a site we now in anticipation call buildings. They are so called because they are made by a process of building-construction. Of what sort this making—building—must be, however, we find out only after we have first given thought to the essence of those things that of themselves require building as the process by which they are made. These things are locales that allow a site for the fourfold, a site that in each case provides for a space. The relation between locale and space lies in the essence of these things as locales, but so does the relation of the locale to the man who lives there. Therefore we shall now try to clarify the essence of these things that we call buildings by the following brief consideration.

For one thing, what is the relation between locale and space? For another, what is the relation between man and space?

The bridge is a locale. As such a thing, it allows a space into which earth and sky, divinities and mortals are admitted. The space allowed by the bridge contains many places variously near or far from the bridge. These places, however, may be treated as mere positions between which there lies a measurable distance; a distance, in Greek *stadion*, always has room made for it, and indeed by bare positions. The space that is thus made by positions is space of a peculiar sort. As distance or "stadion" it is what the same word, *stadion*, means in Latin, a *spatium*, an intervening space or interval. Thus nearness and remoteness between men and things can become mere distance, mere intervals of intervening space. In a space that is represented purely as *spatium*, the bridge now appears as a mere something at some position, which can be occupied at any time by something else or replaced by a mere marker. What is more, the mere dimensions of height, breadth, and depth can be abstracted from space as intervals. What is so abstracted we represent as the pure manifold of the three dimensions. Yet the room made by this manifold is also no longer determined by distances; it is no longer a *spatium*, but now no more than *extensio*—extension. But from space as *extensio* a further abstraction can be made, to analytic-algebraic relations. What these relations make room for is the possibility of the purely mathematical construction of manifolds with an arbitrary number of dimensions. The space provided for in this mathematical manner may be called "space," the "one" space as such. But in this sense "the" space, "space," contains no spaces and no places. We never find in it any locales, that is, things of the kind the bridge is. As against that, however, in the spaces provided for by locales there is always space as interval, and in this interval in turn there is space as pure extension. *Spatium* and *extensio* afford at any time the possibility of measuring things and what they make room for, according to distances, spans, and directions, and of computing these magnitudes. But the fact that they are *univer-*

*sally* applicable to everything that has extension can in no case make numerical magnitudes the *ground* of the essence of spaces and locales that are measurable with the aid of mathematics. How even modern physics was compelled by the facts themselves to represent the spatial medium of cosmic space as a field-unity determined by body as dynamic center cannot be discussed here.*

The spaces through which we go daily are provided for by locales; their essence is grounded in things of the type of buildings. If we pay heed to these relations between locales and spaces, between spaces and space, we get a clue to help us in thinking of the relation of man and space.

When we speak of man and space, it sounds as though man stood on one side, space on the other. Yet space is not something that faces man. It is neither an external object nor an inner experience. It is not that there are men, and over and above them *space*; for when I say "a man," and in saying this word think of a being who exists in a human manner—that is, who dwells—then by the name "man," I already name the stay within the fourfold among things. Even when we relate ourselves to those things that are not in our immediate reach, we are staying with the things themselves. We do not represent distant things merely in our mind—as the textbooks have it—so that only mental representations of distant things run through our minds and heads as substitutes for the things. If all of us now think, from where we are right here, of the old bridge in Heidelberg, this thinking toward that locale is not a mere experience inside the persons present here; rather, it belongs to the essence of our thinking *of* that bridge that *in itself* thinking *persists through* [durchsteht] the distance to that locale. From this spot right here, we are there at the bridge—we are by no means at some representational content in our consciousness. From right here we

*For a discussion of "thing" and "space" in modern physics, see Reading VI. For a criticism of Cartesian "space" and the analysis of the "spatiality" of Dasein, see *Being and Time*, sections 19–24.—ED.

may even be much nearer to that bridge and to what it makes room for than someone who uses it daily as an indifferent river crossing. Spaces, and with them space as such—"space"—are always provided for already within the stay of mortals. Spaces open up by the fact that they are let into the dwelling of man. To say that mortals *are* is to say that *in dwelling* they persist through spaces by virtue of their stay among things and locales. And only because mortals pervade, persist through, spaces by their very essence are they able to go through spaces. But in going through spaces we do not give up our standing in them. Rather, we always go through spaces in such a way that we already sustain them by staying constantly with near and remote locales and things. When I go toward the door of the lecture hall, I am already there, and I could not go to it at all if I were not such that I am there. I am never here only, as this encapsulated body; rather, I am there, that is, I already pervade the space of the room, and only thus can I go through it.

Even when mortals turn "inward," taking stock of themselves, they do not leave behind their belonging to the fourfold. When, as we say, we come to our senses and reflect on ourselves, we come back to ourselves from things *without ever abandoning* our stay among things. Indeed, the loss of rapport with things that occurs in states of depression would be wholly impossible if even such a state were not still what it is as a human state: that is, a staying *with* things. Only if this stay already characterizes human being can the things among which we are also *fail* to speak to us, *fail* to concern us any longer.

Man's relation to locales, and through locales to spaces, inheres in his dwelling. The relationship between man and space is none other than dwelling, thought essentially.

When we think, in the manner just attempted, about the relation between locale and space, but also about the relation of man and space, a light falls on the essence of the things that are locales and that we call buildings.

The bridge is a thing of this sort. The locale allows the simple onefold of earth and sky, of divinities and mortals, to enter into a site by arranging the site into spaces. The locale makes room for the fourfold in a double sense. The locale *admits* the fourfold and it *installs* the fourfold. The two—making room in the sense of admitting and in the sense of installing—belong together. As a double space-making, the locale is a shelter for the fourfold or, by the same token, a house. Things such as locales shelter or house men's lives. Things of this sort are housings, though not necessarily dwelling-houses in the narrower sense.

The making of such things is building. Its essence consists in this, that it corresponds to the character of these things. They are locales that allow spaces. This is why building, by virtue of constructing locales, is a founding and joining of spaces. Because building produces locales, the joining of the spaces of these locales necessarily brings with it space, as *spatium* and as *extensio*, into the thingly structure of buildings. But building never shapes pure "space." Neither directly nor indirectly. Nevertheless, because it produces things as locales, building is closer to the essence of spaces and to the essential origins of "space" than any geometry and mathematics. Building puts up locales that make space and a site for the fourfold. From the simple oneness in which earth and sky, divinities and mortals belong together, building *receives the directive* for its erecting of locales. Building *takes over* from the fourfold the standard for all the traversing and measuring of the spaces that in each case are provided for by the locales that have been founded. The edifices guard the fourfold. They are things that in their own way preserve the fourfold. To preserve the fourfold, to save the earth, to receive the sky, to await the divinities, to initiate mortals—this fourfold preserving is the simple essence of dwelling. In this way, then, do genuine buildings give form to dwelling in its essence, and house this essential unfolding.

Building thus characterized is a distinctive letting-dwell. Whenever it *is* such in fact, building already *has* responded to the summons of the fourfold. All planning remains grounded on this responding, and planning in turn opens up to the designer the precincts suitable for his designs.

As soon as we try to think of the essence of constructive building in terms of a letting-dwell, we come to know more clearly what that process of making consists in by which building is accomplished. Usually we take production to be an activity whose performance has a result, the finished structure, as its consequence. It is possible to conceive of making in that way; we thereby grasp something that is correct, and yet never touch its essence, which is a producing that brings something forth. For building brings the fourfold *hither* into a thing, the bridge, and brings *forth* the thing as a locale, out into what is already present, room for which is only now made *by* this locale.

The Greek for "to bring forth or to produce" is *tiktō*. The word *technē*, technique, belongs to the verb's root, *tec*. To the Greeks *technē* means neither art nor handicraft but, rather, to make something appear, within what is present, as this or that, in this way or that way. The Greeks conceive of *technē*, producing, in terms of letting appear. *Technē* thus conceived has been concealed in the tectonics of architecture since ancient times. Of late it still remains concealed, and more resolutely, in the technology of power machinery. But the essence of the erecting of buildings cannot be understood adequately in terms either of architecture or of engineering construction, nor in terms of a mere combination of the two. The erecting of buildings would not be suitably defined *even if* we were to think of it in the sense of the original Greek *technē* as *solely* a letting-appear, which brings something made, as something present, among the things that are already present.

The essence of building is letting dwell. Building accomplishes its essential process in the raising of locales by the joining of their spaces. *Only if we are capable of dwelling, only then can we build.* Let us think for a while of a farmhouse in the Black Forest, which was built some two hundred years ago by the dwelling of peasants.

Here the self-sufficiency of the power to let earth and sky, divinities and mortals enter *in simple oneness* into things ordered the house. It placed the farm on the wind-sheltered mountain slope, looking south, among the meadows close to the spring. It gave it the wide overhanging shingle roof whose proper slope bears up under the burden of snow, and that, reaching deep down, shields the chambers against the storms of the long winter nights. It did not forget the altar corner behind the community table; it made room in its chamber for the hallowed places of childbed and the "tree of the dead"—for that is what they call a coffin there: the *Totenbaum*—and in this way it designed for the different generations under one roof the character of their journey through time. A craft that, itself sprung from dwelling, still uses its tools and its gear as things, built the farmhouse.

Only if we are capable of dwelling, only then can we build. Our reference to the Black Forest farm in no way means that we should or could go back to building such houses; rather, it illustrates by a dwelling that *has been* how it was able to build.

Dwelling, however, is *the basic character* of Being, in keeping with which mortals exist. Perhaps this attempt to think about dwelling and building will bring out somewhat more clearly that building belongs to dwelling and how it receives its essence from dwelling. Enough will have been gained if dwelling and building have become *worthy of questioning* and thus have remained *worthy of thought*.

But that thinking itself belongs to dwelling in the same sense as building, although in a different way, may perhaps be attested to by the course of thought here attempted.

Building and thinking are, each in its own way, inescapable for dwelling. The two, however, are also insufficient for dwelling so long as each busies itself with its own affairs in separation, instead of listening to the other. They are able to listen if both—building and thinking—belong to dwelling, if they remain within their limits and realize that the one as much as the other comes from the workshop of long experience and incessant practice.

We are attempting to trace in thought the essence of dwelling. The next step on this path would be the question: What is the state of dwelling in our precarious age? On all sides we hear talk about the housing shortage, and with good reason. Nor is there just talk; there is action too. We try to fill the need by providing houses, by promoting the building of houses, planning the whole architectural enterprise. However hard and bitter, however hampering and threatening the lack of houses remains, the *proper plight of dwelling* does not lie merely in a lack of houses. The proper plight of dwelling is indeed older than the world wars with their destruction, older also than the increase of the earth's population and the condition of the industrial workers. The proper dwelling plight lies in this, that mortals ever search anew for the essence of dwelling, that they *must ever learn to dwell*. What if man's homelessness consisted in this, that man still does not even think of the *proper* plight of dwelling as *the* plight? Yet as soon as man *gives thought* to his homelessness, it is a misery no longer. Rightly considered and kept well in mind, it is the sole summons that *calls* mortals into their dwelling.

But how else can mortals answer this summons than by trying on *their* part, on their own, to bring dwelling to the fullness of its essence? This they accomplish when they build out of dwelling, and think for the sake of dwelling.

# IX

# WHAT CALLS FOR THINKING?
## (from *What Is Called Thinking?*)

*We never come to thoughts.*
*They come to us.*

*Being and Time* begins by conceding that the question it wants to think about has been forgotten. Forgottenness, or oblivion, is the kind of concealment that fails to safeguard a thing from the harsh light of the obvious, that neglects the unconcealment of things and so remains blind to the essence of truth. Such oblivion constitutes the danger that threatens man and world in the age of technology. Most thought-provoking for Heidegger is the thoughtlessness—the radical failure of remembrance—characteristic of these times in which we hardly know what to think.

What is called thinking? What calls for thinking? Both questions try to translate the title of Heidegger's 1951–52 lecture course *Was heisst Denken?* The course is divided into two parts; the selections included here appear at the beginning of each part.

What do we call thinking? What does "thinking" mean? These forms of the question are not difficult to answer. We give the name "thinking" to calculating, reckoning, figuring, planning, and problem solving, and also in less earnest moods to whimsical reverie and day-dreaming. Thinking is having ideas or pictures before the mind. "Think" is what you do before and while you "do." Of course if you think too much you never do anything: Hamlet brooding is also what we call thinking. But in its other form the question is not so easily answered. What calls for thinking? What calls on us to think? These questions give us pause. Here we must assert less, listen more. Here we do less problem solving and pay more attention to the way the problem poses itself. Our logical and technological training does not prepare us well for such . . . thinking.

During the first half of the course Heidegger helps his hearers to "unlearn" habitual responses to the question *Was heisst Denken?* by raising the issues of learning and teaching. He contrasts the simplicity of craftsmanship in thinking to the complexity of technical ratiocination. He introduces several lines from the poetry of Hölderlin and asks about their relationship to thinking. The implication is that calculative kinds of thinking, however vital to the conduct of the sciences, do not fulfill all the requirements of man's thinking nature. Poets demand of us another kind of thinking—less exact but no less strict. It is an elusive sort of thinking, whose object steadily withdraws, and which can be of consequence only if it pays heed to its own movement and direction. This leads to the major theme of the first half of the course, memory, which Heidegger calls "the gathering of thought."

During the second half Heidegger concentrates on the "call" that compels us to think about what is most thought-provoking. Our response to this call and the unassertive language of our response increasingly occupy Heidegger's own thinking. (See Reading X, which tries to find a way to a language that can let beings *be,* that is, let them *show themselves.*) We are unable to be self-assertive in our response to the call because it makes demands on us to which we are not equal. It puts us in question. (Cf. Reading II.) What is it that enjoins us to think and so puts us in question? Asked in such a way this question points back to that of the meaning of Being and forward to the task of thought on presence. What is presence? "That remains to be thought about," Heidegger observes, for it alone gives us to think and calls for remembrance.

A recent thinker who has set his hand to pursuing traces of "presence" in the history of metaphysics, Jacques Derrida, confirms the unsettling nature of the questions "What is called thinking?" and "What calls for thinking?" At the very end of Part One of *De la grammatologie* (Minuit, 1967, p. 142) he affirms: "Thinking is that which we already know we have not yet begun to do. . . ."

# WHAT CALLS FOR THINKING?

We come to know what it means to think when we ourselves are thinking. If our attempt is to be successful, we must be ready to learn thinking.

As soon as we allow ourselves to become involved in such learning we have admitted that we are not yet capable of thinking.

Yet man is called the being who can think, and rightly so. Man is the rational animal. Reason, *ratio*, evolves in thinking. Being the rational animal, man must be capable of thinking if he really wants to. Still, it may be that man wants to think, but can't. Ultimately he wants too much when he wants to think, and so can do too little. Man can think in the sense that he possesses the possibility to do so. This possibility alone, however, is no guarantee to us that we are capable of thinking. For we are capable of doing only what we are inclined to do. And again, we truly incline toward something only when it in turn inclines toward us, toward our essential being, by appealing to our essential being as what holds us there. To hold genuinely means to heed protectively, for example, by letting a herd graze at pasture. What keeps us in our essential being holds us only so long, however, as we for our part keep holding on to what holds us. And we keep holding on to it by not letting it out of our memory. Memory is the gathering of thought. To what? To what holds us, in that we give it thought precisely because it remains what must be thought about. What is thought is the gift given in thinking back,

---

Martin Heidegger, *What Is Called Thinking?* translated by Fred D. Wieck and J. Glenn Gray (New York: Harper & Row, 1968), pp. 3–18; 113–121. The German text is Martin Heidegger, *Was heisst Denken?* (Tübingen: Max Niemeyer Verlag, 1954), pp. 1–8, 48–52, and 79–86.

given because we incline toward it. Only when we are so inclined toward what in itself is to be thought about, only then are we capable of thinking.

In order to be capable of thinking, we need to learn it. What is learning? Man learns when he disposes everything he does so that it answers to whatever addresses him as essential. We learn to think by giving heed to what there is to think about.

For example, what is essential in a friend is what we call "friendliness." In the same sense we now call what in itself is to be thought about "the thought-provoking." Everything thought-provoking *gives* us to think. But it always gives that gift just so far as the thought-provoking matter already *is* intrinsically what must be thought about. From now on, we will call "most thought-provoking" what remains to be thought about always, because it is so at the beginning and before all else. What is most thought-provoking? How does it show itself in our thought-provoking time?

*Most thought-provoking is that we are still not thinking*—not even yet, although the state of the world is becoming constantly more thought-provoking. True, this course of events seems to demand rather that man should act without delay, instead of making speeches at conferences and international conventions and never getting beyond proposing ideas on what ought to be, and how it ought to be done. What is lacking, then, is action, not thought.

And yet—it could be that prevailing man has for centuries now acted too much and thought too little. But how dare anyone assert today that we are still not thinking, today when there is everywhere a lively and constantly more audible interest in philosophy, when almost everybody claims to know what philosophy is all about! Philosophers are *the* thinkers *par excellence*. They are called thinkers precisely because thinking properly takes place in philosophy.

Nobody will deny that there is an interest in philosophy today. But—is there anything at all left today in which man does not take an interest, in the sense in which he understands "interest"?

Interest, *interesse*, means to be among and in the midst of things, or to be at the center of a thing and to stay with it. But today's interest accepts as valid only what is interesting. And interesting is the sort of thing that can freely be regarded as indifferent the next moment, and be displaced by something else, which then concerns us just as little as what went before. Many people today take the view that they are doing great honor to something by finding it interesting. The truth is that such a judgment has already relegated the interesting thing to the ranks of what is indifferent and soon boring.

It is no evidence of any readiness to think that people show an interest in philosophy. There is, of course, serious preoccupation everywhere with philosophy and its questions. The learned world is expending commendable efforts in the investigation of the history of philosophy. These are useful and worthy tasks, and only the best talents are good enough for them, especially when they present to us models of great thinking. But even if we have devoted many years to the intensive study of the treatises and writings of the great thinkers, that fact is still no guarantee that we ourselves are thinking, or even are ready to learn thinking. On the contrary—preoccupation with philosophy more than anything else may give us the stubborn illusion that we are thinking just because we are incessantly "philosophizing."

Even so, it remains strange, and seems presumptuous, to assert that what is most thought-provoking in our thought-provoking time is that we are still not thinking. Accordingly, we must prove the assertion. Even more advisable is first to explain it. For it could be that the demand for a proof collapses as soon as enough light is shed on what the assertion says. It runs:

*Most thought-provoking in our thought-provoking time is that we are still not thinking.*

It has been suggested earlier how the term "thought-provoking" is to be understood. Thought-provoking is what gives us to think.

Let us look at it closely, and from the start allow each word its proper weight. Some things are food for thought in themselves, intrinsically, so to speak, innately. And some things make an appeal to us to give them thought, to turn toward them in thought: to think them.

What is thought-provoking, what gives us to think, is then not anything that we determine, not anything that only we are instituting, only we are proposing. According to our assertion, what of itself gives us most to think about, what is most thought-provoking, is this—that we are still not thinking.

This now means: We have still not come face to face with, have not yet come under the sway of, what intrinsically desires to be thought about in an essential sense. Presumably the reason is that we human beings do not yet sufficiently reach out and turn toward what desires to be thought. If so, the fact that we are still not thinking would merely be a slowness, a delay in thinking or at most a neglect on man's part. Such human tardiness could then be remedied in human ways by the appropriate measures. Human neglect would give us food for thought—but only in passing. The fact that we are still not thinking would be thought-provoking, of course, but being a momentary and curable condition of modern man, it could never be called the one most thought-provoking matter. Yet that is what we call it, and we suggest thereby the following: that we are still not thinking is by no means only because man does not yet turn sufficiently toward that which, by origin and innately, wants to be thought about since in its essence it remains what must be thought about. Rather, that we are still not thinking stems from the fact that what is to be thought about turns away from man, has turned away long ago.

We will want to know at once when that event took place. Even before that, we will ask still more urgently how we could possibly know of any such event. And finally, the problems that here lie in wait come rushing at us when we add still further: that which properly gives us food for thought did not turn away from man at some time or other that can be fixed in history—no, what properly must be thought keeps itself turned away from man since the beginning.

On the other hand, in our era man has always thought in some way; in fact, man has thought the profoundest thoughts, and entrusted them to memory. By thinking in that way he did and does remain related to what must be thought. And yet man is not capable of thinking properly as long as that which must be thought about withdraws.

If we, as we are here and now, will not be taken in by empty talk, we must retort that everything said so far is an unbroken chain of hollow assertions, and state besides that what has been presented here has nothing to do with scientific knowledge.

It will be well to maintain as long as possible such a defensive attitude toward what has been said: only in that attitude do we keep the distance needed for a running start by which one or the other of us may succeed in making the leap into thinking. For it is true that what was said so far, and the entire discussion that is to follow, have nothing to do with scientific knowledge, especially not if the discussion itself is to be a thinking. This situation is grounded in the fact that science itself does not think, and cannot think—which is its good fortune, here meaning the assurance of its own appointed course. Science does not think. This is a shocking statement. Let the statement be shocking, even though we immediately add the supplementary statement that nonetheless science always and in its own fashion has to do with thinking. That fashion, however, is genuine and consequently fruitful only after the gulf has become visible that lies between thinking and the sciences, lies there unbridgeably. There is no bridge here—only the leap. Hence there is nothing but mischief in all the makeshift ties and asses' bridges by which men today would set up a comfortable commerce between thinking and the sciences. Hence we, those of us who come from the sciences, must endure what is shocking and strange about thinking—assuming we are ready to learn thinking. To learn means to make everything we do answer to whatever addresses itself to us as essential.

In order to be capable of doing so, we must get under way. It is important above all that on the way on which we set out when we learn to think we do not deceive ourselves and rashly bypass the pressing questions; on the contrary, we must let ourselves be admitted into questions that seek what no inventiveness can find. Especially we moderns can learn only if we always unlearn at the same time. Applied to the matter before us: we can learn thinking only if we radically unlearn what thinking has been traditionally. To do that, we must at the same time come to know it.

We said: man still does not think, and this because what must be thought about turns away from him; by no means only because man does not sufficiently reach out and turn to what is to be thought.

What must be thought about turns away from man. It withdraws from him. But how can we have the least knowledge of something that withdraws from the beginning, how can we even give it a name? Whatever withdraws, refuses arrival. But—withdrawing is not nothing. Withdrawal is an event. In fact, what withdraws may even concern and claim man more essentially than anything present that strikes and touches him. Being struck by actuality is what we like to regard as constitutive of the actuality of the actual. However, in being struck by what is actual, man may be debarred precisely from what concerns and touches him—touches him in the surely mysterious way of escaping him by its withdrawal. The event of withdrawal could be what is most present throughout the present, and so infinitely exceed the actuality of everything actual.

What withdraws from us draws us along by its very withdrawal, whether or not we become aware of it immediately, or at all. Once we are drawn into the withdrawal, we are—albeit in a way quite different from that of migratory birds—caught in the draft of what draws, attracts us by its withdrawal. And once we, being so attracted, are drawing toward what draws us, our essential being already bears the stamp of that "draft." As we are drawing toward what withdraws, we ourselves point toward it. We are who we are by pointing in that direction—not like an incidental adjunct but as follows: this "being in the draft of" is in itself an essential and therefore constant pointing toward what withdraws. To say "being in the draft of" is to say "pointing toward what withdraws."

To the extent that man *is* in this draft, he *points* toward what withdraws. *As* he is pointing that way, man *is* the pointer. Man here is not first of all man, and then also occasionally someone who points. No. Drawn into what withdraws, drawn toward it and thus pointing into the withdrawal, man first *is* man. His essential being lies in being such a pointer. Something which in itself, by its essential being, is pointing, we call a sign. As he draws toward what withdraws, man is a sign. But since this sign points toward what draws *away,* it points not so much at *what* draws away as into the withdrawal. The sign remains without interpretation.

In a draft to one of his hymns Hölderlin writes:

> We are a sign that is not read.

He continues with these two lines:

> We feel no pain, we almost have
> Lost our tongue in foreign lands.

The several drafts of that hymn—besides bearing such titles as "The Serpent," "The Sign," "The Nymph"—also include the title "Mnemosyne." This Greek word may be translated: Memory. . . . Hölderlin uses the Greek word *Mnemosyne* as the name of a Titaness. According to the myth, she is the daughter of Sky and Earth. Myth means the telling word. For the Greeks, to tell is to lay bare and let appear—both the appearance and what has its essence in the appearance, its epiphany. *Mythos* is what has its essence in its telling—what appears in the unconcealment of its appeal. The *mythos* is that appeal of foremost and radical concern to all human beings which lets man think of what appears, what unfolds. *Logos* says the same; *mythos* and *logos* are not, as our current historians of philosophy claim, placed into opposition by philosophy as such; on the contrary, the early Greek thinkers (Parmenides, fragment

VIII) are precisely the ones to use *mythos* and *logos* in the same sense. *Mythos* and *logos* become separated and opposed only at the point where neither *mythos* nor *logos* can keep to its pristine essence. In Plato's work this separation has already taken place. Historians and philologists, by virtue of a prejudice modern rationalism adopted from Platonism, imagine that *mythos* was destroyed by *logos*. But nothing religious is ever destroyed by logic; it is destroyed only by the god's withdrawal.

Mnemosyne, daughter of Sky and Earth and bride of Zeus, in nine nights becomes the mother of the nine Muses. Drama and music, dance and poetry are of the womb of Mnemosyne, Memory. It is plain that the word means something else than merely the psychologically demonstrable ability to retain a mental representation of something that is past. Memory thinks back to something thought. But when it is the name of the Mother of the Muses, "Memory" does not mean just any thought of anything that can be thought. Memory is the gathering of thought upon what everywhere demands to be thought about first of all. Memory is the gathering of recollection, thinking back. It safely keeps and keeps concealed within it that to which at any given time thought must first be given in everything that essentially unfolds, appealing to us as what has being and has been in being. Memory, Mother of the Muses—the thinking back to what is to be thought—is the source and ground of poesy. This is why poesy is the water that at times flows backward toward the source, toward thinking as a thinking back, a recollection. Surely, as long as we take the view that logic gives us insight into what thinking is, we shall never be able to think how much all poesy rests upon thinking back, recollection. Poetry wells up only from devoted thought thinking back, recollecting.

Under the heading *Mnemosyne*, Hölderlin says:

> We are a sign that is not read . . .

We? Who? We the men of today, of a today that has lasted since long ago and will still last for a long time, so long that no calendar in history can give its measure. The same hymn, "Mnemosyne," says: "Long is / The time"—the time in which we are an uninterpreted sign. And this, that we are a sign, indeed an uninterpreted one, does this not give enough food for thought? What the poet says in these words, and those that follow, may have a part in showing us what is most thought-provoking: precisely what the assertion about our thought-provoking time attempts to think of. And that assertion, provided only we discuss it adequately, may throw some light upon the poet's word; Hölderlin's word, in turn, because it is a word of poesy, may summon us with a larger appeal, and hence greater allure, upon a way of thought that tracks in thought what is most thought-provoking. Even so, it is as yet obscure what purpose this reference to the words of Hölderlin is supposed to serve. It is still questionable with what right we, by way of an attempt to think, make mention of a poet, this poet in particular. And it is also still unclear upon what ground, and within what limits, our reference to the poetic must remain. . . .

By way of this series of lectures we are attempting to learn thinking. The way is long. We dare take only a few steps. If all goes well, they will take us to the foothills of thought. But they will take us to places that we must explore to reach the point where only the leap will help further. The leap alone takes us into the neighborhood where thinking resides. We therefore shall take a few practice leaps right at the start, though we will not notice it at once, nor need to.

In contrast to a steady progress, where we move unawares from one thing to the next and everything remains alike, the leap takes us abruptly to a place where everything is different, so different that it strikes us as strange. Abrupt means the sudden sheer descent or rise that marks the chasm's edge. Though we may not founder in such a leap, what the leap takes us to will confound us.

It is quite in order, then, that we receive notice from the very start of what will confound us. But all would not be well if the strangeness were due only to the fact that you, the listeners, are not yet listening closely enough. If that were the case, you would be bound to overlook completely the strangeness that lies in the matter itself. The matter of thinking is always confounding—all the more in proportion as we keep clear of prejudice. To keep clear of prejudice, we must be ready and willing to listen. Such readiness allows us to surmount the boundaries in which all customary views are confined, and to reach a more open terrain. In order to encourage such readiness, I shall insert here some transitional remarks, which will also apply to all subsequent lectures.

In universities especially the danger is still very great that we misunderstand what we hear of thinking, particularly if the immediate subject of the discussion is scientific. Is there any place compelling us more forcibly to rack our brains than the research and training institutions pursuing scientific work? Now, everyone admits unreservedly that the arts and the sciences are totally different from each other, though in official oratory they are still mentioned jointly. But if a distinction is made between thinking and the sciences, and the two are contrasted, that is immediately considered a disparagement of science. There is the fear even that thinking might open hostilities against the sciences, and becloud the seriousness and spoil the joy of scientific work.

But even if those fears were justified, which is emphatically not the case, it would still be both tactless and tasteless to take a stand against science upon the very rostrum that serves scientific education. Tact alone ought to prevent all polemics here. But there is another consideration as well. Any kind of polemics fails from the outset to assume the attitude of thinking. The role of thinking is not that of an opponent. Thinking is thinking only when it pursues whatever speaks *for* a matter. Everything said here defensively is always intended exclusively to protect the matter. When we speak of the sciences as we pursue our way, we shall be speaking not against but for them, for clarity concerning their essential being. This alone implies our conviction that the sciences are in themselves positively essential. However, their essence is frankly of a different sort than what our universities today still fondly imagine it to be. In any case, we still seem afraid of facing the exciting fact that today's sciences belong in the realm of the essence of modern technology, and nowhere else. Note that I am saying "in the realm of the *essence* of technology," and not simply "in technology." A fog still surrounds the essence of modern science. That fog, however, is not produced by individual investigators and scholars in the sciences. It is not produced by man at all. It arises from the region of what is most thought-provoking—that we are still not thinking; none of us, including me who speaks to you, me first of all.

This is why we are here attempting to learn thinking. We are all on the way together, and are not reproving each other. To learn means to make everything we do answer to whatever addresses us as essential. Depending on the kind of essentials, depending on the realm from which they address us, the answer and with it the kind of learning differs.

A cabinetmaker's apprentice, someone who is learning to build cabinets and the like, will serve as an example. His learning is not mere practice, to gain facility in the use of tools. Nor does he merely gather knowledge about the customary forms of the things he is to build. If he is to become a true cabinetmaker, he makes himself answer and respond above all to the different kinds of wood and to the shapes slumbering within wood—to wood as it enters into man's dwelling with all the hidden riches of its essence. In fact, this relatedness to wood is what maintains the whole craft. Without that relatedness, the craft will never be anything but empty busywork, any occupation with it will be determined exclusively by business concerns. Every handicraft, all human dealings, are constantly in that danger. The writing of poetry is no more exempt from it than is thinking.

Whether or not a cabinetmaker's apprentice, while he is learning, will come to respond to wood and wooden things depends obviously on the presence of some teacher who can teach the apprentice such matters.

True. Teaching is even more difficult than learning. We know that; but we rarely think about it. And why is teaching more difficult than learning? Not because the teacher must have a larger store of information, and have it always ready. Teaching is more difficult than learning because what teaching calls for is this: to let learn. Indeed, the proper teacher lets nothing else be learned than—learning. His conduct, therefore, often produces the impression that we really learn nothing from him, if by "learning" we now automatically understand merely the procurement of useful information. The teacher is ahead of his apprentices in this alone, that he has still far more to learn than they—he has to learn to let them learn. The teacher must be capable of being more teachable than the apprentices. The teacher is far less sure of his material than those who learn are of theirs. If the relation between the teacher and the learners is genuine, therefore, there is never a place in it for the authority of the know-it-all or the authoritative sway of the official. It still is an exalted matter, then, to become a teacher—which is something else entirely than becoming a famous professor. That nobody wants any longer to become a teacher today, when all things are downgraded and graded from below (for instance, from business), is presumably because the matter is exalted, because of its altitude. And presumably this disinclination is linked to that most thought-provoking matter which gives us to think. We must keep our eyes fixed firmly on the true relation between teacher and taught—if indeed learning is to arise in the course of these lectures.

We are trying to learn thinking. Perhaps thinking, too, is just something like building a cabinet. At any rate, it is a craft, a "handicraft." The hand is something altogether peculiar. In the common view, the hand is part of our bodily organism. But the hand's essence can never be determined, or explained, by its being an organ that can grasp. Apes, too, have organs that can grasp, but they do not have hands. The hand is infinitely different from all the grasping organs—paws, claws, or fangs—different by an abyss of essence. Only a being who can speak, that is, think, can have hands and can handily achieve works of handicraft.

But the craft of the hand is richer than we commonly imagine. The hand does not only grasp and catch, or push and pull. The hand reaches and extends, receives and welcomes—and not just things: the hand extends itself, and receives its own welcome in the hands of others. The hand holds. The hand carries. The hand designs and signs, presumably because man is a sign. Two hands fold into one, a gesture meant to carry man into the great oneness. The hand is all this, and this is the true handicraft. Everything is rooted here that is commonly known as handicraft, and commonly we go no further. But the hand's gestures run everywhere through language, in their most perfect purity precisely when man speaks by being silent. And only when man speaks, does he think—not the other way around, as metaphysics still believes. Every motion of the hand in every one of its works carries itself through the element of thinking, every bearing of the hand bears itself in that element. All the work of the hand is rooted in thinking. Therefore, thinking itself is man's simplest, and for that reason hardest, handiwork, if from time to time it would be accomplished properly.

We must learn thinking, because our being able to think, and even gifted for it, is still no guarantee that we are capable of thinking. To be capable we must before all else incline toward what addresses itself to thought—and that is what of itself gives food for thought. What gives us this gift, the gift of what must properly be thought about, is what we call most thought-provoking.

Our answer to the question as to what the most thought-provoking thing might be is the assertion: most thought-provoking for our thought-provoking time is that we are still not thinking.

The reason is never exclusively or primarily that we human beings do not sufficiently reach out and turn toward what properly gives food for thought; the reason is that this most thought-provoking thing turns away from us, in fact has long since turned away from man.

And what withdraws in such a manner keeps and develops its own incomparable nearness.

Once we are so related and drawn to what withdraws, we are drawing into what withdraws, into the enigmatic and therefore mutable nearness of its appeal. Whenever man is properly drawing that way, he is thinking—even though he may still be far away from what withdraws, even though the withdrawal may remain as veiled as ever. All through his life and right into his death, Socrates did nothing else than place himself into this draft, this current, and maintain himself in it. This is why he is the purest thinker of the West. This is why he wrote nothing. For anyone who begins to write out of thoughtfulness must inevitably be like those people who run to seek refuge from any draft too strong for them. An as yet hidden history still keeps the secret why all great Western thinkers after Socrates, with all their greatness, had to be such fugitives. Thinking entered into literature. And literature has decided the fate of Western science, which, by way of the *doctrina* of the Middle Ages, became the *scientia* of modern times. In this form all the sciences have sprung from the womb of philosophy, in a twofold manner. The sciences come out of philosophy, because they have to part with her. And now that they are so apart they can never again, by their own power as sciences, make the leap back into the source from whence they have sprung. Henceforth they are remanded to a realm of essence where only thinking can find them, provided thinking is capable of doing what is its own to do.

When man is drawing into what withdraws, he points into what withdraws. As we are drawing that way we are a sign, a pointer. But we are pointing then at something that has not, not yet, been transposed into the language that we speak. It remains uncomprehended. We are an uninterpreted sign.

In his draft for the hymn "Mnemosyne," Hölderlin says:

> We are a sign that is not read,
> We feel no pain, we almost have
> Lost our tongue in foreign lands.

And so, on our way toward thinking, we hear a word of poesy. But the question to what end and with what right, upon what ground and within what limits, our attempt to think allows itself to get involved in a dialogue with poesy, let alone with the poetry of this poet—this question, which is inescapable, we can discuss only after we ourselves have taken the path of thinking. . . .

*What is called thinking?* The question sounds definite. It seems unequivocal. But even a slight reflection shows it to have more than one meaning. No sooner do we ask the question than we begin to vacillate. Indeed, the ambiguity of the question foils every attempt to push toward the answer without some further preparation.

We must, then, clarify the ambiguity. The ambiguousness of the question "What is called thinking?" conceals several possible ways of dealing with it. Looking ahead, we may stress four ways in which the question can be posed.

"What is called thinking?" says for one thing, and in the first place: what is it we call "thought" and "thinking," what do these words signify? What is it to which we give the name "thinking"?

"What is called thinking?" says also, in the second place: how does traditional doctrine conceive and define what we have named thinking? What is it that for two and a half thousand years has been regarded as the basic characteristic of thinking? Why does the traditional doctrine of thinking bear the curious title "logic"?

"What is called thinking?" says further, in the third place: what are the prerequisites we need so that we may be able to think with essential rightness? What is called for on our part in order that we may each time achieve good thinking?

"What is called thinking?" says finally, in the fourth place: what is it that calls us, as it were, commands us to think? What is it that calls us into thinking?

These are four ways in which we can ask the question and bring it closer to an answer by corresponding analyses. These four ways of asking the question are not just superficially strung together. They are all interrelated. What is disturbing about the question therefore lies less in the multiplicity of its possible meanings than in the single meaning toward which all four ways are pointing. We must consider whether only one of the four ways is the right one, while the others prove to be incidental and untenable; or whether all four of them are equally necessary because they are unified and of a piece. But how are they unified, and by what unity? Is oneness added to the multiplicity of the four ways as a fifth piece, like a roof to four walls? Or does one of the four ways of asking the question take precedence? Does this precedence establish a hierarchy within the group of questions? Does the hierarchy exhibit a structure by which the four ways are coordinated and yet subordinated to the one that is decisive?

The four ways we have mentioned, in which the question "What is called thinking?" may be asked, do not stand side by side, separate and unrelated. They belong together by virtue of a union that is enjoined by one of the four ways. However, we must go slow, one step at a time, if we are to become aware how this is so. We must therefore begin our attempt with a statement that will at first remain a mere assertion.

The meaning of the question that we noted in the fourth place tells us how the question would want to be asked first in the decisive way: "What calls for thinking?" Properly understood, the question asks what it is that commands us to enter into thought, that calls on us to think. The turn of phrase "What calls for thinking on our part?" could of course intend no more than "What does the term 'thinking' signify to us?" But the question, asked properly, "What calls for thinking on our part?," means something else. . . . It means: What is it that directs us into thought and gives us directives for thinking?

Accordingly, does the question ask what it is that gives us the impetus to think on each occasion and with regard to a particular matter? No. The directives that come from what directs us into thought are much more than merely the given impetus to do some thinking.

That which directs us to think gives us directives in such a way that we first become capable of thinking, and thus *are* as thinkers, only by virtue of its directive. It is true, of course, that the question

"What calls for thinking?," in the sense of "What calls on us to think?," is foreign to the common understanding. But we are all the less entitled simply to overlook the fact that the question "What is called thinking?" presents itself at first quite innocently. It sounds as if, and we unknowingly take it as if, the question merely asked for more precise information about what is supposedly meant when we speak of such a thing as thinking. Thinking here appears as a theme with which one might deal as with any other. Thus thinking becomes the object of an investigation. The investigation considers a process that occurs in man. Man takes a special part in the process, in that he performs the thinking. Yet this fact, that man is naturally the performer of thinking, need not further concern the investigation of thinking. The fact goes without saying. Being irrelevant, it may be left out of our reflection on thinking. Indeed, it must be left out. For the laws of thought are after all valid independently of the one who performs the individual acts of thinking.

But if the question "What calls for thinking?" is asking what it is that first of all directs us to think, then we are asking for something that concerns ourselves because it calls upon us, upon our essence. It is we ourselves to whom the question "What is called thinking—what calls for thinking?" is addressed directly. We ourselves are in the text and texture of the question. The question "What calls on us to think?" has already drawn us into the issue in question. We ourselves are, in the strict sense of the word, put in question by the question. The question "What calls on us to think?" strikes us directly, as a lightning bolt. Asked in this way, the question "What calls for thinking?" does more than merely struggle with an object, in the manner of a scientific problem.

This other formulation of the question, which strikes us as strange, is open to the following immediate objection. The new meaning of the question "What calls for thinking?" has been obtained here by arbitrarily forcing on the question a signification totally different from the one that all the world would attach to it on hearing or reading it. This trick is easily exposed. It obviously relies on a mere play with words. And the victim of the play is the word that, as the verb of the question, sustains the sentence "What is called thinking?" We are playing with the verb "to call."

One might ask, for instance: "What do you call that village up there on the hill?" We want to know the name of the village. Or we may ask: "What shall we call the child?" That says: What name shall it bear? "To call" means in that sense to be named and to name. "What is called thinking?" means, then, what idea shall we form about the process which has been given the name "thinking?" This is how we understand the question if we take it simply and naturally.

But if we are to hear the question in a sense that asks for what it is that directs us to think, we find ourselves suddenly compelled to accept the verb "to call" in a signification that is strange to us, or at least no longer familiar.

We are now supposed to use the word "to call" in a signification that one might paraphrase approximately with the verbs summon, demand, instruct, direct. We call on someone who is in our way to give way, to make room. But the "call" does not necessarily imply demand, still less command; it rather implies an anticipatory reaching out for something that is reached by our call, through our calling.

In the widest sense, "to call" means to set in motion, to get something under way—which may be done in a gentle and therefore unobtrusive manner, and in fact is most readily done that way. In the New Testament, Matthew 8:18, we read, *Videns autem Jesus turbas multas circum se, iussit ire trans fretum.* ["But seeing a large crowd about him, Jesus 'commanded' them to go across the sea."] Luther translates, *Und da Jesus viel Volks um sich sah, hiess er hinüber jenseit des Meeres fahren.* ["And when Jesus saw many people around him he called them to go over across the sea."] To call [*heissen*] here corresponds to the Latin *iubēre* of the Vulgate, which properly means to wish that something might happen. Jesus "called" them to go over: he did not give a command or issue an order. What *heissen* in this passage means comes to light more clearly if we keep to the older Greek version of the Gospel. Here we read, *Idōn de ho Iēsous ochlon peri auton ekeleusin apelthein eis to peran* ["Seeing a large crowd around him, Jesus called to them to go to the other side"]. The Greek verb *keleuein* properly means to get something on the road, to get it under way. The Greek noun *keleuthos* means way. And that the old word "to call" means not so much a command as a letting-reach, that therefore the "call" has an assonance of helpfulness and complaisance, is shown by the fact that the same word in Sanskrit means something like "to invite."

The meaning of the word "call" which we have described is thus not altogether unfamiliar to us. It still is unaccustomed as we encounter it in the question "What is called thinking—what calls for thinking?" When we hear that question, the meaning of "call" in the sense of instruct, demand, allow to reach, get on the way, convey, provide with a way, does not immediately occur to us. We are not so much at home with these meanings of the word that we hear them at first, let alone first of all. We do not have the habit, or only just barely have it, of using the word "call" in this sense. And so it remains unfamiliar to us. Instead, we follow the habitual signification of the verb "to call," and mostly stay within it, not giving it much thought. "To call" simply means to bestow this or that name. In that signification the word is current among us. And why do we prefer the customary meaning, even unknowingly? Presumably because the unaccustomed and apparently uncustomary signification of the word "to call" is its proper one: the one that is innate to the word, and thus remains the only one—for from its native realm stem all the others.

In short, "to call" means "to command," provided we hear this word too in its native, telling sense. For "to command" basically means, not to give commands and orders, but to commend, entrust, give into safekeeping, to shelter. To call is to appeal commendingly, to direct and so let something be reached. To promise [*Verheissung*] means to respond to an entreaty in such a way that what is spoken here is spoken *to* and spoken *for.* To call means to appeal, and so to let something arrive and come to presence. It means to speak to something by addressing it.

Accordingly, when we hear our question, "What is called thinking?" in the sense that it asks, "What is it that claims us so that we must think?" we then are asking: "What is it that enjoins our essential being to think, and thus lets it arrive in thinking, there to shelter it?"

When we ask in this way we do, of course, use the word "to call" in a rather unfamiliar signification. But it is unhabitual not because our spoken speech has never yet been at home in it, but rather because *we* are no longer at home with this telling word, because we no longer really live in it.

We turn back to the original and vital significance of the word "to call" and ask: "What is it that calls on us to think?"

Is this return a whim, or is it to play games? Neither one nor the other. If we may talk here of playing games at all, it is not we who play with words; rather, the essence of language plays with us, not only in this case, not only now, but long since and always. For language plays with our speech—it likes to let our speech drift away into the more obvious meanings of words. It is as though man had to make an effort to live properly with language. It is as though such a dwelling were especially prone to succumb to the danger of commonness.

The place of language properly inhabited, and of its habitual words, is usurped by common terms. The common speech becomes the current speech. We meet it on all sides, and since it is common to all, we now accept it as the only standard. Anything that departs from this commonness, in order to inhabit the formerly habitual proper speaking of language, is at once considered a violation of the standard. It is branded as a frivolous whim. All this is in fact quite in order, as soon as we regard the common as the only legitimate standard, and become generally incapable of fathoming the commonness of the common. This floundering in a commonness that we have placed under the protection of so-called natural common sense is not accidental, nor are we free to deprecate it. This floundering in commonness is part of the high and dangerous game and gamble in which, by the essence of language, we are the stakes.

Is it playing with words when we attempt to give heed to this play of language and to hear what language really says when it speaks? If we succeed in hearing such play, then it may happen—provided we proceed carefully—that we get more truly to the matter that is expressed in any telling and asking.

We give heed to the proper signification of the word "to call," and accordingly ask our question, "What calls for thinking?" in this way: what is it that directs us into thinking, that calls on us to think? But after all, the word "to call" means also, and commonly, to give a name to something or to be named. The current meaning of the word cannot simply be pushed aside in favor of the rare one, even though the rare signification may still be the proper one. That would be an open violation of language. Besides, the presently more current signification of the word "call" is not totally unconnected and unrelated to the proper one. On the contrary, the presently customary signification is rooted in the other, original, decisive one. For what is it that the word "to name" tells us?

When we name a thing, we furnish it with a name. But what about this furnishing? After all, the name is not just draped over the thing. On the other hand, no one will deny that the name is coordinated with the thing as an object. If we conceive the situation in this way, we turn the name, too, into an object. We represent the relation between name and thing as the coordination of two objects. The coordination in turn is by way of an object, which we can see and conceive and deal with and describe according to its various possibilities. The relation between what is named and its name can always be conceived as a coordination. The only question is whether this correctly conceived coordination will ever allow us, will allow us at all, to give heed to what constitutes the peculiar character of the name.

To name something—that is to call it by name. More fundamentally, to name is to call something into its word. What is so called is then at the call of the word. What is called appears as what is present, and in its presence it is secured, commanded, called into the calling word. So called by name, called into presencing, it in turn calls. It is named, has the name. By naming, we call on what is present to arrive. Arrive where? That remains to be thought about. In any case, all naming and all being named is the familiar "to call" only because naming itself consists essentially in proper calling, in the call to come, in a commending and a command.

What is called thinking? At the outset we mentioned four ways to ask the question. We said that the way listed in the fourth place is the first, first in the sense of being highest in rank, since it sets the standard. When we understand the question "What is called thinking?" in the sense that it is a question about what calls upon us to think, we then have understood the word "to call" in its proper significance. That is to say also: we now ask the question as it properly wants to be asked. Presumably we shall now almost automatically get to the three remaining ways to ask the question. It will therefore be advisable to explicate the proper question a little more clearly. It runs: "What is it that calls on us to think?" What makes a call upon us that we should think and, by thinking, be who we are?

That which calls us to think in this way presumably can do so only insofar as the calling itself, on its own, needs thought. What calls us to think, and thus commands, that is, brings our essential being into the keeping of thought, needs thinking because what calls us wants itself to be thought about according to its essence. What calls on us to think demands for itself that it be tended, cared for, husbanded in its own essential being, by thought. What calls on us to think gives us food for thought.

What gives us food for thought we call thought-provoking. But what is thought-provoking not just occasionally, and not just in some given limited respect, but rather giving food for thought inherently and hence from the start and always—is that which is thought-provoking *per se*. This is what we call most thought-provoking. And what it gives us to think about, the gift it gives to us, is nothing less than itself—itself, which calls on us to enter into thinking.

The question "What calls for thinking?" asks for what wants to be thought about in the preeminent sense: it does not just give us something to think about, nor only itself, but it first gives thought and thinking to us, it entrusts thought to us as our essential destiny, and thus first joins and appropriates us to thought.

# X

# THE WAY TO LANGUAGE

*What is spoken is never, in any language, what is said.*

Early and late, Heidegger remained on the trail of language. If *being, time,* and *truth* constitute the motto on his escutcheon, it is nonetheless true that these things, whatever else they may be, are *words*. Heidegger never lost sight of that fact. Virtually every other text in these *Basic Writings* thematizes language, however briefly; in the present essay the question of language receives its most intensive treatment in Heidegger's oeuvre. Here his thought goes to encounter that of many others in our century—one thinks of Russell and Wittgenstein, Carnap, Quine, and Austin, to mention only a few—for whom language is *the* matter for thinking. Yet the way it goes to encounter them is unfamiliar and even uncanny.

Heidegger seeks *a way to* language. He does not come on the scene already outfitted with a program and a procedure, a methodology and a prescription *for* language. He does not run an analytical vacuum cleaner over language in order to tidy it up; he does not put it through the wringer of formalization in order to make it fit to occupy the House of Science. He does not even formulate arguments concerning language, spin out a theory of it, or concoct a meta-language that would allow him to say impossible things *about* language. His search is less impressive than all that. Indeed, there is an undeniable simplicity about "The Way to Language," which is doubtless why it is the most difficult of these *Basic Writings*. A word now about the gestation of the piece, followed by a brief discussion of a few of the decisive turns on Heidegger's simple way to language.

In January of 1959 Heidegger joined a group of distinguished colleagues in a lecture series sponsored jointly by the Bavarian and Berlin art academies. The series' unadorned title: "Language." The contributions varied widely in subject-matter and approach. Carl von Weizsäcker spoke on cybernetics and information theory; Thrasybulos Georgiades recounted the importance of traditionally set rhythms in and for ancient Greek diction. Heidegger took the opportunity to summarize the whole of his later thinking on language, which is also a thinking of *Ereignis* or "propriation." ("Later" here means from 1935 onward; see especially the remarks on language in Readings IV, V, and IX.)

In *Being and Time* Heidegger had emphasized the primary importance of discourse or talk (*die Rede*) for language, and the secondary or "derivative" character of assertions and propositions—the discourses of science and philosophy, but also of journalism, politics, and culture generally. He also stressed the importance of our listening to and heeding speech, suggesting that the silence that enables us to listen is more significant than all the noise of signification. (His remarks on silence in *Being and Time* receive a noteworthy qualification in the present essay.)

"The Way to Language" too takes its orientation from the spoken rather than the written word. After all, a long line of thinkers from Aristotle to Wilhelm von Humboldt set their written seal of approval on the primacy of speech for language. Language speaks. The Romantic writer and thinker, Novalis (Friedrich von Hardenberg, 1772–1801), who contributes the phrase that opens Heidegger's "Way," tells us that language "concerns itself purely with itself alone." If language speaks, its speech is a "monologue." Heidegger's own way to language begins with this dual inheritance. Language *speaks*. *Language* speaks. Yet its monologue is not a self-absorbed mumble. Language says something when it speaks, and such saying (*sagen, die Sage*) will be Heidegger's major concern in "The Way to Language." As simple as that sounds.

By saying something, language addresses people and things in the world; it points to them, as it were, showing them to be matters of concern. The showing and pointing (*zeigen, die Zeige*) that language perform constitute the very essence of language. They delineate its profile, its rift-design (*der Auf-Riss:* see Reading IV). Through its saying, showing, and pointing, language lets people and things be there for us, allows them to come into their own and radiate in presence. Monologue never simply upstages the things. It owns up to the fact that its saying becomes telling only when it lets a being come into its own.

What about this "owning up" and its "owning"? Perhaps the most hazardous turns on Heidegger's way to language involve the words *own* and *owning* in their many cognate forms. "Own" is *eigen* in German, and it is the root of a whole series of resonant words for Heidegger: *eigentlich,* the crucial epithet of Heidegger's analysis of

Dasein, meaning "appropriate," applied in this essay to language "proper"; *eignen,* to own or possess, especially in the form *an-eignen,* "to appropriate," a word Heidegger often employs, though not in this essay; *eigens,* meaning "expressly" or "explicitly," as when Heidegger tries to say explicitly what language on its own *is; das Eigene,* whatever is a thing's "own," that is, whatever shows itself when language lets a being advene under its own power, or lets it withdraw into concealment and abide on its own; and finally, *das Eigentümliche,* what is "peculiar" to language proper. By far the most important and complex of these words is *Ereignis,* often written *Er-eignis,* and its verb *sich ereignen.* Customarily translated as "event," *Ereignis* is here rendered as "propriation" in an effort to save the sense of "ownness," Latin *proprius,* French *propre.* Yet we should keep an eye on all such renderings. (See p. 414, below.)

It will not be possible to say quickly why *owning* and *propriating* become key words (along with *saying* and *showing*) of an essay on the way to language. For that would be to ignore the final turn of Heidegger's tripartite essay—his recognition that the way to language is never finished, never put behind us, but is itself always under way. Perhaps two remarks on owning and propriation are in order. First, a warning. The most treacherous turn on the way to language occurs when we first hear talk of propriation. Because propriation smacks of property and appropriation, we can easily misunderstand it as one aspect of man's assault on being—as an element of the aggrandizing essence of technology. To be sure, propriation does bear a special relation to the *essence* of technology. Yet propriation is not subject to human calculation; it is rather what is sent as the historical destiny of mortals. The hardest lesson to learn is that the owning is not ours, except perhaps in one sense. Here—second and last remark—the English word *to own* offers food for thought.

To own is not only to appropriate, but also to recognize and acknowledge an other, to declare or make manifest one's acceptance or affirmation of some other thing, to confess or profess something as true, and even as holding sway over us. Such owning would involve not a commandeering of language but a responding to it. While under way to it.

# THE WAY TO LANGUAGE

At the outset we shall hear some words of Novalis. They stand in a text he entitled *Monologue*. The title directs us to the mystery of language: language speaks solely and solitarily with itself. One sentence in the text goes as follows: "Precisely what is peculiar to language—that it concerns itself purely with itself alone—no one knows."

If we grasp what we shall now try to say as a sequence of assertions about language, it will remain a concatenation of unverified and scientifically unverifiable claims. If on the contrary we experience the way to language in terms of what transpires with the way while we are under way on it, then a kind of surmise could awaken, a surmise by which language would henceforth strike us as exceedingly strange.

The way to language: it sounds as though language lay far afield, at some place toward which we would first of all have to set out on our way. However, do we really need a way *to* language? According to an ancient pronouncement, we ourselves are those creatures who can speak and who thus already possess language. Nor is the capacity to speak merely *one* capability of human beings, on a par with the remaining ones. The capacity to speak distinguishes the human being as a human being. Such a distinguishing mark bears in itself the very design of the human essence. Man would not be man if it were denied him to speak—ceaselessly, ubiquitously, with respect

---

Reading X is a new translation of the final essay of Heidegger's *Unterwegs zur Sprache* (Pfullingen: G. Neske, 1959), "Der Weg zur Sprache," made especially for these *Basic Writings* by the editor. I am grateful to have had the earlier translation by Peter Hertz, in *On the Way to Language* (New York: Harper & Row, 1971), for purposes of comparison.

to all things, in manifold variations, yet for the most part tacitly—by way of an "*It is*." Inasmuch as language grants this very thing, the essence of man consists in language.

Thus we are within language, at home in language, prior to everything else. A way to it is superfluous. Moreover, the way to language is impossible, if indeed we are already at the place to which it is supposed to lead us. Yet are we there? Are we within language in such a way that we experience its essence, thinking it as language by apprehending and listening to what is proper to it? Do we already linger in nearness to language, without our having to take any trouble concerning it at all? Or does the way to language as language constitute the farthest stretch for our thought? Not only the farthest, but also one that is bestrewn with obstacles, obstacles that arise from language itself the moment we try to suspend every type of diversion and follow its trail into what is purely its own?

In this regard we shall risk something strange, something we might adumbrate in the following way: *To bring language as language to language*. That sounds like a formula. It is to serve us as a guideline on the way to language. The formula employs the word *language* three times; each time it says something different, though nonetheless selfsame. The selfsame is what conjoins all that is held apart, conjoins it on the basis of that one thing in which the peculiarity of language consists. To be sure, the formula directs us in the first place to a weft of relations in which we ourselves are already interwoven. Our proposed way to language is woven into a speaking that would like to liberate nothing else than language, liberate it in order to present it, giving utterance to it as something represented—which straightway testifies to the fact that language itself has woven us into its speaking.

The weft announced by our path's formula designates the predetermined realm in which not only this lecture series but also the whole of linguistics, all theory of language and philosophy of language, and every attempt to follow the trail of language must reside.

A weft compresses, tightens, and thus obstructs any straightforward view into its mesh. Yet at the same time the weft designated by our path's formula is language, language for its own sake. We therefore dare not divert our gaze from this weft, even if it seems to draw everything together into an inextricable tangle. Rather, the formula must compel our meditation to try, not of course to eliminate the weft, but to loosen it in such a way that it grants a view upon the unconstrained cohesion of the various elements designated in the formula. Perhaps the weft is permeated by a bond that unbinds language to what is peculiar to it, albeit in a way that is passing strange. It is a matter of experiencing that unbinding bond in the weft of language.

The lecture which undertook to think language as information, and which in turn had to think information as language, called this self-reverting relation a circle, indeed an unavoidable though meaningful circle.* The circle is a special case of the weft to which we have referred. The circle possesses meaning because the direction and the manner of language's circling are determined by language itself; that is, by a movement within language. We would like to experience the character and scope of this movement in terms of language itself by seeking an entry into the weft.

How might such an effort succeed? By means of a relentless pursuit of whatever it is that our path's formula indicates when it says: To bring language as language to language.

The more clearly language itself shows itself in what is its own, the more significant the way to language becomes for itself while under way, and the more decisively the sense of the formula is transformed. It loses its formulaic flavor, imperceptibly passing over into a soundless intimation, an intimation that enables us to hear the faint ring of what is peculiar to language.

*See the bibliographical reference at the end of the book. [In these *Basic Writings*, see the Introduction to this essay—Ed.]. In the lecture series mentioned there, Carl Friedrich von Weizsäcker spoke on the theme "Language as Information."

I

Language: by it we mean speech, something we know as an activity of our own, an activity we are confident we can perform. Nevertheless, speech is not a secure possession. A human being may be speechless with astonishment or terror. He is altogether astonished, thunderstruck. He no longer speaks: he is silent. Someone else has an accident and loses the power of speech. He no longer speaks. Nor is he silent. He remains mute. Speech implies the creation of articulated sounds, whether we produce these, by speaking, or refrain from doing so, in silence, or are incapable of doing so, due to loss of speech. The creation of articulated sounds by the voice pertains to speech. In speech, language shows itself to be activation of the phonic instruments that we possess: mouth, lips, the "barricade of the teeth,"* tongue, and larynx. That language has since ancient times been immediately represented in terms of these phenomena is evident in the very names Western languages have bestowed on language: *glōssa*, *lingua*, *langue*, *language*. Language is tongue, and it works by word of mouth.

At the outset of a treatise later given the title *Peri hermēneias*, *De interpretatione*, or *On Utterance*, Aristotle says the following:

> Ἔστι μὲν οὖν τὰ ἐν τῇ φωνῇ τῶν ἐν τῇ ψυχῇ παθημάτων σύμβολα, καὶ τὰ γραφόμενα τῶν ἐν τῇ φωνῇ. καὶ ὥσπερ οὐδὲ γράμματα πᾶσι τὰ αὐτά, οὐδὲ φωναὶ αἱ αὐταί· ὧν μέντοι ταῦτα σημεῖα πρώτων, ταὐτὰ πᾶσι παθήματα τῆς ψυχῆς, καὶ ὧν ταῦτα ὁμοιώματα πράγματα ἤδη ταὐτά.

Only a meticulous interpretation would permit an adequate translation of the text. Here a makeshift must suffice. Aristotle says:

> Now, whatever it is [that transpires] in the creation of sound by the voice is a showing of whatever affections there may be in the soul, and the written is a

*Das "Gehege der Zähne." Presumably a reference to the familiar Homeric epithet, *herkos odontōn*. See, for example, *The Odyssey*, I, 64; V, 22; X, 328, etc.—Ed.

showing of the sounds of the voice. Hence, just as writing is not identical among all [human beings], so too the sounds of the voice are not identical. However, that of which these [sounds and writing] are in the first place a showing are among all [human beings] the identical affections of the soul; and the matters of which these [the affections] form approximating presentations are likewise identical.

Our translation consistently understands the *sēmeia* (that which shows), the *symbola* (that which holds together), and the *homoiōmata* (that which approximates) in terms of showing; it understands showing in the sense of letting appear, which for its part depends on the ruling sway of revealing (*alētheia*). And yet our translation neglects the variety in the modes of showing that the text introduces.

Aristotle's text contains the confident, sober saying that marks the classical construction, the construction that harbors language as speech. Letters show sounds; sounds show affections in the soul; affections show the matters that impinge on us.

The braces and supports of the construction are shaped and borne aloft by showing. In manifold ways, by unveiling or veiling, showing brings something to appear, lets what appears be apprehended, and enables what is apprehended to be thoroughly discussed (so that we can act on it). However, the kinship of the showing with what it shows never unfolds purely in terms of the kinship itself and its provenance. In subsequent periods, the kinship is transformed into the conventional relationship between a sign and its signified. Greek civilization at its acme experiences the sign on the basis of showing, the sign having been coined by showing for showing. From the Hellenistic (and Stoic) period onward, as the convention becomes sheer stipulation, the sign comes to be an instrument for designating; by means of such designation, representation is coordinated and directed from one object to another. Designation is no longer a showing in the sense that it lets something appear. The alteration of the sign—from that which shows to that which designates—is based on a transformation in the essence of truth.*

Ever since the age of the Greeks, beings have been experienced as what comes to presence. Inasmuch as language *is*, coming as speech again and again on the scene, it pertains to what comes to presence. One represents language, having taken one's departure from speech, with a view to articulated sounds as bearers of meanings. Speaking is one form of human activity.

The representation of language that we have sketched here in rough outline has remained throughout manifold transformations the guiding and supporting one in Western European thought over the centuries. This way of looking at language, having commenced in Greek antiquity and ramifying along many different paths, gathers to a kind of summit in Wilhelm von Humboldt's meditation on language. That meditation assumes final form in the magnificent Introduction to his work on the Kawi language of Java. A year after his death, his brother, Alexander von Humboldt, published the Introduction separately under the title, *On the Diversity of the Structure of Human Language and Its Influence on the Intellectual Development of Mankind* (Berlin, 1836).† Since that date, down to the present day, this treatise has shaped all subsequent linguistics and philosophy of language, whether tacitly or explicitly, whether through advocacy or refutation.

Every listener who is present at the lecture series we are attempting here would have to have thought through and have in mind the astonishing but scarcely penetrable treatise by Wilhelm von Humboldt. It is a treatise that vacillates in obscurity whenever it is a matter of fundamental concepts but that nonetheless never fails to

*See "Plato's Doctrine of Truth," 1947, first published in *Geistige Überlieferung*, vol. II, 1942, pp. 96–124. [See *Wegmarken*, 1967, pp. 109–44. However, on this important matter see also Heidegger's later qualification, in these *Basic Writings* on p. 446.—Ed.]

†The following quotations derive from the anastatic reprint of von Humboldt's text, edited by E. Wasmuth, 1936.

stimulate. If that prerequisite were met, a shared vantage point for our view upon language would be made available to us all. Such a prerequisite is lacking. We shall have to make our peace with that lack. It will be enough if we avoid forgetting it.

"Articulated sound" is, according to Wilhelm von Humboldt, "the basis and the essence of all speech" (*On the Diversity*, section 10, p. 65). In section 8 of his treatise (p. 41), Humboldt coins those statements that are often cited but seldom considered, that is to say, seldom considered solely with a view to the manner in which they define Humboldt's *way to language*. The statements run as follows:

> *Language*, grasped in its actual essence, is perpetually and at every moment something *transitory*. Even its preservation through writing is always a merely incomplete preservation, a kind of mummification, which is necessary if we are to try to render once again the delivery of the living word. Language itself is not a work (*ergon*), but an activity (*energeia*). Its true definition can thus only be a genetic one. For language is the eternally self-repeating *labor of spirit* to make *articulated sound* capable of being an expression of *thought*. Taken strictly and directly, this is the definition of every instance of *speaking*; but in the true and essential sense, one can also regard the totality of such speech only as an approximation to language.

Here Humboldt says that he sees the essential element of language in speech. Does he thereby also say what language viewed in this way is, as language? Does he bring speech as language to language? We leave the question deliberately without reply, but observe the following points.

Humboldt represents language as a particular "labor of spirit." Guided by this view of the matter, he pursues the sort of thing language shows itself to be, that is to say, what it is. Such what-being is called the *essence*. Now, as soon as we approach and delineate the labor of spirit with a view to its linguistic achievements, the essence of language thus conceived has to stand out in bolder relief. However, spirit lives—in Humboldt's sense as well—also in other activities and achievements. Yet if language is reckoned to be but one among them, speech is not experienced on its own—in terms of language—but is oriented in that very view to something else. Nevertheless, this "something else" is too significant for us who are meditating on language to be permitted to overlook it. What activity does Humboldt have in view when he conceives of language as the labor of spirit? Several statements at the outset of section 8 supply the answer:

> One must not regard *language* as a lifeless *product*. It is fár more like a *reproducing*. One must endeavor more keenly to abstract from the things it achieves by way of designating objects and mediating the understanding. As opposed to that, one must go back more meticulously to its origin, so tightly interwoven with the inner activity of spirit, and to their influence upon one another.

Humboldt here refers to the "inner linguistic form" described in section 11, a notion quite difficult to define in terms of his own conceptual apparatus. We get a bit closer to it when we ask: What is speech as the expression of thought; what is speech when we ponder it in accord with its provenance from the inner activity of spirit? The answer lies in a statement (section 20, p. 205) whose adequate interpretation would require a separate discussion: "Whenever the feeling truly awakens in the soul that language is not merely a medium of exchange for the sake of mutual understanding, but a true *world*, which *spirit* must posit between itself and *objects* by the inner labor of its own force, then it is on the true way to finding more and more in language and to investing more and more in it." According to the doctrine of modern idealism, the labor of spirit is positing. Because spirit is grasped as subject and thus represented in the subject-object schema, positing (thesis) must be the synthesis between the subject and its objects. What is posited in this way affords a view upon the totality of objects. What the force of the subject elaborates, what it posits by means of labor between itself and the objects, Humboldt calls a "world." In such a "view upon the world" a form of humanity brings itself to expression.

Yet why does Humboldt envisage language as world and view upon the world? Because *his* way to language is not so much determined by language as language; rather, it strives to depict by means of a history the entire historical-spiritual development of mankind as a whole, but also at the same time in its prevailing individuality. In a fragment toward an autobiography from the year 1816 Humboldt writes, "Precisely what I am striving for is a conception of the world in its individuality and totality."

Now, a conception of the world that sets out in this fashion can draw from various wells, inasmuch as the force of spirit expressing itself is active in manifold ways. Humboldt recognizes and selects language as one of the principal sources. Language is of course not the only form of that view upon the world which human subjectivity elaborates; but to its prevailing imprinting power one must attribute a special status, as the standard by which the historical development of humanity can be measured. The title of Humboldt's treatise now speaks more clearly with regard to his way to language.

Humboldt treats of "the diversity of the structure of human language" to the extent that "the intellectual development of mankind" stands under "its influence." Humboldt brings language to language as *one* form and variety of the view upon the world that is elaborated by human subjectivity.

To what sort of language? To a series of assertions that speak the language of the metaphysics of his age. The philosophy of Leibniz contributes a definitive word to this language. This is most clearly announced in the fact that Humboldt defines the essence of language as *energeia*, understanding it however in a way that is foreign to the Greeks; he takes it in the sense of an activity of the subject, as Leibniz's *Monadology* takes it. Humboldt's way to language goes in the direction of man, passing through language on its way to something else: demonstration and depiction of the intellectual development of the human race.

However, the essence of language conceived in terms of such a view does not of itself show language in its essence: it does not show the way in which language essentially unfolds as language; that is, the way it perdures; that is, the way it remains gathered in what it grants itself on its own as language.

## II

If we are on the trail of language as language, we have already abandoned the procedures that have long prevailed in linguistic study. We can no longer root about for general notions like energy, activity, labor, force of spirit, view upon the world, or expression, under which we might subsume language as a particular instance of this or that universal. Instead of explaining language as this or that, and thus fleeing from it, the way to language wants to let language be experienced as language. True, in the essence of language, language is grasped conceptually; but it is caught in the grip of something other than itself. If on the contrary we pay heed only to language as language, it demands of us that we begin by bringing to the fore all those things that pertain to language as language.

Yet it is one thing to collate the multiplicity of elements that show themselves in the essence of language, and another to gather one's gaze to what of itself unifies the coherent elements, unifies them insofar as its uniting grants to the essence of language the unity that is appropriate to it.

The way to language will now try to advance more strictly along the guidelines spelled out in our formulation—to bring language as language to language. It is a matter of getting closer to what is peculiar to language. Here too language initially shows itself as our speech. For the moment we shall heed all the things that speak along with us in our speech, always from the outset and in accord with the selfsame measure, whether we are aware of it or not.

To speech belong the speakers, but not as cause to effect. Rather, in speech the speakers have their presencing. Where to? Presencing to the wherewithal of their speech, to that by which they linger, that which in any given situation already matters to them. Which is to say, their fellow human beings and the things, each in its own way; everything that makes a thing a thing and everything that sets the tone for our relations with our fellows. All this is referred to, always and everywhere, sometimes in one way, at other times in another. As what is referred to, it is all talked over and thoroughly discussed; it is spoken of in such a way that the speakers speak to and with one another, and also to themselves. Meanwhile, what is spoken remains multifaceted. It is often only what is spelled out in so many words, something that quickly evanesces or in some way is retained. What is spoken can be long gone, but it can also be what has long gone on, as what is addressed.

What is spoken derives in manifold ways from the unspoken, whether in the form of the not yet spoken or of what has to remain unspoken—in the sense that it is denied speech. Thus the bizarre impression arises that what in manifold ways is spoken is cut off from speech and from speakers, and does not belong to them; whereas it alone holds up to speech and to the speakers those things to which they attend, no matter how they reside in the spoken elements of the unspoken.

In the essence of language a multiplicity of elements and relations shows itself. We enumerated these, but did not put them in proper sequence. In running through them—which is to say, in original counting, which is not a reckoning in numbers—a certain coherence announced itself. Counting is a recounting. It previews the unifying power in cohesion, but cannot yet bring it to the fore.

The incapacity of our way of seeing things that is here coming to light, the inability of our thought to experience the unifying unity in the essence of language, has a long provenance. That is why the unifying unity has received no name. The traditional names for what one means under the rubric *language* name this unity always only in one or other respect, as the essence of language proffers them.

Let the unity in the essence of language that we are seeking be called the rift-design.* The name calls upon us to descry more clearly what is proper to the essence of language. *Riss* [rift] is the same word as *ritzen* [to notch, carve]. We often come across the

**Der Aufriss*. See Reading IV, esp. pp. 188–89.—Ed.

word *Riss* in the purely pejorative form, for example, as a crack in the wall. Today when farmers speak in dialect about plowing a field, drawing furrows through it, they still say *aufreissen* or *umreissen* [literally, to tear up, to rend or rive, to turn over]. They open up the field, that it may harbor seed and growth. The rift-design is the totality of traits in the kind of drawing that permeates what is opened up and set free in language. The rift-design is the drawing of the essence of language, the well-joined structure of a showing in which what is addressed enjoins the speakers and their speech, enjoins the spoken and its unspoken.

Yet the rift-design in the essence of language remains veiled even in its most approximate adumbration as long as we fail to pay explicit attention to the sense in which we have been speaking all along of speech and the spoken.

Speech is, of course, the creation of sounds. It can also be taken as an activity of human beings. Both are correct representations of language as speech. Both will remain outside our purview here, although we do not intend to forget how long the sounding of language has been waiting for its fitting definition. For the phonetic, acoustic, physiological explanation of such sounding does not experience the provenance of sounding from the ringing of stillness; even less does it experience the attunement of the sounding in that stillness.

Yet how have speech and what is spoken been thought in our earlier, quite brief recounting of the essence of language? They showed themselves as the sort of thing through which and in which something comes to language, that is to say, comes to the fore *whenever something is said*. Saying and speaking are not identical. One can speak, speak endlessly, and it may all say nothing. As opposed to that, one can be silent, not speak at all, and in not speaking say a great deal.

Yet what is it we call *saying?* To experience this, we shall hold to what our language itself calls on us to think in this word. *Sagan* means to show, to let something appear, let it be seen and heard.*

What we are saying here becomes obvious, though hardly pondered in its full scope, when we indicate the following. To speak to one another means to say something to one another; it implies a mutual showing of something, each person in turn devoting himself or herself to what is shown.† To speak with one another means that together we say something about something, showing one another the sorts of things that are suggested by what is addressed in our discussion, showing one another what the addressed allows to radiate of itself. The unspoken is not merely what is deprived of sound; rather, it is the unsaid, what is not yet shown, what has not yet appeared on the scene. Whatever has to remain unspoken will be held in reserve in the unsaid. It will linger in what is concealed as something unshowable. It is mystery. The addressed speaks as a pronouncement, in the sense of something allotted; its speech need not make a sound.

As saying, speech belongs to the rift-design in the essence of language. Various modes of saying and the said permeate the rift-design, modes in which what is present or absent says something about itself, affirms or denies itself—shows itself or withdraws. What pervades the rift-design in the essence of language is a richly configured saying, from various provenances. With a view to the concatenations of saying, we shall call the essence of language as a whole *the saying* [die Sage]. Even so, we have to admit that the unifying element in these concatenations is not yet in sight.

We are now accustomed to using the word *Sage* [saying, saga], like many other words in our language, for the most part in a dis-

*It is more difficult to show the connection in English between "saying" (*Sagen*) and "showing" (*Zeigen*). Yet the Latin *dico* brings both senses together: "I say" originally means "I show through words."—ED.

†The German text here (*Unterwegs zur Sprache*, 1959, p. 253, ll. 2–3) is marred by two typographical errors that disrupt the sense. The lines should read as follows: *Zueinandersprechen heisst: einander etwas sagen, gegenseitig etwas zeigen, wechselweise sich dem Gezeigten zutrauen.*—ED.

paraging sense. A saying is taken to be sheer hearsay, as someone's say-so, which may or may not hold water and which therefore leaves us incredulous. That is not the way we are thinking *die Sage* here. Nor are we referring to the admittedly essential sense that is intended when one invokes the "sagas of gods and heroes." But perhaps we are thinking it as Georg Trakl's "venerable saying of the blue font" ["*die ehrwürdige Sage des blauen Quells*"]. In accord with the word's oldest usage, we understand the saying in terms of "to say" in the sense of "to show." In order to name the saying on which the essence of language depends, we shall use an old, well-testified, but archaic word: *die Zeige* [the pointing]. What Latin grammar calls the "demonstrative pronoun" is often translated as "the little indicator" ["*Zeigewörtlin*"]. Jean Paul calls the phenomena of nature "the spiritual index finger" ["*den geistigen Zeigefinger*"].

*What unfolds essentially in language is saying as pointing.* Its showing does not culminate in a system of signs. Rather, all signs arise from a showing in whose realm and for whose purposes they can be signs.

However, in view of the well-joined structure of the saying, we dare not attribute showing either exclusively or definitively to human doing. Self-showing as appearing characterizes the coming to presence or withdrawal to absence of every manner and degree of thing present. Even when showing is accomplished by means of our saying, such showing or referring is preceded by a thing's letting itself be shown.

Only when we ponder our saying in this regard do we arrive at an adequate determination of what essentially unfolds in all speech. We know speech to be the articulate vocalization of thought by means of the instruments of speech. However, speech is simultaneously hearing. Speaking and hearing are customarily set in opposition to one another: one person speaks, the other hears. Yet hearing does not merely accompany and encompass speaking, such as we find it in conversation. That speaking and hearing occur simultaneously means something more. Speech, taken on its own, is hearing. It is listening to the language we speak. Hence speaking is not *simultaneously* a hearing, but is such *in advance*. Such listening to language precedes all other instances of hearing, albeit in an altogether inconspicuous way. We not only *speak* language, we speak *from out of* it. We are capable of doing so only because in each case we have already listened to language. What do we hear there? We hear language speaking.

But then does language itself speak? How should it manage to do so, when it is not even equipped with the instruments of voice? Nevertheless, it is *language* that speaks. What language properly pursues, right from the start, is the essential unfolding of speech, of saying. Language speaks by saying; that is, by showing. Its saying wells up from the once spoken yet long since unspoken saying that permeates the rift-design in the essence of language. Language speaks by pointing, reaching out to every region of presencing, letting what is present in each case appear in such regions or vanish from them. Accordingly, we listen to language in such a way that we let it tell us its saying. No matter what other sorts of hearing we engage in, whenever we hear something we find ourselves caught up in a hearing that *lets itself be told*, a hearing that embraces all apprehending and representing. In speech, as listening to language, we reiterate the saying we have heard. We let its soundless voice advance, requesting the sound that is already held in reserve for us, calling for it, reaching out to it in a way that will suffice. With that, at least one trait in the rift-design of the essence of language announces itself more clearly, a trait that allows us to descry how language as speech is brought home into its own, thus speaking as language.

If speech as listening to language lets itself be told the saying, such letting can be given only insofar—and in so near—as our own essence is granted entry into the saying. We hear it only because we belong to it. However, the saying grants those who belong to it their listening to language and hence their speech. Such granting perdures in the saying; it lets us attain the capacity of speech. What unfolds essentially in language depends on the saying that grants in this way.

And the saying itself? Is it something separate from our speech, something to which we must first span a bridge? Or is the saying the stream of stillness that conjoins its own two banks—the saying and our reiterating—by forming them both? Our customary representations of language hardly go so far. The saying: when we try to think the essence of language in terms of it, are we not in danger of hypostasizing language to a phantasm, a self-subsistent essence that is nowhere to be found as long as we remain sober and follow hard upon the trail of language? Language does remain unmistakably bound up with human speech. Certainly. However, of what sort is this binding? Whence and in what way does such binding hold sway? Language needs human speech and is nonetheless not the mere contrivance of our speech activities. On what does the essence of language rest; in what is it grounded? Perhaps when we search for grounds we pass on by the essence of language.

Might not the saying itself be what does the "resting," what grants the repose of cohesion to those elements that belong to the well-joined structure of the essence of language?

Before we think any further in this direction, let us once again pay heed to the way to language. By way of introduction we suggested that the more clearly language as such comes to the fore, the more decisively the way to it is transformed. Heretofore the way had the character of a passage that would lead us as we set out to follow the trail of language, a passage into that curious weft designated by our path's formula. We took our orientation from speech, in the company of Wilhelm von Humboldt, and tried first to represent the essence of language, then to ground it. Accordingly, it was a matter of recounting the elements that pertain to the rift-design in the essence of language. On the trail of the rift-design, we arrived at language as the saying.

## III

With our recounting elucidation of the essence of language as the saying, the way to language has arrived at language as language and thus reached its goal. Our commemorative thought has left the way to language behind. So it seems, and so it is, as long as one considers the way to language to be the passage of a thinking that is on the trail of language. In truth, however, commemorative thought merely finds itself confronting the *way to language* that it seeks, and is but barely tracing it. For in the meantime something has shown itself in the essence of language, and it says: In language as the saying, something like a way unfolds essentially.

What is a way? The way lets us get somewhere. Here it is the saying that lets us get to the speaking of language, provided we listen to the saying.

The way to speech unfolds essentially in language itself. The way to language in the sense of speech is language as the saying. What is peculiar to language thus conceals itself on the way, the way by which the saying lets those who listen to it get to language. We can be those listeners only if we belong to the saying. The way to speech, which lets us arrive, itself derives from a letting-belong to the saying. Such letting-belong harbors what properly can be said to unfold essentially on the way to language. Yet how does the saying unfold essentially, so that it is capable of letting someone belong? If the essential unfolding of language is to announce itself explicitly at all, it should do so as soon as we have heeded with greater determination the things already yielded by the foregoing elucidation.

The saying is a showing. In everything that appeals to us; in everything that strikes us by way of being spoken or spoken of; in everything that addresses us; in everything that awaits us as unspoken; but also in every speaking of *ours*—showing holds sway. It lets what is coming to presence shine forth, lets what is withdrawing into absence vanish. The saying is by no means the supplementary linguistic expression of what shines forth; rather, all shining and fading depend on the saying that shows. It liberates what comes to presence to its particular presencing, spirits away what is withdrawing into absence to its particular kind of absence. The saying joins and pervades the open space of the clearing which every shining must seek, every evanescence abandon, and to which every presencing and absencing must expose itself and commit itself.

The saying is a gathering that joins every shining of a showing. The showing, for its part, is multiple; everywhere it lets what is shown stand on its own.

Whence does the showing arise? Our question asks too much, and too quickly. It suffices if we heed what it is that bestirs itself in showing and brings its stirrings to a culmination. Here we need not search forever. The simple, abrupt, unforgettable and therefore ever-renewed gaze toward what is familiar to us suffices, although we can never try to know it, much less cognize it in the appropriate way. This unknown but familiar thing, every showing of the saying, with regard to what it stirs and excites in each coming to presence or withdrawing into absence, is the dawn, the daybreak, with which the possible alternation of day and night first commences. It is at once the earliest and the oldest. We can only name it, because it will deign no discussion. For it is the place [*Ortschaft*] that encompasses all locales and time-play-spaces. We shall name it by using an old word. We shall say:

*What bestirs in the showing of saying is owning.*

Owning conducts what comes to presence and withdraws into absence in each case into its own. On the basis of owning, these things show themselves, each on its own terms, and linger, each in its own manner. Let us call the owning that conducts things in this way—the owning that bestirs the saying, the owning that points in any saying's showing—the propriating. Propriating dispenses the open space of the clearing into which what is present can enter for a while, and from which what is withdrawing into absence can depart, retaining something of itself while all the while in withdrawal. What the propriating yields through the saying is never the effect of a cause, nor the consequence of a reason. The owning that conducts, the propriating, grants more than any effecting, making, or grounding can grant. What propriates is propriation itself—and nothing besides.* Propriation, espied in the showing of the saying, can be represented neither as an event nor as a happening; it can only be experienced in the showing of the saying as that which grants. There is nothing else to which propriation reverts, nothing in terms of which it might even be explained. Propriating is not an outcome or a result of something else; it is the bestowal whose giving reaches out in order to grant for the first time something like a "There is / It gives," which "being" too needs if, as presencing, it is to come into its own.†

Propriation gathers the rift-design of the saying and unfolds it in such a way that it becomes the well-joined structure of a manifold showing. Propriation is the most inconspicuous of inconspicuous things, the simplest of simple things, the nearest of things near and most remote of things remote, among which we mortals reside all our lives.

The propriation that rules in the saying is something we can name only if we say: It—propriation—owns. When we say this, we are speaking in what is already our own spoken language. We hear some of Goethe's lines, lines that use the verbs *eignen* and *sich eignen* [to own, to own itself] in proximity to *sich zeigen* and *be-*

*See *Identität und Differenz*, 1957, pp. 28ff. [Even though Heidegger does not draw our attention to other similar wordings, it would be interesting to compare this formulation—"and nothing besides," "nothing else"—to Reading II, esp. p. 95, above.—Ed.]

†See *Being and Time*, 1927, section 44. [Discussed in the Introduction to Reading III. The pages of section 44 that are most relevant here are 226–30. On the phrase, "There is / It gives," *Es gibt*, see Reading XI, esp. p. 449.—Ed.]

*zeichnen* [to show itself, to designate], although not with a view to the essence of language. Goethe says:

> Von Aberglauben früh und spät umgarnt:
> Es eignet sich, es zeigt sich an, es warnt.
>
> Wrapped then as now in superstition's yarns:
> It owns itself, it shows itself, it warns.*

Elsewhere, in a somewhat altered fashion, he says:

> Sei auch noch so viel bezeichnet,
> Was man fürchtet, was begehrt,
> Nur weil es dem Dank sich eignet,
> Ist das Leben schätzenswert.
>
> Designate all else into the scheme
> Of things that make you fear or dream;
> Only when it owns itself to thanking
> Is life held in esteem.†

Propriation bestows on mortals residence in their essence, such that they can be the ones who speak. If by "law" we mean the gathering of what lets everything come to presence on its own and cohere with all that belongs to it, then propriation is the most candid and most gentle of laws, gentler still than the law acknowledged by Adalbert Stifter to be "the gentle law." To be sure, propriation is not a law in the sense of a norm that hovers over us somewhere; it is not an ordinance that orders and regulates a certain course of events.

Propriation is *the* law, inasmuch as it gathers mortals in such a way that they own up to their own essence. It gathers them and holds them there.

Because the showing of the saying is an owning, our being able to hear the saying, our belonging to it, also depends on propriation.

**Faust*, Part II, Act V, "Midnight." [Note that what here is owned, announces itself, and warns is *Sorge*, "Care," the name that Heidegger in *Being and Time* chose as the existential-ontological designation of human existence.—ED.]

†"For Grand Duke Karl August, New Year's, 1828."

In order to catch a glimpse of this state of affairs in its full enormity, we would have to think the essence of mortals, in all its sundry connections, in a sufficiently comprehensive way. And of course, above all else, we would have to think propriation as such. Here a mere reference must suffice.*

Propriation propriates the mortals by envisaging the essence of man.† It does so by remanding mortals to that which in the saying advances from all sides in order to converge on the concealed, which thus becomes *telling* for man.‡ The remanding of human beings, the ones who hear, to the saying is distinctive in that it releases the essence of man into its own. Yet it does so only in order

*See *Vorträge und Aufsätze*, 1954, as follows: "The Thing," pp. 163ff. [in the translation by Albert Hofstadter in *Poetry, Language, Thought*, pp. 163–86]; "Building Dwelling Thinking," pp. 145ff. [see Reading VIII]; "The Question Concerning Technology," pp. 13ff. [see Reading VII]. Today, when half-baked thoughts, or things scarcely thought at all, are rushed into print in one form or another, many readers may be incredulous about the fact that the author has used the word *Ereignis* [propriation] in his manuscripts for the matter thought here for more than twenty-five years. This matter, albeit simple in itself, remains at first recalcitrant to thought. For thought must wean itself from the habit of lapsing into the view that here "Being" ["*das Sein*"] is being thought as propriation. Yet propriation is essentially other, other because richer than every possible metaphysical determination of Being. On the contrary, Being lets itself be thought—with a view to its essential provenance—from out of propriation.

†*Das Ereignis ereignet in seinem Er-äugen des Menschenwesens die Sterblichen dadurch*. . . . The homophony and homology of *Er-eignen/Er-äugen* is lost in translation. Once again Goethe provides the fundamental clue. Where one would expect to find *ereignen* in *Faust* (e.g. ll. 5917 and 7750) one finds instead *sich eräugnen*, containing the root *Auge*, "eye." Although the relation to *eignen*, "to own," cannot be denied, *Ereignis* also has to do with "bringing something before the eyes, showing." *Ereignis* is as much related to envisagement (Old High German *irougen*, Middle High German *eröugen*) as to enownment.—Ed.

‡Continuing the above phrase: . . . *dadurch, dass es sie dem vereignet, was sich dem Menschen in der Sage von überall her auf Verborgenes hin zu-sagt*. The verb *vereignen*, here rendered as "to remand," is a neologism whose sense is extremely difficult to hear. *Ver-* has no fewer than seven different functions as a verbal prefix in modern German. The two that seem most relevant are these: *vereignen* could either be an enhancement and intensification of *eignen* or a negation, distortion, or transformation of it. That *concealment* here becomes telling somehow suggests *both* enhancement and negation of owning and eyeing. The reflexive *sich zu-sagen*, here rendered as "*telling*," more literally suggests that in the saying things are "said to" man, affirmed (*Zusage* means "acceptance"), precisely as concealed.—Ed.

that human beings—the ones who speak, and that means, the ones who say—go to encounter the saying; indeed, encounter it on the basis of what is proper to it. The latter is the sounding of the word. When mortals say, and thus encounter, they respond. Every spoken word is already a response—a reply, a saying that goes to encounter, and listens. The remanding of mortals to the saying releases the essence of man to that usage by which man is needed—needed in order to bring the soundless saying into the resonance of language.

In the remanding to usage, propriation lets the saying arrive at speech. The way to language pertains to the saying that is determined by propriation. On this way, which pertains to the essence of language, what is peculiar to language conceals itself. The way is propriating.

To clear a way—for instance, across a snowfield—is still today in the Alemannic-Swabian dialect called *wëgen* [literally, "waying"]. This transitive verb suggests creating a way, giving shape to it and keeping it in shape. *Be-wëgen* (*Be-wëgung*) [cf. *bewegen*, *Bewegung*, to move, motion], thought in this way, no longer means merely transporting something on a way that is already at hand; rather, it means rendering the way to . . . in the first place, thus being the way.

Propriation propriates human beings for itself, propriates them into usage. Propriating showing as owning, propriation is thus the saying's way-making movement toward language.

Such way-making brings language (the essence of language) as language (the saying) to language (to the resounding word). Our talk concerning the way to language no longer means exclusively or even preeminently the course of our thought on the trail of language. While under way, the way to language has transformed itself. It has transposed itself from being some deed of ours to the propriated essence of language. Except that the transformation of the way to language looks like a transposition that has just now been effected only for us, only with respect to us. In truth, the way to language has its sole place always already in the essence of language itself. However, this suggests at the same time that the way to language as we first intended it is not superfluous; it is simply that it becomes possible and necessary only by virtue of the way proper, the way-making movement of propriation and usage. Because the essence of language, as the saying that shows, rests on the propriation that delivers us human beings over to releasement toward unconstrained hearing, the saying's way-making movement toward speech first opens up the path on which we can follow the trail of the proper way to language.

Our path's formula—*to bring language as language to language*—no longer merely encapsulates a directive for us who ponder over language. Rather, it betells the *forma*, the configuration of the well-enjoined structure within which the essence of language, which rests on propriation, makes its way.

If we do not think about it, but merely string along with the string of words, then the formula expresses a weft of relations in which language simply entangles itself. It seems as though every attempt to represent language needs the learned knack of dialectic in order to master the tangle. However, such a procedure, which the formula formidably provokes, bypasses the possibility that by remaining on the trail—that is to say, by letting ourselves be guided expressly into the way-making movement—we may yet catch a glimpse of the essence of language in all its simplicity, instead of wanting to represent language.

What looks more like a tangle than a weft loosens when viewed in terms of the way-making movement. It resolves into the liberating motion that the way-making movement exhibits when propriated in the saying. It unbinds the saying for speech. It holds open the way for speech, the way on which speaking as hearing, hearing the saying, registers what in each case is to be said, elevating what it receives to the resounding word. The saying's way-making movement to language is the unbinding bond, the bond that binds by propriating.

Thus freed to its own open space, language can concern itself solely with itself alone. That resembles the talk one hears about egoistic solipsism. Yet language does not insist on itself, is not a self-mirroring that forgets everything else because it is so enamored of itself. As the saying, the essence of language is the propriating showing that in fact disregards itself in order to liberate what is shown into its own, into its appearance.

Language, which speaks by saying, is concerned that our speech, heeding the unspoken, corresponds to what language says. Hence silence too, which one would dearly like to subtend to speech as its origin, is already a corresponding.* Silence corresponds to the noiseless ringing of stillness, the stillness of the saying that propriates and shows. The saying that rests on propriation is, as showing, the most proper mode of propriating. Propriation is telling [*sagend*]. Accordingly, language speaks after the manner of the given mode in which propriation reveals itself as such or withdraws. A thinking that thinks back to *propriation* can just barely surmise it, and yet can already experience it in the *essence* of modern technology, an essence given the still odd-sounding name *Ge-Stell* ["enframing"].† The enframing, because it sets upon human beings—that is, challenges them—to order everything that comes to presence into a technical inventory, unfolds essentially after the manner of propriation; at the same time, it distorts propriation, inasmuch as all ordering sees itself committed to calculative thinking and so speaks the language of enframing. Speech is challenged to correspond to the ubiquitous orderability of what is present.

Speech, when posed in this fashion, becomes information.‡ It informs itself concerning itself, in order to establish securely, by means of information theories, its own procedure. Enframing, the essence of modern technology that holds sway everywhere, ordains

*See *Being and Time*, 1927, section 34. [This section, "Dasein and Discourse; Language," in fact argued strongly that speech, talk, or discourse is "grounded in" silence, so that silence—not speech—is primordial. That thesis is not dropped here, but altered: not silence as such but *Ent-sprechen*, a corresponding that is quite literally an "un-speaking," is the focal point of "The Way to Language."—ED.]

†See *Vorträge und Aufsätze*, 1954, pp. 31–32. [In these *Basic Writings*, see Reading VII, esp. pp. 324–28, including the explanatory note.—ED.]

‡See *Hebel—Friend of the Household*, (Pfullingen: G. Neske, 1957), pp. 34ff.

for itself a formalized language—that kind of informing by virtue of which man is molded and adjusted into the technical-calculative creature, a process by which step-by-step he surrenders his "natural language." Even when information theory has to concede that formalized language must again and again revert to "natural language," in order by means of nonformalized language to bring to language what the technological inventory has to say, this happenstance represents—according to the current self-interpretation of information theory—merely a transitional stage. For the "natural language" that perforce must be invoked here is posited from the outset as a language that, while not yet formalized, has already been ordained to formalization. Formalization, the calculative orderability of saying, is the goal and the standard. What is "natural" in language, whose existence the will to formalization finds itself compelled as it were to concede for the time being, is not experienced with a view to the originary nature of language. Such a nature is *physis*, which in turn rests on propriation, out of which the saying bestirs itself and surges upward. Information theory conceives of the natural as a shortfall in formalization.

Yet even if a long path should lead us to the insight that the essence of language can never be dissolved into a formalism and then tabulated as such; even if we should accordingly have to say that "natural language" is *not* formaliz*able* language; even then "natural language" would still be defined purely negatively; that is to say, against the backdrop of the possibility or impossibility of formalization.

However, what if "natural language," which for information theory remains but a disturbing remnant, drew its nature—that is, the essential unfolding of the essence of language—from the saying? What if the saying, instead of merely disturbing the devastation that is information, had already surpassed information on the basis of a propriation that is not subject to our ordering? What if propriation—when and how, no one knows—were to become a *penetrating gaze* [Ein-Blick], whose clearing lightning strikes what is and what the being is held to be? What if propriation by its entry withdrew every present being that is subject to sheer orderability and brought that being back into its own?

Every language that human beings possess propriates in the saying. Every language is, as such, in the strict sense of the word, language proper, allowing for variations in the measure of its nearness to propriation. Every proper language, because it is allotted to human beings through the way-making movement of the saying, is sent, hence fateful.

There is no such thing as a natural language, a language that would be the language of a human nature at hand in itself and without its own destiny. Every language is historical, also in cases where human beings know nothing of the discipline of history in the modern European sense. Nor is language as information *the sole* language in itself. Rather, it is historical in the sense of, and written within the limits set by, the current age. Our age begins nothing new, but only brings to utter culmination something quite old, something already prescribed in modernity.

What is peculiar to language depends on the propriative provenance of the word; that is, on the provenance of human speech from the saying.

Let us at the end remember as we did at the outset these words of Novalis: "Precisely what is peculiar to language—that it concerns itself purely with itself alone—no one knows." Novalis understands the word *peculiar* in the sense of the particularity that makes language exceptional. Through the experience of the essence of language as the saying, a saying whose showing rests on propriation, what is *peculiar* [*das* Eigen*tümliche*] comes into the proximity of *owning* [Eignen] and *propriating* [Ereignen]. There the peculiar receives its birth certificate, as it were; but this is not the place for us to think back to the primordial determination of such peculiarity.

The peculiar character of language, which is determined on the basis of propriation, lets itself be known even less than the particularity of language, if "knowing" means having by circumspection seen something in the entirety of its essence. The essence of language does not submit to our circumspection, inasmuch as we—we who can say only by reiterating the saying—ourselves belong within the saying. The monological character of the essence of language has its well-joined structure in the rift-design of the saying. The rift-design does not and cannot coincide with the *Monologue* that Novalis was thinking of, because he represents language dialectically in terms of subjectivity and within the purview of absolute idealism.

Yet language *is* monologue. This now says something twofold: it is language *alone* that properly speaks; and it speaks *in solitude*. Yet only one who is *not* alone can be solitary; not alone, that is to say, not in separation and isolation, not devoid of all kinship. On the contrary, precisely in the solitary [*Im Einsamen*] there unfolds essentially the lack of what is in common [*der Fehl des Gemeinsamen*], as the most binding relation *to* what is in common. The suffix *-sam* is the Gothic *sama*, the Greek *hama*. *Einsam* suggests the selfsame, in the unifying of things that belong to one another. The saying that shows opens the way for language to the speech of human beings. The saying needs to resound in the word. Yet man can speak only by listening to the saying, belonging to it; only by means of reiteration is he able to say a word. Such needing and reiterating rest on that lack mentioned above, which is neither a mere shortcoming nor anything negative at all.

We human beings, in order to be who we are, remain within the essence of language to which we have been granted entry. We can therefore never step outside it in order to look it over circumspectly from some alternative position. Because of this, we catch a glimpse of the essence of language only to the extent that we ourselves are envisaged by it, remanded to it. That we cannot know the essence of language—according to the traditional concept of knowledge, defined in terms of cognition as representation—is certainly not a defect; it is rather the advantage by which we advance to an exceptional realm, the realm in which we dwell as the *mortals*, those who are needed and used for the speaking of language.

The saying will not allow itself to be captured in any assertion. It demands of us a telling silence as regards the propriative, way-making movement in the essence of language, without any talk *about* silence.

The saying that rests on propriation is, as showing, the most proper mode of propriating. That sounds like an assertion. If we hear only that, it does not say what is to be thought. The saying is the mode in which propriation speaks. Yet mode is meant here not so much in the sense of *modus* or "kind"; it is meant in the musical sense of the *melos*, the song that says by singing. For the saying that propriates brings what comes to presence out of its propriety to a kind of radiance; it lauds what comes to presence; that is, allows it in its own essential unfolding. At the beginning of the eighth stanza of *Friedensfeier* ["The Celebration of Peace"], Hölderlin sings as follows:

> Viel hat von Morgen an,
> Seit ein Gespräch wir sind und hören voneinander,
> Erfahren der Mensch; bald sind aber Gesang (wir).
>
> Much, from morning onward,
> Since we became a conversation and hear from one another,
> Have human beings undergone; but soon (we) will be song.

Language was once called the "house of Being."* It is the guardian of presencing, inasmuch as the latter's radiance remains entrusted to the propriative showing of the saying. Language is the house of Being because, as the saying, it is propriation's mode.

In order to think back to the essence of language, in order to reiterate what is its own, we need a transformation of language, a transformation we can neither compel nor concoct. The transformation does not result from the fabrication of neologisms and novel

*See "Letter of Humanism," 1947. [In *Wegmarken*, 1967, see pp. 188–89; in these *Basic Writings*, see Reading V, p. 223.—Ed.]

phrases. The transformation touches on our relation to language. That relation is determined in accordance with the sending that determines whether and in what way we are embraced in propriation by the essence of language, which is the original pronouncement of propriation. For propriation—owning, holding, keeping to itself—is the relation of all relations. For this reason, *our* saying, as answering, constantly remains relational. The relation [*Das Verhältnis*, literally, our "being held"] is here thought always and everywhere in terms of propriation, and is no longer represented in the form of a mere relationship. Our relation to language is defined by the mode according to which we belong to propriation, we who are needed and used by it.

Perhaps we can in some slight measure prepare for the transformation in our kinship with language. The following experience might awaken: Every thinking that is on the trail of something is a poetizing, and all poetry a thinking. Each coheres with the other on the basis of the saying that has already pledged itself to the unsaid, the saying whose thinking is a thanking.

That the possibility of an appropriate transformation of language emerged in the complex of Wilhelm von Humboldt's thought receives eloquent testimony in his treatise, *On the Diversity of the Structure of Human Language.* Wilhelm von Humboldt worked on this treatise, as his brother writes in the Preface, "in solitude, in nearness to a *grave*," until his death. Wilhelm von Humboldt, whose deeply dark insight into the essence of language should never cease to astonish us, says:

> The *application* of already available phonetic forms to the inner purposes of the language . . . may be considered a possibility during the central periods of *language formation*. A people could through inner illumination and propitious external circumstances devise such a different form for the language it has inherited that it would thereby become a wholly different language, a new language.
>
> (Section 10, p. 84)

In a later passage we find the following:

> Without changing the language phonetically, much less changing its forms and laws, *time* often introduces into it an enhanced power of thought and a more penetrating sensibility than it possessed hitherto, and it does so through the burgeoning development of ideas. It is as though a variant sense occupies the old husk, something different is given in the unaltered coinage, and a differently scaled sequence of ideas is intimated according to unchanged syntactical laws. Here we have one of the bounteous fruits of a people's *literature*, and, preeminent in this domain, their *poetry* and *philosophy*.
>
> (Section 11, p. 100)

# XI

# THE END OF PHILOSOPHY AND THE TASK OF THINKING

*We may venture the step back*
*out of philosophy into the*
*thinking of Being as soon as*
*we have grown familiar with*
*the provenance of thinking.*

The title is provocative. It wants to provoke an "immanent criticism" of *Being and Time,* composed some forty years earlier, which is to say, to inquire into the "basic experience" underlying that book and the aptness of its "formulations" without abandoning the perspective of the *question* of Being. Heidegger has exercised such criticism before, for example in his "Letter on Humanism" (Reading V), and in fact has done so continually since 1927. As a result of this latest reappraisal the key words of Heidegger's project change. Instead of "Being and Time" (*Sein und Zeit*) he now speaks of "Clearing and Presence" (*Lichtung und Anwesenheit*). (Readers should recall that the word *Lichtung,* although cognate with "lighting," has been translated throughout as "clearing.") But Heidegger's alteration is not so much a change in terminology as a transformation of thinking. To what extent this transformation is already envisaged in earlier texts, for example in section 44 of *Being and Time* or in "On the Essence of Truth" (Reading III), is an arresting question.

In the French edition of this essay, the "end" of philosophy is translated as *achèvement*. In the *Vollendung* of philosophy Heidegger accentuates the "full" rather than the "ending" by analyzing the full consequences of the dissolution of philosophy into the specialized sciences. The completion of philosophy, the most extreme possibility or "place" for metaphysics, is a world civilization based on the Western technological model. This model is the Platonic *idea* ostensibly drained of all ontological content and become a mere cipher, a monadic carrier of information, a unit of cybernetic science. In the present essay, which appears here in its entirety, Heidegger asks whether a kind of thinking different from the calculative sort, a reflection that is neither scientific nor metaphysical, is possible. Against the background of the Hegelian and Husserlian phenomenologies Heidegger recounts clearly and decisively what his own thinking wants to accomplish. Neither a "system of science" grounded in the absolute identity-within-difference of substance and subject, nor a "rigorous science" that appeals to an incorrigible source of ultimate evidence, but something less grand and less influential is the matter for whose sake Heidegger thinks and writes.

Goethe's *Urphänomen* or primal phenomenon—that beings become present—provides a clue in this respect. Heidegger invites thought on the free or open space where things appear, linger, endure, and disappear. He calls this *die Lichtung des Seins,* the clearing of Being. In colloquial German, *eine Lichtung* has the sense of a forest "clearing" where the pines have been thinned out and the woods made "lighter," more "open." With the word *Lichtung* Heidegger wants to designate that unencumbered place for the presencing (*Anwesen,* Being) of things. Metaphysics, which stresses the "natural light" of the thinking subject who casts his beam on "objects," has not attended to the clearing or lighting of Being, the opening that precedes all natural and divine light. Such attendance Heidegger names "the task of thinking." It requires a creative return to early Greek thinking—creative because even the Greeks did not secure the clearing for thought and save it from oblivion.

Heidegger questions the early words of Parmenides regarding "well-rounded *alētheia,*" unconcealment thought as the *Lichtung* of presence. He now declines to translate *alētheia* as "truth." Citing a passage from *Being and Time* (section 44) that had already sketched the salient features of *alētheia,* Heidegger criticizes his later use of such expressions as "the truth of Being." (He often used this phrase in the 1940s: see for example the "Letter on Humanism," above.) Note that this criticism has nothing to do with Heidegger's "turn" as it is normally interpreted. Indeed Heidegger is here turning away from certain aspects of his post–*Being and Time* writings *toward* the initial project and insights of *Being and Time* itself. Hence the task of thinking at the end of philosophy, at least so far as Heidegger's own career is concerned, is to deepen meditation "On the Essence of Truth" in such a way that this title too would have to change.

That for the sake of which thought gets under way is the *Lichtung* or clearing in which beings come to presence. Thought must pursue the mystery of this clearing: the need of unconcealment for self-concealing; the need of self-showing or upsurgence for reticence or hiding; the need of gathering for sheltering. Most mysterious is the reciprocal play of *Lēthē* and *Alētheia* in the clearing. Whatever the origins of that insatiable need for self-concealment, it is essential that at the end of philosophy—no matter how that "end" may be understood, whether as the achievement of absolute knowing or science (Hegel), the consummation of nihilism (Nietzsche), the closure of the metaphysics of presence and/or the foundering of every apocalyptic invocation of "ends" (Derrida)—our thinking remember the task Heraclitus and Parmenides assigned it: *to protect* the interplay of unconcealment and concealment in the *Lichtung des Seins*. Such protection Socrates called "wonder," whose daughter is iridescent speech (*Theaetetus* 155 d, *Cratylus* 408 b).

# THE END OF PHILOSOPHY AND THE TASK OF THINKING

The title designates the attempt at a reflection that persists in questioning. Questions are paths toward an answer. If the answer could be given it would consist in a transformation of thinking, not in a propositional statement about a matter at stake.

The following text belongs to a larger context. It is the attempt undertaken again and again ever since 1930 to shape the question of *Being and Time* in a more primordial fashion. This means to subject the point of departure of the question in *Being and Time* to an immanent criticism. Thus it must become clear to what extent the *critical* question as to what the matter of thinking is necessarily and continually belongs to thinking. Accordingly, the name of the task of *Being and Time* will change.

We are asking:

1. To what extent has philosophy in the present age entered into its end?
2. What task is reserved for thinking at the end of philosophy?

---

Martin Heidegger, "The End of Philosophy and the Task of Thinking," appears in Martin Heidegger, *On Time and Being*, translated by Joan Stambaugh (New York: Harper & Row, 1972), pp. 55–73. The essay first appeared in a French translation by Jean Beaufret and François Fédier in *Kierkegaard vivant* (Paris: Gallimard, 1966). The German text appears in Martin Heidegger, *Zur Sache des Denkens* (Tübingen: Max Niemeyer Verlag, 1969), pp. 61–80. I have altered the translation slightly here.

## I
## To what extent has philosophy in the present age entered into its end?

Philosophy is metaphysics. Metaphysics thinks beings as a whole—the world, man, God—with respect to Being, with respect to the belonging together of beings in Being. Metaphysics thinks beings as beings in the manner of a representational thinking that gives grounds. For since the beginning of philosophy, and with that beginning, the Being of beings has shown itself as the ground (*archē*, *aition*, principle). The ground is that from which beings as such are what they are in their becoming, perishing, and persisting as something that can be known, handled, and worked upon. As the ground, Being brings beings in each case to presencing. The ground shows itself as presence. The present of presence consists in the fact that it brings what is present each in its own way to presence. In accordance with the given type of presence, the ground has the character of grounding as the ontic causation of the actual, the transcendental making possible of the objectivity of objects, the dialectical mediation of the movement of absolute spirit and of the historical process of production, and the will to power positing values.

What characterizes metaphysical thinking, which seeks out the ground for beings, is the fact that metaphysical thinking, starting from what is present, represents it in its presence and thus exhibits it as grounded by its ground.

What is meant by the talk about the end of philosophy? We understand the end of something all too easily in the negative sense as mere cessation, as the lack of continuation, perhaps even as decline and impotence. In contrast, what we say about the end of philosophy means the completion of metaphysics. However, completion does not mean perfection, as a consequence of which philosophy would have to have attained the highest perfection at its end. Not only do we lack any criterion that would permit us to evaluate the perfection of an epoch of metaphysics as compared with any other epoch; the right to this kind of evaluation does not exist. Plato's thinking is no more perfect than Parmenides'. Hegel's philosophy is no more perfect than Kant's. Each epoch of philosophy has its own necessity. We simply have to acknowledge the fact that a philosophy is the way it is. It is not for us to prefer one to the other, as can be the case with regard to various *Weltanschauungen*.

The old meaning of the word "end" means the same as place: "from one end to the other" means from one place to the other. The end of philosophy is the place, that place in which the whole of philosophy's history is gathered in its uttermost possibility. End as completion means this gathering.

Throughout the entire history of philosophy, Plato's thinking remains decisive in its sundry forms. Metaphysics is Platonism. Nietzsche characterizes his philosophy as reversed Platonism. With the reversal of metaphysics that was already accomplished by Karl Marx, the uttermost possibility of philosophy is attained. It has entered into its end. To the extent that philosophical thinking is still attempted, it manages only to attain an epigonal renaissance and variations of that renaissance. Is not then the end of philosophy after all a cessation of its way of thinking? To conclude this would be premature.

As a completion, an end is the gathering into the uttermost possibilities. We think in too limited a fashion as long as we expect only a development of new philosophies in the previous style. We forget that already in the age of Greek philosophy a decisive characteristic of philosophy appears: the development of the sciences within the field that philosophy opened up. The development of the sciences is at the same time their separation from philosophy and the establishment of their independence. This process belongs to the completion of philosophy. Its development is in full swing today in all regions of beings. This development looks like the mere dissolution of philosophy, yet in truth is precisely its completion.

It suffices to refer to the independence of psychology, sociology, anthropology as cultural anthropology, or to the role of logic as symbolic logic and semantics. Philosophy turns into the empirical science of man, of all that can become for man the experiential object of his technology, the technology by which he establishes himself in the world by working on it in the manifold modes of making and shaping. All of this happens everywhere on the basis of and according to the criterion of the scientific discovery of the individual areas of beings.

No prophecy is necessary to recognize that the sciences now establishing themselves will soon be determined and regulated by the new fundamental science that is called cybernetics.

This science corresponds to the determination of man as an acting social being. For it is the theory of the regulation of the possible planning and arrangement of human labor. Cybernetics transforms language into an exchange of news. The arts become regulated-regulating instruments of information.

The development of philosophy into the independent sciences that, however, interdependently communicate among themselves ever more markedly, is the legitimate completion of philosophy. Philosophy is ending in the present age. It has found its place in the scientific attitude of socially active humanity. But the fundamental characteristic of this scientific attitude is its cybernetic, that is, technological character. The need to ask about modern technology is presumably dying out to the same extent that technology more decisively characterizes and directs the appearance of the totality of the world and the position of man in it.

The sciences will interpret everything in their structure that is still reminiscent of their provenance from philosophy in accordance with the rules of science, that is, technologically. Every science understands the categories upon which it remains dependent for the articulation and delineation of its area of investigation as working hypotheses. Not only is their truth measured in terms of the effect that their application brings about within the progress of research, scientific truth is also equated with the efficiency of these effects.

The sciences are now taking over as their own task what philosophy in the course of its history tried to present in certain places, and even there only inadequately, that is, the ontologies of the various regions of beings (nature, history, law, art). The interest of the sciences is directed toward the theory of the necessary structural concepts of the coordinated areas of investigation. "Theory" means now supposition of the categories, which are allowed only a cybernetic function, but denied any ontological meaning. The operational and model-based character of representational-calculative thinking becomes dominant.

However, the sciences still speak about the Being of beings in the unavoidable supposition of their regional categories. They only do not say so. They can deny their provenance from philosophy, but never dispense with it. For in the scientific attitude of the sciences the certification of their birth from philosophy still speaks.

The end of philosophy proves to be the triumph of the manipulable arrangement of a scientific-technological world and of the social order proper to this world. The end of philosophy means the beginning of the world civilization that is based upon Western European thinking.

But is the end of philosophy in the sense of its evolving into the sciences also already the complete actualization of all the possibilities in which the thinking of philosophy was posited? Or is there a *first* possibility for thinking apart from the *last* possibility that we characterized (the dissolution of philosophy in the technologized sciences), a possibility from which the thinking of philosophy would have to start, but which as philosophy it could nevertheless not expressly experience and adopt?

If this were the case, then a task would still have to be reserved for thinking in a concealed way in the history of philosophy from its beginning to its end, a task accessible neither to philosophy as metaphysics nor, even less, to the sciences stemming from philosophy. Therefore we ask:

## II
## What task is reserved for thinking at the end of philosophy?

The mere thought of such a task of thinking must sound strange to us. A thinking that can be neither metaphysics nor science?

A task that has concealed itself from philosophy since its very beginning, even in virtue of that beginning, and thus has withdrawn itself continually and increasingly in the times that followed?

A task of thinking that—so it seems—includes the assertion that philosophy has not been up to the matter of thinking and has thus become a history of mere decline?

Is there not an arrogance in these assertions which desires to put itself above the greatness of the thinkers of philosophy?

This suspicion obtrudes. But it can easily be quelled. For every attempt to gain insight into the supposed task of thinking finds itself moved to review the whole history of philosophy. Not only that. It is even forced to think the historicity of that which grants a possible history to philosophy.

Because of this, the thinking in question here necessarily falls short of the greatness of the philosophers. It is less than philosophy. Less also because the direct or indirect effect of this thinking on the public in the industrial age, formed by technology and science, is decisively less possible for this thinking than it was for philosophy.

But above all, the thinking in question remains unassuming, because its task is only of a preparatory, not of a founding character. It is content with awakening a readiness in man for a possibility whose contour remains obscure, whose coming remains uncertain.

Thinking must first learn what remains reserved and in store for it, what it is to get involved in. It prepares its own transformation in this learning.

We are thinking of the possibility that the world civilization that is just now beginning might one day overcome its technological-scientific-industrial character as the sole criterion of man's world sojourn. This may happen, not of and through itself, but in virtue of the readiness of man for a determination which, whether heeded or not, always speaks in the destiny of man, which has not yet been decided. It is just as uncertain whether world civilization will soon be abruptly destroyed or whether it will be stabilized for a long time. Such stabilization, however, will not rest in something enduring, but establish itself in a sequence of changes, each presenting the latest novelty.

The preparatory thinking in question does not wish and is not able to predict the future. It only attempts to say something to the present that was already said a long time ago, precisely at the beginning of philosophy and for that beginning, but has not been explicitly thought. For the time being, it must be sufficient to refer to this with the brevity required. We shall take a directive that philosophy offers as an aid in our undertaking.

When we ask about the task of thinking, this means in the scope of philosophy to determine that which concerns thinking, is still controversial for thinking, and is the controversy. This is what the word *Sache* [matter] means in the German language. It designates that with which thinking has to do in the case at hand, in Plato's language, *to pragma auto* (See "The Seventh Letter," 341c 7).

In recent times, philosophy has of its own accord expressly called thinking "to the things themselves." Let us mention two cases that receive particular attention today. We hear this call "to the things themselves" in the Preface that Hegel placed at the front of the work he published in 1807, *System of Science,* First Part: The Phenomenology of Spirit*. This preface is not the preface to the *Phenomenology*, but to the *System of Science*, to the whole of philosophy. The call "to

**Wissenschaft, scientia*, body of knowledge, not "science" in the present use of that word. For German Idealism, science is the name for philosophy.—Tr.

the things themselves" refers ultimately—and that means according to the matter, primarily—to the *Science of Logic*.

In the call "to the things themselves" the emphasis lies on the "themselves." Heard superficially, the call has the sense of a rejection. The inadequate relations to the matter of philosophy are rejected. Mere talk about the purpose of philosophy belongs to these relations, but so does mere reporting about the results of philosophical thinking. Neither is ever the actual whole of philosophy. The whole shows itself only in its becoming. This occurs in the developmental presentation of the matter. In the presentation, theme and method coincide. For Hegel, this identity is called the idea. With the idea, the matter of philosophy "itself" comes to appear. However, this matter is historically determined as subjectivity. With Descartes's *ego cogito*, says Hegel, philosophy steps on firm ground for the first time, where it can be at home. If the *fundamentum absolutum* is attained with the *ego cogito* as the distinctive *subiectum*, this means the subject is the *hypokeimenon* transferred to consciousness, is what truly presences; and this, vaguely enough, is called "substance" in traditional terminology.

When Hegel explains in the Preface (ed. Hoffmeister, p. 19), "The true (in philosophy) is to be understood and expressed, not as substance, but, just as much, as subject," then this means: the Being of beings, the presence of what is present, is manifest and thus complete presence only when it becomes present as such for itself in the absolute idea. But since Descartes, *idea* means *perceptio*. Being's coming to itself occurs in speculative dialectic. Only the movement of the idea, the method, is the matter itself. The call "to the thing itself" requires a philosophical method appropriate to its matter.

However, what the matter of philosophy should be is presumed to be decided from the outset. The matter of philosophy as metaphysics is the Being of beings, their presence in the form of substantiality and subjectivity.

A hundred years later, the call "to the thing itself" again is heard in Husserl's treatise *Philosophy as Rigorous Science*. It was published in the first volume of the journal *Logos* in 1910–11 (pp. 289ff.). Again, the call has at first the sense of a rejection. But here it aims in another direction than Hegel's. It concerns naturalistic psychology, which claims to be the genuine scientific method of investigating consciousness. For this method blocks access to the phenomena of intentional consciousness from the very beginning. But the call "to the thing itself" is at the same time directed against historicism, which gets lost in treatises about the standpoints of philosophy and in the ordering of types of philosophical *Weltanschauungen*. About this Husserl says in italics (ibid., p. 340): "*The stimulus for investigation must start, not with philosophies, but with issues* [Sachen] *and problems*."

And what is the matter at stake in philosophical investigation? In accordance with the same tradition, it is for Husserl as for Hegel the subjectivity of consciousness. For Husserl, the *Cartesian Meditations* were not only the topic of the Paris lectures in February of 1929. Rather, from the time following the *Logical Investigations*, their spirit accompanied the impassioned course of his philosophical investigations to the end. In its negative and also in its positive sense, the call "to the matter itself" determines the securing and elaborating of method. It also determines the procedure of philosophy, by means of which the matter itself can be demonstrated as a datum. For Husserl, "the principle of all principles" is first of all not a principle of content but one of method. In his work published in 1913, *Ideas toward a Pure Phenomenology and Phenomenological Philosophy*, Husserl devoted a special section (24) to the determination of "the principle of all principles." "No conceivable theory can upset this principle," says Husserl.

"The principle of all principles" reads:

> . . . Every originarily giving intuition [is] *a source of legitimation for knowledge; everything* that presents itself to us *in the 'Intuition' originarily* (in its bodily actuality, so to speak) [is] simply to be *accepted as it gives itself*, but also *only within the limits in which it gives itself there*. . . .

"The principle of all principles" contains the thesis of the precedence of method. This principle decides what matter alone can suffice for the method. "The principle of principles" requires absolute subjectivity as the matter of philosophy. The transcendental reduction to absolute subjectivity gives and secures the possibility of grounding the objectivity of all objects (the Being of these beings) in their valid structure and consistency, that is, in their constitution, in and through subjectivity. Thus transcendental subjectivity proves to be "the sole absolute being" (*Formal and Transcendental Logic*, 1929, p. 240). At the same time, transcendental reduction as the method of "universal science" of the constitution of the Being of beings has the same mode of Being as this absolute being, that is, the manner of the matter most native to philosophy. The method is not only directed toward the matter of philosophy. It does not merely belong to the matter as a key does to a lock. Rather, it belongs to the matter because it is "the matter itself." If one wished to ask: Where does "the principle of all principles" get its unshakable right? the answer would have to be: from transcendental subjectivity, which is already presupposed as the matter of philosophy.

We have chosen a discussion of the call "to the matter itself" as our directive. It was to bring us to the path that leads us to a determination of the task of thinking at the end of philosophy. Where are we now? We have arrived at the insight that for the call "to the matter itself" what concerns philosophy as its matter is established from the outset. From the perspective of Hegel and Husserl—and not only from their perspective—the matter of philosophy is subjectivity. It is not the matter as such that is controversial for the call, but rather the presentation by which the matter itself becomes present. Hegel's speculative dialectic is the movement in which the matter as such comes to itself, comes to its own presence [*Präsenz*]. Husserl's method is supposed to bring the matter of philosophy to its ultimate originary givenness, and that means to its own presence [*Präsenz*].

The two methods are as different as they could possibly be. But the matter that they are to present as such is the same, although it is experienced in different ways.

But of what help are these discoveries to us in our attempt to bring the task of thinking to view? They do not help us at all as long as we do not go beyond a mere discussion of the call. Rather, we must ask what remains unthought in the call "to the matter itself." Questioning in this way, we can become aware that something that it is no longer the matter of philosophy to think conceals itself precisely where philosophy has brought its matter to absolute knowledge and to ultimate evidence.

But what remains unthought in the matter of philosophy as well as in its method? Speculative dialectic is a mode in which the matter of philosophy comes to appear of itself and for itself, and thus becomes present [*Gegenwart*]. Such appearance necessarily occurs in luminosity. Only by virtue of some sort of brightness can what shines show itself, that is, radiate. But brightness in its turn rests upon something open, something free, which it might illuminate here and there, now and then. Brightness plays in the open and strives there with darkness. Wherever a present being encounters another present being or even only lingers near it—but also where, as with Hegel, one being mirrors itself in another speculatively—there openness already rules, the free region is in play. Only this openness grants to the movement of speculative thinking the passage through what it thinks.

We call this openness that grants a possible letting appear and show "clearing." In the history of language the German word *Lichtung* is a translation derived from the French *clairière*. It is formed in accordance with the older words *Waldung* [foresting] and *Feldung* [fielding].

The forest clearing [*Lichtung*] is experienced in contrast to dense forest, called *Dickung* in our older language. The substantive *Lichtung* goes back to the verb *lichten*. The adjective *licht* is the same word as "light." To lighten something means to make it light, free and open, e.g., to make the forest free of trees at one place. The free space thus originating is the clearing. What is light in the sense of being free and open has nothing in common with the adjective "light" which means "bright," neither linguistically nor materially. This is to be observed for the difference between clearing and light.* Still, it is possible that a material relation between the two exists. Light can stream into the clearing, into its openness, and let brightness play with darkness in it. But light never first creates the clearing. Rather, light presupposes it. However, the clearing, the open region, is not only free for brightness and darkness but also for resonance and echo, for sound and the diminishing of sound. The clearing is the open region for everything that becomes present and absent.

It is necessary for thinking to become explicitly aware of the matter here called clearing. We are not extracting mere notions from mere words, e.g., *Lichtung*, as it might easily appear on the surface. Rather, we must observe the unique matter that is named with the name "clearing" in accordance with the matter. What the word designates in the connection we are now thinking, free openness, is a "primal phenomenon" [*Urphänomen*], to use a word of Goethe's. We would have to say a "primal matter" [*Ursache*]. Goethe notes (*Maxims and Reflections*, no. 993): "Look for nothing behind phenomena: they themselves are what is to be learned." This means the phenomenon itself, in the present case the clearing, sets us the task of learning from it while questioning it, that is, of letting it say something to us.

Accordingly, we may suggest that the day will come when we will not shun the question whether the clearing, free openness, may not be that within which alone pure space and ecstatic time and

*"Light" is also two adjectives in English, each having its own origin. "Light" in the sense of having little weight derives from the Sanskrit *laghu* and the Greek *elaphros*, *elachus* (slight, small); in the sense "bright, shining, luminous" it derives from the Indo-Germanic *leuk-* (white) and Sanskrit *ruc* (to shine). Yet already in Old English, though not yet in Old High German, the words take the same form; during the history of both languages they increasingly converge. The verb *lichten*, "to lighten," also has two senses: to illuminate and to alleviate. Heidegger emphasizes the less familiar second sense—to make less dense and heavy, for example, to lighten a ship by dispatching "lighters" to it to relieve it of cargo—see Whitman, "Crossing Brooklyn Ferry," lines 47–48 and 92.—Ed.

everything present and absent in them have the place that gathers and protects everything.

In the same way as speculative dialectical thinking, originary intuition and its evidence remain dependent upon openness that already holds sway, the clearing. What is evident is what can be immediately intuited. *Evidentia* is the word that Cicero uses to translate the Greek *enargeia*, that is, to transform it into the Roman. *Enargeia*, which has the same root as *argentum* (silver), means that which in itself and of itself radiates and brings itself to light. In the Greek language, one is not speaking about the action of seeing, about *vidēre*, but about that which gleams and radiates. But it can radiate only if openness has already been granted. The beam of light does not first create the clearing, openness, it only traverses it. It is only such openness that grants to giving and receiving and to any evidence at all the free space in which they can remain and must move.

All philosophical thinking that explicitly or inexplicitly follows the call "to the matter itself" is in its movement and with its method already admitted to the free space of the clearing. But philosophy knows nothing of the clearing. Philosophy does speak about the light of reason, but does not heed the clearing of Being. The *lumen naturale*, the light of reason, throws light only on the open. It does concern the clearing, but so little does it form it that it needs it in order to be able to illuminate what is present in the clearing. This is true not only of philosophy's *method*, but also and primarily of its *matter*, that is, of the presence of what is present. To what extent the *subiectum*, the *hypokeimenon*, that which already lies present, thus what is present in its presence is constantly thought also in subjectivity, cannot be shown here in detail. (Refer to Heidegger, *Nietzsche*, vol. 2 [1961], pages 429ff.)*

We are concerned now with something else. Whether or not what is present is experienced, comprehended, or presented, presence as

*This material appears in English in Martin Heidegger, *The End of Philosophy*, trans. Joan Stambaugh (New York: Harper & Row, 1973), pp. 26ff.—Ed.

lingering in the open always remains dependent upon the prevalent clearing. What is absent, too, cannot be as such unless it presences in the *free space of the clearing.*

All metaphysics, including its opponent, positivism, speaks the language of Plato. The basic word of its thinking, that is, of its presentation of the Being of beings, is *eidos, idea:* the outward appearance in which beings as such show themselves. Outward appearance, however, is a manner of presence. No outward appearance without light—Plato already knew this. But there is no light and no brightness without the clearing. Even darkness needs it. How else could we happen into darkness and wander through it? Still, the clearing as such as it prevails through Being, through presence, remains unthought in philosophy, although it is spoken about in philosophy's beginning. Where does this occur and with which names? Answer:

In Parmenides' thoughtful poem which, as far as we know, was the first to reflect explicitly upon the Being of beings, which still today, although unheard, speaks in the sciences into which philosophy dissolves. Parmenides listens to the claim:

> . . . χρεὼ δέ σε πάντα πυθέσθαι
> ἠμὲν Ἀληθείης εὐκυκλέος ἀτρεμὲς ἦτορ
> ἠδὲ βροτῶν δόξας, ταῖς οὐκ ἔνι πίστις ἀληθής.
>
> Fragment I, 28ff.

> . . . but you should learn all:
> the untrembling heart of unconcealment, well-rounded,
> and also the opinions of mortals
> who lack the ability to trust what is unconcealed.

*Alētheia,* unconcealment, is named here. It is called well-rounded because it is turned in the pure sphere of the circle in which beginning and end are everywhere the same. In this turning there is no possibility of twisting, distortion, and closure. The meditative man is to experience the untrembling heart of unconcealment. What does the phrase about the untrembling heart of unconcealment mean? It means unconcealment itself in what is most its own, means the place of stillness that gathers in itself what first grants unconcealment. That is the clearing of what is open. We ask: openness for what? We have already reflected upon the fact that the path of thinking, speculative and intuitive, needs the traversable clearing. But in that clearing rests possible radiance, that is, the possible presencing of presence itself.

What prior to everything else first grants unconcealment is the path on which thinking pursues one thing and perceives it: *hopōs estin . . . einai:* that presencing presences. The clearing grants first of all the possibility of the path to presence, and grants the possible presencing of that presence itself. We must think *alētheia*, unconcealment, as the clearing that first grants Being and thinking and their presencing to and for each other. The quiet heart of the clearing is the place of stillness from which alone the possibility of the belonging together of Being and thinking, that is, presence and apprehending, can arise at all.

The possible claim to a binding character or commitment of thinking is grounded in this bond. Without the preceding experience of *alētheia* as the clearing, all talk about committed and noncommitted thinking remains without foundation. Whence does Plato's determination of presence as *idea* have its binding character? With regard to what is Aristotle's interpretation of presencing as *energeia* binding?

Strangely enough, we cannot even ask these questions, always neglected in philosophy, as long as we have not experienced what Parmenides had to experience: *alētheia*, unconcealment. The path to it is distinguished from the lane along which the opinion of mortals wanders. *Alētheia* is nothing mortal, just as little as death itself.

It is not for the sake of etymology that I stubbornly translate the name *alētheia* as unconcealment, but for the sake of the matter that must be considered when we think adequately that which is called Being and thinking. Unconcealment is, so to speak, the element in which Being and thinking and their belonging together exist. *Alētheia* is named at the beginning of philosophy, but afterward it is not explicitly thought as such by philosophy. For since Aristotle it has become the task of philosophy as metaphysics to think beings as such ontotheologically.

If this is so, we have no right to sit in judgment over philosophy, as though it left something unheeded, neglected it and was thus marred by some essential deficiency. The reference to what is unthought in philosophy is not a criticism of philosophy. If a criticism is necessary now, then it rather concerns the attempt, which is becoming more and more urgent ever since *Being and Time*, to ask about a possible task of thinking at the end of philosophy. For the question now arises, late enough: Why is *alētheia* not translated with the usual name, with the word "truth"? The answer must be:

Insofar as truth is understood in the traditional "natural" sense as the correspondence of knowledge with beings, demonstrated in beings; but also insofar as truth is interpreted as the certainty of the knowledge of Being; *alētheia*, unconcealment in the sense of the clearing, may not be equated with truth. Rather, *alētheia*, unconcealment thought as clearing, first grants the possibility of truth. For truth itself, like Being and thinking, can be what it is only in the element of the clearing. Evidence, certainty in every degree, every kind of verification of *veritas*, already moves *with* that *veritas* in the realm of the clearing that holds sway.

*Alētheia*, unconcealment thought as the clearing of presence, is not yet truth. Is *alētheia* then less than truth? Or is it more, because it first grants truth as *adaequatio* and *certitudo*, because there can be no presence and presenting outside the realm of the clearing?

This question we leave to thinking as a task. Thinking must consider whether it can even raise this question at all as long as it thinks philosophically, that is, in the strict sense of metaphysics, which questions what is present only with regard to its presence.

In any case, one thing becomes clear: to raise the question of *alētheia*, of unconcealment as such, is not the same as raising the question of truth. For this reason, it was immaterial and therefore misleading to call *alētheia*, in the sense of clearing, "truth."[1] The talk about the "truth of Being" has a justified meaning in Hegel's *Science of Logic*, because here truth means the certainty of absolute knowledge. And yet Hegel, as little as Husserl, as little as all metaphysics, does not ask about Being as Being, that is, does not raise the question as to how there can be presence as such. There is presence only when clearing holds sway. Clearing is named with *alētheia*, unconcealment, but not thought as such.

The natural concept of truth does not mean unconcealment, not in the philosophy of the Greeks either. It is often and justifiably pointed out that the word *alēthes* is already used by Homer only in the *verba dicendi*, in statements, thus in the sense of correctness and reliability, not in the sense of unconcealment. But this reference means only that neither the poets nor everyday linguistic usage, nor even philosophy, see themselves confronted with the task of asking how truth, that is, the correctness of statements, is granted only in the element of the clearing of presence.

In the scope of this question, we must acknowledge the fact that *alētheia*, unconcealment in the sense of the clearing of presence, was originally experienced only as *orthotēs*, as the correctness of representations and statements. But then the assertion about the essential transformation of truth, that is, from unconcealment to correctness, is also untenable. Instead we must say: *alētheia*, as clearing of presence and presentation in thinking and saying, immediately comes under the perspective of *homoiōsis* and *adaequatio*, that is, the perspective of adequation in the sense of the correspondence of representing with what is present.

But this process inevitably provokes another question: How is it that *alētheia*, unconcealment, appears to man's natural experience and

1. How the attempt to think a matter can for a time stray from what a decisive insight has already shown is demonstrated by a passage from *Being and Time*, 1927 (p. 219): "The translation [of the word *alētheia*] by means of the word 'truth,' and even the very theoretical-conceptual determinations of this expression [truth], cover up the meaning of what the Greeks established as basically 'self-evident' in the prephilosophical understanding of their terminological employment of *alētheia*."

speech *only* as correctness and dependability? Is it because man's ecstatic sojourn in the openness of presencing is turned only toward what is present and the presentation of what is present? But what else does this mean than that presence as such, and together with it the clearing that grants it, remains unheeded? Only what *alētheia* as clearing grants is experienced and thought, not what it is as such.

This remains concealed. Does that happen by chance? Does it happen only as a consequence of the carelessness of human thinking? Or does it happen because self-concealing, concealment, *lēthē*, belongs to *a-lētheia*, not as a mere addition, not as shadow to light, but rather as the heart of *alētheia*? Moreover, does not a sheltering and preserving rule in this self-concealing of the clearing of presence, from which alone unconcealment can be granted, so that what is present can appear in its presence?

If this were so, then the clearing would not be the mere clearing of presence, but the clearing of presence concealing itself, the clearing of a self-concealing sheltering.

If this were so, then only with these questions would we reach the path to the task of thinking at the end of philosophy.

But is not all this unfounded mysticism or even bad mythology, in any case a ruinous irrationalism, the denial of *ratio*?

I ask in return: What does *ratio*, *nous*, *noein*, apprehending, mean? What do ground and principle and especially principle of all principles mean? Can this ever be sufficiently determined unless we experience *alētheia* in a Greek manner as unconcealment and then, above and beyond the Greek, think it as the clearing of self-concealing? As long as *ratio* and the rational still remain questionable in what is their own, talk about irrationalism is unfounded. The technological-scientific rationalization ruling the present age justifies itself every day more surprisingly by its immense results. But this says nothing about what first grants the possibility of the rational and the irrational. The effect proves the correctness of technological-scientific rationalization. But is the manifest character of what *is* exhausted by what is demonstrable? Does not the insistence on what is demonstrable block the way to what is?

Perhaps there is a thinking that is more sober-minded than the incessant frenzy of rationalization and the intoxicating quality of cybernetics. One might aver that it is precisely this intoxication that is extremely irrational.

Perhaps there is a thinking outside of the distinction of rational and irrational, more sober-minded still than scientific technology, more sober-minded and hence removed, without effect, yet having its own necessity. When we ask about the task of this thinking, then not only this thinking but also the question concerning it is first made questionable. In view of the whole philosophical tradition this means:

We all still need an education in thinking, and first of all, before that, knowledge of what being educated and uneducated in thinking means. In this respect Aristotle gives us a hint in Book IV of his *Metaphysics* (1006aff.): ἔστι γὰρ ἀπαιδευσία τὸ μὴ γιγνώσκειν τίνων δεῖ ζητεῖν ἀπόδειξιν καὶ τίνων οὐ δεῖ. "For it is uneducated not to have an eye for when it is necessary to look for a proof and when this is not necessary."

This sentence demands careful reflection. For it is not yet decided in what way that which needs no proof in order to become accessible to thinking is to be experienced. Is it dialectical mediation, or originarily giving intuition, or neither of the two? Only the peculiar quality of what demands of us above all else to be granted entry can decide about that. But how is this to make the decision possible for us when we have not yet granted it? In what circle are we moving here, indeed, inevitably?

Is it the *eukukleos Alētheiē*, well-rounded unconcealment itself, thought as the clearing?

Does the title for the task of thinking then read, instead of *Being and Time:* Clearing and Presence?

But where does the clearing come from and how is it given? What speaks in the "There is / It gives"?

The task of thinking would then be the surrender of previous thinking to the determination of the matter for thinking.

# SUGGESTIONS FOR FURTHER STUDY

The following list provides a selection of Martin Heidegger's works available in English translation. It is by no means a complete inventory. A catalogue of Heidegger's works, along with translations, and a thorough survey of the secondary literature can be found in Hans-Martin Sass, *Martin Heidegger: Bibliography and Glossary* (Bowling Green State University, Ohio: Philosophy Documentation Center, 1982). Readers are advised that in addition to the volumes already published by Indiana University Press (translations of volumes in the *Martin Heidegger Gesamtausgabe*) further volumes will be appearing in the future. HarperCollins too will be updating its list constantly, so that readers should examine these publishers' lists regularly.

Among the works by Martin Heidegger available in English are:

"Art and Space." Translated by Charles H. Seibert. *Man and World*, vol. 6, no. 1 (1973), pp. 3–5. [*Kunst und Raum*, 1969.]

*Basic Problems of Phenomenology*. Translated by Albert Hofstadter. Bloomington: Indiana University Press, 1982. [*Die Grundprobleme der Phänomenologie*, 1975 (1927).]

*Being and Time*. Translated by John Macquarrie and Edward Robinson. New York: Harper & Row, 1962. [*Sein und Zeit*, 1927.] A new translation by Joan Stambaugh of this work is now in preparation.

*Discourse on Thinking*. Translated by John M. Anderson and E. Hans Freund. New York: Harper & Row, 1966. [*Gelassenheit*, 1959.]

*Early Greek Thinking*. Translated by David Farrell Krell and Frank A. Capuzzi. New York: Harper & Row, 1975. ["Der Spruch des Anaximander" from Holzwege, 1950, pp. 296–343; "Logos (Heraklit, Fragment B 50)," "Moira (Parmenides VIII, 34–41)," and "Aletheia (Heraklit, Fragment B 16)" from *Vorträge und Aufsätze*, 1954, pp. 207–82.]

*The End of Philosophy*. A new edition of this volume, as a supplement to Heidegger's *Nietzsche*, is now in preparation by HarperCollins. ["Die Metaphysik als Geschichte des Seins," "Entwürfe zur Geschichte des Seins als

Metaphysik," and "Die Erinnerung in die Metaphysik" from *Nietzsche*, 1961, vol. II, pp. 399–490; "Überwindung der Metaphysik" from *Vorträge und Aufsätze*, 1954, pp. 71–99.]

*The Essence of Reasons*. A bilingual edition. Translation by Terrence Malick. Evanston, IL: Northwestern University Press, 1969. [*Vom Wesen des Grundes*, 1929.]

*Existence and Being*. Edited, with introduction, by Werner Brock. Chicago: Henry Regnery Company, 1949. [In addition to other translations of Readings II and III in this anthology, the volume contains translations of "Heimkunft: An die Verwandten" and "Hölderlin und das Wesen der Dichtung" from *Erläuterungen zu Hölderlins Dichtung*, 1951, pp. 9–45.]

*Hegel's Concept of Experience*. Translated by J. Glenn Gray and Fred D. Wieck. New York: Harper & Row, 1970. ["Hegels Begriff der Erfahrung," *Holzwege*, 1950, pp. 105–92.]

*Hegel's Phenomenology of Spirit*. Translated by Parvis Emad and Kenneth Maly. Bloomington: Indiana University Press, 1988. [*Hegels Phänomenologie des Geistes*, 1988 (1930–31).]

*Heraclitus Seminar, 1966–1967*. With Eugen Fink. Translated by Charles H. Seibert. University, AL: University of Alabama Press, 1979. [*Heraklit*, 1970.]

*History of the Concept of Time: Prologomena*. Translated by Theodore Kisiel. Bloomington: Indiana University Press, 1985. [*Prologomena zur Geschichte des Zeitbegriffs*, 1979 (1925).]

*Identity and Difference*. A bilingual edition. Translated by Joan Stambaugh. New York: Harper & Row, 1969. [*Identität und Differenz*, 1957.]

*An Introduction to Metaphysics*. Translated by Ralph Manheim. Garden City, NY: Doubleday-Anchor Books, 1961. [*Einführung in die Metaphysik*, 1953.]

*Kant and the Problem of Metaphysics*. Translated by Richard Taft. Bloomington: Indiana University Press, 1990. [*Kant und das Problem der Metaphysik*, 1929.]

*Metaphysical Foundations of Logic*. Translated by Michael Heim. Bloomington: Indiana University Press, 1984. [*Metaphysische Anfangsgründe der Logik im Ausgang von Leibniz*, 1984 (1928).]

*Nietzsche*. Four volumes. Edited by David Farrell Krell. New York: Harper & Row, 1979–87. Published as two paperback volumes in 1991. [*Nietzsche*, 2 vols., 1961.]

*On the Way to Language*. Translated by Peter D. Hertz and Joan Stambaugh. New York: Harper & Row, 1971. [*Unterwegs zur Sprache*, 1959. The English edition does not follow the sequence of the essays in the German edition and omits the first essay, which appears in *Poetry, Language, Thought*, listed below.]

*On Time and Being*. Translated by Joan Stambaugh. New York: Harper & Row, 1972. [*Zur Sache des Denkens*, 1969.]

"'Only a God Can Save Us Now': An Interview with Martin Heidegger." Translated by David Schendler. *Graduate Faculty Philosophy Journal*, vol. 6, no.

1 (1977), pp. 5–27. The interview also appears, in a translation by Sister Maria P. Alter and John D. Caputo, in *Philosophy Today*, vol. 20, no. 4 (Winter 1976), pp. 267–84. ["Nur noch ein Gott kann uns retten," *Der Spiegel*, 31 May 1976 (1966).]

*Parmenides*. Translated by André Schuwer and Richard Rojcewicz. Bloomington: Indiana University Press, 1992. [*Parmenides*, 1982 (1942–43).]

*The Piety of Thinking*. Translation, notes, and commentary by James G. Hart and John C. Maraldo. Bloomington: Indiana University Press, 1976. [Contains Heidegger's review of Ernst Cassirer's *Mythical Thought*, 1928, "Grundsätze des Denkens," 1958, and other pieces.]

*Poetry, Language, Thought*. Translated by Albert Hofstadter. New York: Harper & Row, 1971. [*Aus der Erfahrung des Denkens*, 1954; *Der Ursprung des Kunstwerkes* (Reclam), 1960; "Wozu Dichter?" from *Holzwege*, 1950, pp. 248–95; "Bauen Wohnen Denken," "Das Ding," and ". . . Dichterisch wohnet der Mensch . . ." from *Vorträge und Aufsätze*, 1954, pp. 145–204; "Die Sprache" from *Unterwegs zur Sprache*, 1959, pp. 9–33.]

*The Principle of Reason*. Translated by Reginald Lilly. Bloomington: Indiana University Press, 1991. [*Der Satz vom Grund*, 1957.]

*The Question Concerning Technology and Other Essays*. Translated by William Lovitt. New York: Harper & Row, 1977. ["Die Frage nach der Technik" and "Wissenschaft und Besinnung" from *Vorträge und Aufsätze*, 1954, pp. 13–70; "Die Zeit des Weltbildes" and "Nietzsches Wort 'Gott ist tot'" from *Holzwege*, 1950, pp. 69–104 and 193–247.]

*The Question of Being*. A bilingual edition. Translation by William Kluback and Jean T. Wilde. New Haven, CT: College & University Press, 1958. [*Zur Seinsfrage*, 1956.]

*Schelling's Treatise on the Essence of Human Freedom*. Translated by Joan Stambaugh. Athens: Ohio University Press, 1985. [*Schellings Abhandlung über das Wesen der menschlichen Freiheit (1809)*, 1971 (1936–43).]

"The Self-Assertion of the German University." Translated by Karsten Harries. *Review of Metaphysics*, vol. 38 (1985), 467–80. (With "The Rectorate 1933–34: Facts and Thoughts," 481–502.) [*Die Selbstbehauptung der deutschen Universität*, 1983 (1933; 1945).]

*What Is a Thing?* Translated by W. B. Barton, Jr., and Vera Deutsch. Chicago: Henry Regnery Company, 1967. [*Die Frage nach dem Ding*, 1962.]

*What Is Called Thinking?* Translated by Fred D. Wieck and J. Glenn Gray. New York: Harper & Row, 1968. [*Was heisst Denken?* 1954.]

*What Is Philosophy?* A bilingual edition. Translated by William Kluback and Jean T. Wilde. New Haven, CT: College & University Press, 1958. [*Was ist das—die Philosophie?* 1956.]